Reboot Development

Scan to go to this publication online.

Reboot Development

The Economics of a Livable Planet

Richard Damania, Ebad Ebadi, Kentaro Mayr,
Jason Russ, and Esha Zaveri

WORLD BANK GROUP

Contents

PART 2: CITIES AND COMMERCE

PART 3: POLICIES, JOBS, AND SOLUTIONS FOR A LIVABLE PLANET

BOXES

FIGURES

MAPS

TABLES

Acknowledgments

This book was prepared by a World Bank team led by Richard Damania (Chief Economist, Planet Vice Presidency), and a core team comprising Ebad Ebadi (Economist), Kentaro Mayr (Data Scientist), Jason Russ (Senior Economist), and Esha Zaveri (Senior Economist). The book greatly benefited from the strategic guidance and general direction of Axel van Trotsenburg (Senior Managing Director), Juergen Voegele (Vice President, Planet Vice Presidency), and Renaud Seligmann (Director, Strategy and Operations, Planet Vice Presidency).

In addition to research completed by the authors, this work drew on background papers, notes, and analyses prepared by a wide group of internal and external collaborators, for which the authors are grateful.

Internal collaborators included Enrique Aldaz-Carroll (Senior Economist), Paolo Avner (Senior Economist), Seth Ayers (Senior Digital Development Specialist), Brian Blankespoor (Senior Geographer), Claire Chase (Senior Water Economist), Wendy Cunningham (Lead Economist), Susmita Dasgupta (Lead Environmental Economist), Shenghui Feng (Evaluation Analyst), Joshua Gill (Senior Economist), Philip Grinsted (Private Sector Specialist), Nagaraja Rao Harshadeep (Lead Environmental Specialist), Catherine Highet (Consultant), Tejasvi Hora (Consultant), Duong Trung Le (Economist), Ralf Martin (Principal Economist, International Finance Corporation), Nicholas Menzies (Senior Environmental Specialist), Alejandra Guardia Muguraza (Consultant), Jane Park (Consultant), Matias Piaggio (Consultant), Jonah Rexer (Economist), Mark Roberts (Lead Urban Economist), and Bhavya Srivastava (Young Professional).

External collaborators included Rafael Araujo (São Paulo School of Economics), Nandita Basu (University of Waterloo), Maksym Chepeliev (Purdue University), Esther Choi (World Resources Institute), Roman Czebiniak (World Resources Institute), Eyal Frank (University of Chicago), Gargee Goswami (Princeton University), Vinicius Hector (Brazilian School of Economics and Finance), Erik Katovich (University of Connecticut), Ron Milo (Weizmann Institute of Science), Anant Sudarshan (University of Warwick), Farzad Taheripour (Purdue University), and Wen Zhou (World Resources Institute).

The team is grateful to other colleagues from the World Bank and external contributors for their helpful inputs and suggestions at various stages of the book's development, including Elizabeth Ruppert Bulmer, Eileen Burke, German Caruso, Sarah Coll-Black, Jose Antonio Cuesta Leiva, Gabriel Demombynes, Sebastien Dessus, Gabriel Englander,

Maria Julia Granata, Maddalena Honorati, Sandor Karacsony, Julian Lampietti, Maryla Maliszewska, Robert Marty, Megha Mukim, Yogita Mumssen, Helena Naber, Regassa Namara, Dilip Ratha, Ernesto Sanchez-Triana, Harris Selod, Siddharth Sharma, Iryna Sikora, Amal Talbi, Estefania Vergara-Cobos, and Xiao'ou Zhu, as well as Yoshihide Wada, Reshmita Nath, and Yawei Bai (King Abdullah University of Science and Technology).

The team was fortunate to receive invaluable feedback and advice from the following peer reviewers: Kevin Carey (Program Manager), Lorenzo Carrera (Lead Disaster Risk Management Specialist), Andrew Dabalen (Chief Economist, Africa Region), Marianne Fay (Country Director), Arunabha Ghosh (Founder and CEO of the Council on Energy, Environment and Water, India), Chakib Jenane (Regional Practice Director), Paolo Mauro (Director, International Finance Corporation), Giovanni Ruta (Lead Environmental Economist), Stephane Straub (Chief Economist, Infrastructure), Timothy Olalekan Williams (Feed Africa Special Envoy, African Development Bank, and Former Director, CGIAR), and Sergiy Zorya (Lead Agricultural Economist).

The team also benefited greatly from the feedback provided by the report's external advisory committee: Carl Folke (Beijer Institute of Ecological Economics of the Royal Swedish Academy of Sciences and Stockholm Resilience Centre, Stockholm University), Ma Jun (Institute of Public and Environmental Affairs, China), Gretchen Daily (Stanford University), Deepak Mishra (formerly at the Indian Council for Research on International Economic Relations), and Yvonne Aki-Sawyerr OBE (Mayor of Freetown, Sierra Leone).

The authors thank Lucy Southwood, Sandra Gain, and Ann O'Malley for editing and proofreading. Yann Kerblat provided excellent design support and contributed several interior images. Bradley Amburn was responsible for the design and layout of the report's stand-alone Overview. The World Bank communications team, including Ferzina Banaji, Kimberly Versak, Nigina Alieva, and Diana Manevskaya provided valuable guidance on outreach and communications. Cindy Fisher, Amy Lynn Grossman, and Jewel McFadden of the World Bank's publishing program were responsible for managing editing, design, typesetting, printing, and dissemination of the print and digital versions of the report and provided inputs into the production of the stand-alone Overview.

Finally, Desy Adiati provided helpful administrative support, for which the team is grateful.

This work was made possible by the financial contribution of the Korea Green Growth Trust Fund (refer to https://www.wbgkggtf.org) of the Planet Vice Presidency, World Bank Group.

Main Messages

Key findings

Reboot Development: The Economics of a Livable Planet explores how the foundational endowments of land, air, and water can continue enabling human prosperity, supporting health, food, energy, and economic opportunity. In the span of mere generations, much of the world has emerged from the shadows of poverty and hunger to an era of relative abundance. Yet, in reshaping the world for prosperity, humanity is unsettling the very foundations of that progress. The same forces that fueled economic growth—industrial expansion, energy consumption, and unsustainable agriculture—now strain the planet's ability to sustain it. This report argues that maintaining a livable planet is not merely a distant environmental concern but a present economic threat.

Land, air, and water underpin a livable planet. Yet today all are under threat with both global and increasingly severe local impacts.

- Around ninety percent of people *globally* live with degraded land, or polluted air, or water stress.

- Around eighty percent of people in *low-income countries* live with all three environmental stressors—degraded land, polluted air, and water stress (figure MM.1). By contrast, in high-income countries 43 percent of people are not exposed to any of the three stressors.

FIGURE MM.1 **Close to 80 percent of low-income country residents are exposed to poor air quality, unsafe water, and degraded land**

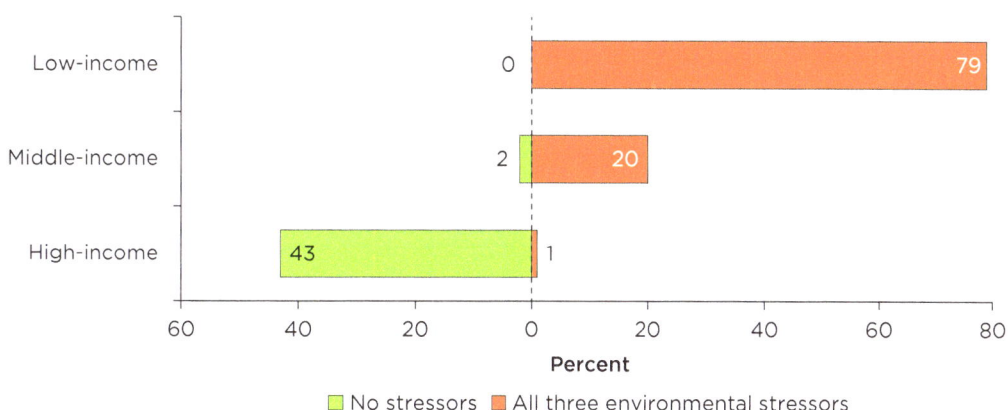

Sources: Original calculations for this report. Water risks, World Resources Institute 2023; air pollution, van Donkelaar et al. 2016; land degradation, Bai et al. 2025.
Note: The figure shows the share of inhabitants in each income group exposed to water risks, air pollution, or land degradation. Definitions of the risks are given in the notes in map 1.1 in chapter 1.

The scale of impacts is so vast that humans have transitioned from being passive beneficiaries of the planet to becoming the dominant force in its transformation.

- Today humans and their livestock account for an astonishing 95 percent of total mammalian biomass (by weight) on Earth, leaving wild mammals a vanishing 5 percent (figure MM.2).

- The Earth has transgressed six of the nine environmental thresholds—termed "Planetary Boundaries"—needed for human progress.

FIGURE MM.2 **Wild mammal biomass has plummeted while human impacts have soared**

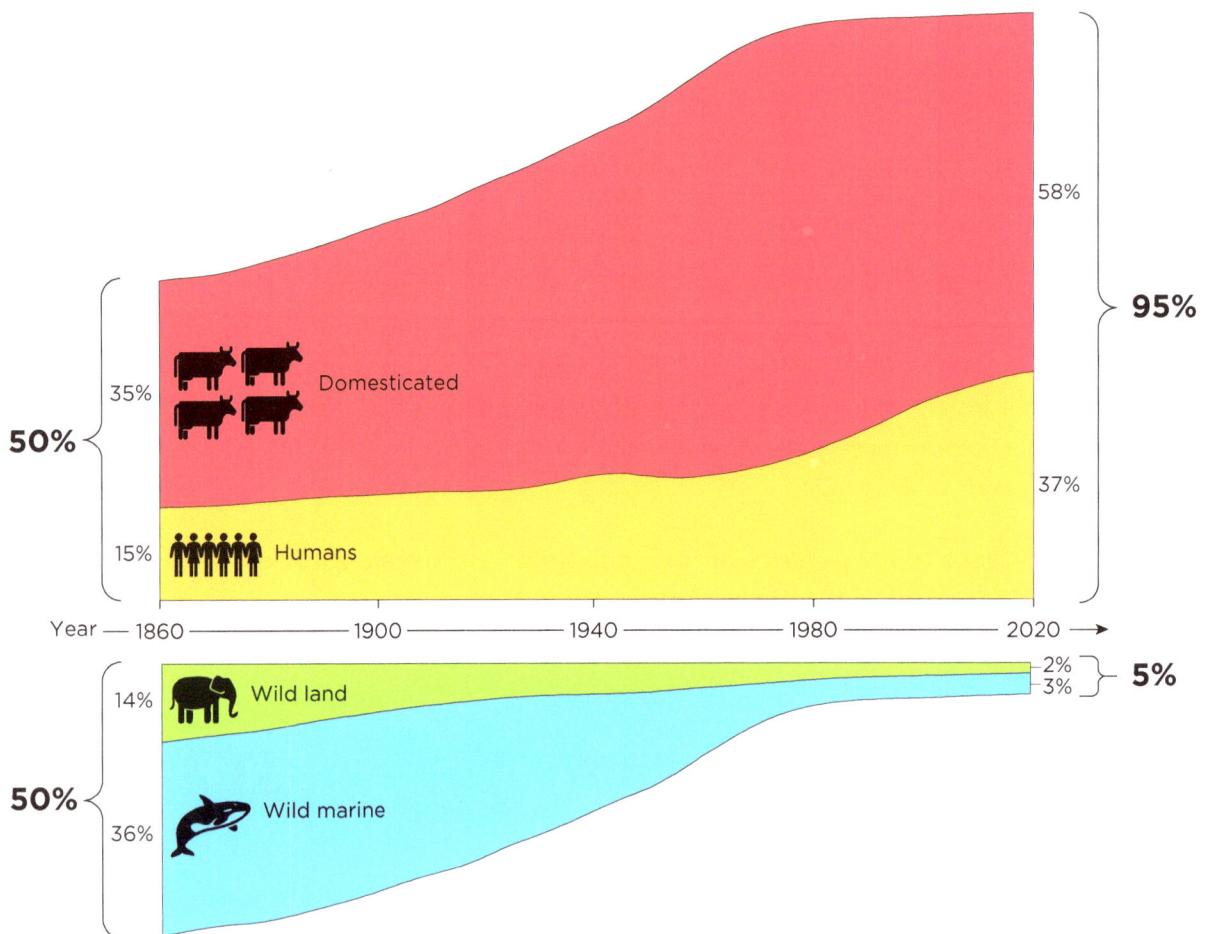

Source: Adapted from Greenspoon et al. (forthcoming)
Note: Refer to chapter 1 for details.

The hidden connections between land and water

Forests are more than hosts of biodiversity, sources of timber, and carbon sinks. They also fuel rainfall and store water.

- Loss of *natural forests* dries out soils and reduces crop yields, *costing the world $379 billion annually*—about 8 percent of global agricultural gross domestic product (GDP).

- *The protective role of natural forests is especially important during droughts.* Natural forests reduce the growth losses from droughts by more than half, unlike plantations and monoculture forests (figure MM.3).

- *Forests are also rainmakers*—nearly half of all rainfall comes from vegetation, most of it driven by forests, sustaining crops and cities alike. Deforestation-induced rainfall loss costs countries in the Amazon $14 billion per year in lost economic growth.

FIGURE MM.3 **Upstream forest cover buffers against droughts**

Loss in growth of GDP (percentage points)

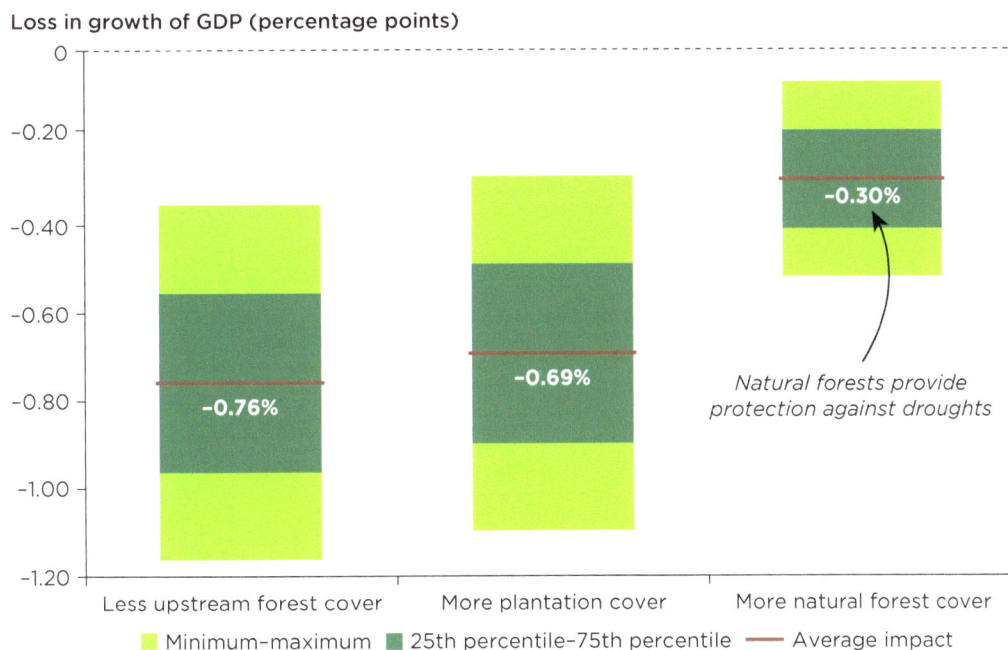

Sources: Original calculations based on data from Defourny 2019; Kummu, Taka, and Guillaume 2018; Lesiv et al. 2022; and Matsuura and Willmott 2018.
Note: Refer to chapter 2 for details. GDP = gross domestic product.

Too much of a good thing can be toxic: The nitrogen legacy

Nitrogen fertilizer has shaped the modern world by more than doubling crop yields. Yet unbalanced nitrogen fertilizer use has caused problems in some regions:

- Excess nitrogen use *harms crop yields*. Half of the global food supply is produced in regions where nitrogen use is so excessive that it diminishes crop yields (figure MM.4).

- Excess nitrogen *pollutes land, air, and water*—costing up to $3.4 trillion a year globally.

The deadliest risk no one sees: Air pollution

Air pollution is a ubiquitous problem and is especially severe in low- and middle-income countries.

- *Outdoor air pollution kills at least 5.7 million people each year*—more than tobacco, malnutrition, or lives lost to wars and violence. Polluted air reduces productivity, increases sick days, and even lowers cognitive function and, as a consequence, significantly lowers GDP growth in affected regions (World Bank 2025).

- *China's "war on pollution" proves that change is possible.* In just a decade, many of the country's most polluted cities transformed their air quality through a strategy that included robust monitoring, incentivizing cleaner energy, and enforcing stricter emissions standards. China's success has been responsible for bending the global air pollution curve downward (figure MM.5).

FIGURE MM.4 **Half of the global food supply is grown where nitrogen does more harm than good to yields**

Increase in yield productivity (%)

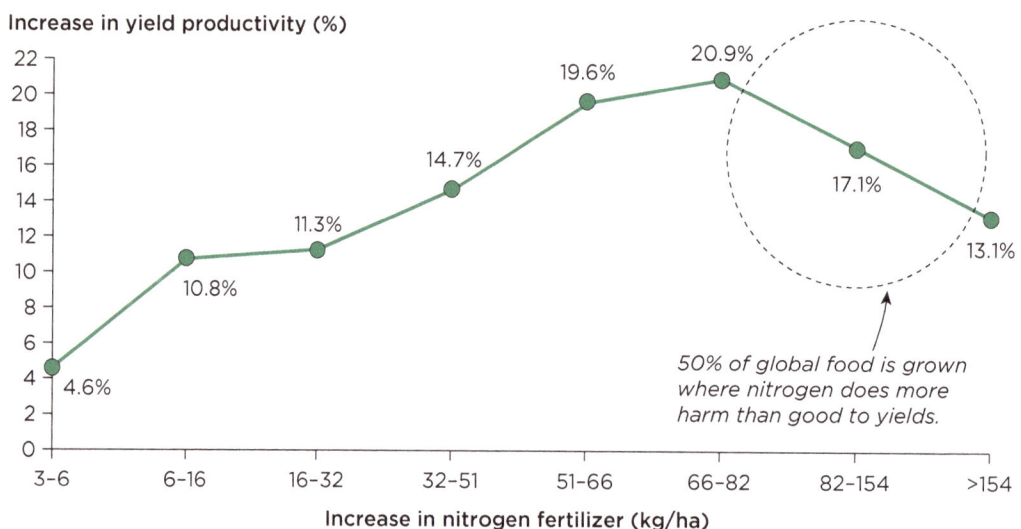

50% of global food is grown where nitrogen does more harm than good to yields.

Increase in nitrogen fertilizer (kg/ha)

Source: Zaveri 2025.
Note: Refer to chapter 4 for details. kg/ha = kilograms per hectare.

FIGURE MM.5 It is possible to reduce air pollution (PM$_{2.5}$) while growing the economy

Population-weighted PM$_{2.5}$ concentrations (indexed to 2000)

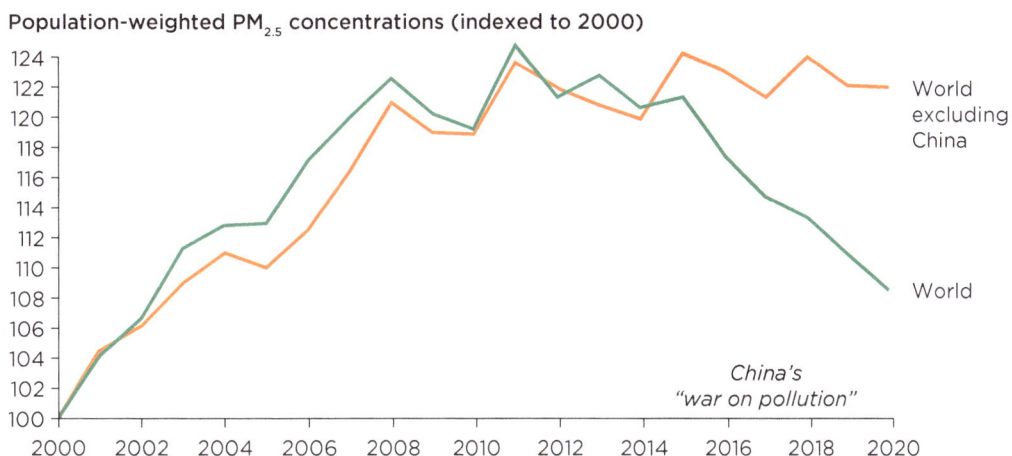

Sources: Original calculations using PM$_{2.5}$ data from van Donkelaar et al. 2016 and population data from CIESIN 2018.
Note: Refer to chapter 5 for details. PM$_{2.5}$ = fine particulate matter.

Pathways to progress

Nature's wealth is a source of growth and comparative advantage, calling for policies that sustain, rather than deplete, the natural assets essential to progress. As the *scale* of economic activity has increased, pollution and resource use have increased. This can be offset by improvements in *efficiency*—producing more with less or the same—and by changing the *composition* of production toward less resource- and pollution-intensive goods. The evidence suggests that:

- High-income countries are starting to *decouple* economic growth from some forms of environmental damage, thanks largely to gains in efficiency.

- *Efficiency improvements* have significantly reduced environmental impacts. They have reduced water use by 50 percent, land use by 69 percent, and air pollution by 59 percent (table MM.1).

- Simply improving efficiency is unlikely to suffice, however. Ultimately, sustaining natural assets will require not just producing things better, but also *producing better things* (such as the switch from the horse and cart to internal combustion engines to electric vehicles).

TABLE MM.1 **As resource use increases, increased efficiency is the main force curbing degradation**

	Water withdrawals	Land use	Air pollution (PM$_{2.5}$)	GHG emissions
Scale (Normalized)	100	100	100	100
Composition	–30	4	–4	–10
Effciency	–50	–69	–59	–52
Net change	20	35	37	38

Source: Original table for this report using the GTAP 11 Data Base.
Note: Refer to spotlight 1 for details. GHG = greenhouse gas; PM$_{2.5}$ = fine particulate matter.

Jobs for a livable future

Economic growth, productivity, and jobs rely on nature's wealth—clean air, healthy soil, and resilient ecosystems. In contrast, a degraded environment undermines livelihoods, drains talent, and weakens economies.

- *Fertile land, forests, and fisheries form the backbone of rural economies.* Agriculture alone supports 3.2 billion people, and fishing employs nearly 62 million globally.

- *Destruction of natural capital*—through pollution or disasters—undermines human health, lowers productivity, and shrinks wages.

- *Investment in less-polluting sectors* on average can create *more jobs per dollar invested* compared to more-polluting ones (figure MM.6).

FIGURE MM.6 **Less-polluting sectors create more jobs**

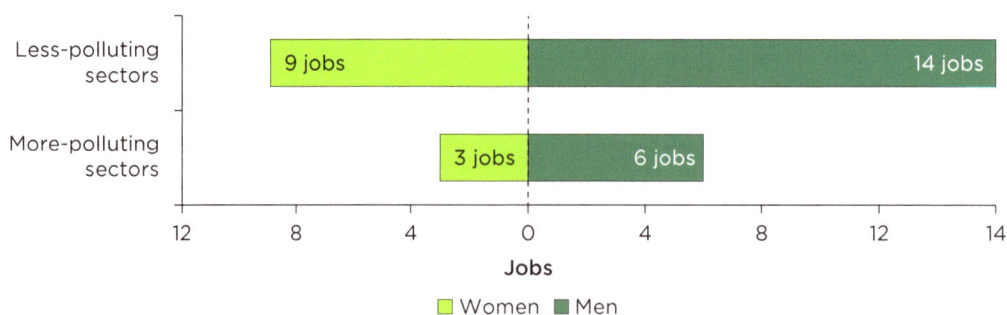

Source: Original calculations based on data from Taheripour et al. 2022.
Note: Refer to chapter 9 for details.

Transition minerals: The bedrock of the transition

Certain "transition minerals," such as lithium, cobalt, nickel, and rare earth elements, are essential for continued innovation and a more livable planet, but strong governance is needed to manage their risks. For resource-rich countries, this demand creates opportunities for industrialization, economic diversification, and job creation.

- Significant reserves exist in high-income countries, while a minor proportion of reserves co-occur in locations with risk factors, including deforestation, biodiversity loss, water contamination, and governance challenges (map MM.1).

- Estimates presented in this report indicate that deforestation rates are higher where these transition minerals are extracted than at conventional mining sites.

- Good governance partly ameliorates the impact. Stronger environmental safeguards, investments in local value chains, and ensuring the transparency of supply chains will be crucial to ensure that mineral wealth becomes a source of prosperity.

MAP MM.1 **"Transition mineral" mines registered in areas that overlap with potential risks**

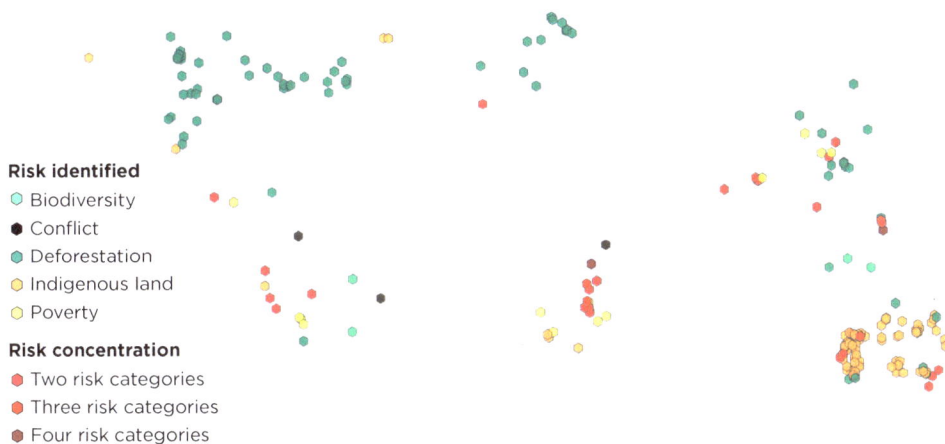

Risk identified
- ○ Biodiversity
- ● Conflict
- ● Deforestation
- ○ Indigenous land
- ○ Poverty

Risk concentration
- ● Two risk categories
- ● Three risk categories
- ● Four risk categories

Source: Katovich and Rexer 2025.
Note: Refer to spotlight 6 for details.

Fix the system, not the symptom

Development cannot succeed on a damaged planet, and policies cannot succeed without considering the whole system. Environmental reforms that ignore how land, water, air, and people interact often backfire. A more effective approach involves a three-pronged policy strategy:

- *Inform.* Information is power. From air pollution monitors to satellite imagery, real-time data are key to targeting problems, empowering citizens, and driving accountability.

- *Enable.* Policies work best when they work together. A systems approach aligns actions across sectors and time, avoiding unintended consequences and trade-offs.

- *Evaluate.* Learning beats guessing. Regular evaluation keeps policies on track, helps scale what works, and ensures that reforms can adapt to shifting realities.

The wealth of nature is a source of prosperity. Protecting ecosystems is not just an environmental concern—it is necessary for improving the resilience and productivity of jobs. Countries that leverage their comparative environmental advantages and build industries that thrive within ecological limits will be the ones that prosper into the future.

References

Bai, Z., J. D. Russ, K. F. Mayr, and D. Dent. 2025. "How Is Gaia Doing? Trends in Global Land Degradation and Improvement." *Ambio*, 1-37.

CIESIN (Center for International Earth Science Information Network). 2018. *Gridded Population of the World, Version 4 (GPWv4): Population Count, Revision 11*. New York: CIESIN, Columbia University. doi:10.7927/H4JW8BX5.

Defourny, P. 2019. "ESA Land Cover Climate Change Initiative (Land_Cover_cci): Global Land Cover Maps, Version 2.0.7." Centre for Environmental Data Analysis, Didcot, UK. https://catalogue.ceda.ac.uk/uuid/b382ebe6679d44b8b0e68ea4ef4b701c.

Greenspoon, L., N. Ramot, U. Moran, et al. Forthcoming. "The Global Biomass of Mammals Since 1850." *Nature Communications*.

Katovich, E., and J. Rexer. 2025. "Critical Mining Contributes to Economic Growth and Forest Loss in High-Corruption Settings." Available at SSRN 5291760. Background paper for this report.

Kummu, M., M. Taka, and J. H. Guillaume. 2018. "Gridded Global Datasets for Gross Domestic Product and Human Development Index over 1990–2015." *Scientific Data* 5 (1): 1–15.

Lesiv, M., D. Schepaschenko, M. Buchhorn, et al. 2022. "Global Forest Management Data for 2015 at a 100 m Resolution." *Scientific Data* 9: 199. doi:10.1038/s41597-022-01332-3.

Matsuura, K., and C. J. Willmott. 2018. "Terrestrial Air Temperature and Precipitation: Monthly and Annual Time Series (1900–2017)." https://climate.geog.udel.edu/html_pages/download.html

Taheripour, F., M. Chepeliev, R. Damania, T. Farole, N. L. Gracia, and J. D. Russ. 2022. "Putting the Green Back in Greenbacks: Opportunities for a Truly Green Stimulus." *Environmental Research Letters* 17 (4): 044067.

van Donkelaar, A., R. V. Martin, M. Brauer, et al. 2016. "Global Estimates of Fine Particulate Matter Using a Combined Geophysical-Statistical Method with Information from Satellites, Models, and Monitors." *Environmental Science & Technology* 50: 3762–72. doi:10.1021/acs.est.5b05833.

World Resources Institute. 2023. "Aqueduct 4.0: Updated Decision-Relevant Global Water Risk Indicators." Technical Note. World Resources Institute, Washington, DC.

World Bank. 2025. *Accelerating Access to Clean Air for a Livable Planet*. Washington, DC: World Bank. https://documents.worldbank.org/curated/en/099032625132535486.

Zaveri, E. 2025. "Fixing Nitrogen: Agricultural Productivity, Environmental Fragility, and the Role of Subsidies." Policy Research Working Paper 11050, World Bank, Washington, DC. https://hdl.handle.net/10986/42737.

Abbreviations

μg/m³	micrograms of pollutant per cubic meter of air
4Rs	right nutrients, at the right rate, at the right time, and in the right place
AI	artificial intelligence
AIIB	Asian Infrastructure Investment Bank
ANR	assisted natural regeneration
ASM	artisanal and small-scale mining
BISP	Benazir Income Support Program [Pakistan]
BLS	Bureau of Labor Statistics [United States]
CAC	command-and-control
CIESIN	Center for International Earth Science Information Network
CLC	Center for Livable Cities
CLRTAP	Geneva Convention on Long-Range Transboundary Air Pollution
CO_2	carbon dioxide
CTAs	contingent trade agreements
CYGNSS	Cyclone Global Navigation Satellite System
EEA	European Environment Agency
EECs	environmental Engel curves
EITI	Extractive Industries Transparency Initiative
ERIA	Energy Research Institute for ASEAN and East Asia
ESMAP	Energy Sector Management Assistance Program
EVs	electric vehicles
EWS	early warning systems
FAO	Food and Agriculture Organization of the United Nations
FARs	floor-area ratios
FFI	Framework for Financial Incentives
GBIF	Global Biodiversity Information Facility
GCI	Green Complexity Index
GDP	gross domestic product
GGW	Great Green Wall
GHG	greenhouse gas
GPSC	Global Platform for Sustainable Cities

GPT-3	Generative Pre-trained Transformer 3
GROW	Global and Regional Opportunities Window
GSMA	Groupe Spécial Mobile Association
GTAP	Global Trade Analysis Project
G-WAN	government wide area network
HICs	high-income countries
IBRD	International Bank for Reconstruction and Development
ICCT	International Council on Clean Transportation
IDA	International Development Association
IEA	International Energy Agency
IGF	Intergovernmental Forum on Mining, Minerals, Metals and Sustainable Development
IIASA	International Institute for Applied Systems Analysis
ILO	International Labour Organization
IPBES	Intergovernmental Science-Policy Platform on Biodiversity and Ecosystem Services
IRENA	International Renewable Energy Agency
ITDP	Institute for Transportation & Development Policy
ITU	International Telecommunication Union
km	kilometer
L/pkm	liters required to transport one passenger one kilometer
LICs	low-income countries
LMDI	Logarithmic Mean Divisia Index
LMICs	lower-middle-income countries
LPG	liquefied petroleum gas
MBI	market-based instrument
MICs	middle-income countries
MOSAP3	Smallholder Agricultural Transformation Project [Angola]
MPAs	marine protected areas
$MtCO_2eq$	million tons of carbon dioxide equivalent
NADRA	National Database and Registration Authority [Pakistan]
NASA	National Aeronautics and Space Administration [United States]
NbS	nature-based solutions
NH_3	ammonia
NJILA	Angola Strengthening Governance for Enhanced Service Delivery Project [Angola]
NSER	National Socio-Economic Registry [Pakistan]

NUE	nitrogen use efficiency
OECD	Organisation for Economic Co-operation and Development
OP	operation manual
PES	payment for ecosystem services
$PM_{2.5}$	fine particulate matter with a diameter of 2.5 micrometers or less
PM_{10}	particulate matter with a diameter of 10 micrometers or less
PMAY-U	Pradhan Mantri Awas Yojana-Urban [India]
PWP	public works program
QUARG	Quality of Urban Air Review Group
R&D	research and development
RCA	revealed comparative advantage
RNPA	National Registry of Agricultural Producers [Angola]
RZSM	root-zone soil moisture
SAWAP	Sahel and West Africa Program in Support of the Great Green Wall
SRC	Seismic Research Center [University of the West Indies]
SRCA	symmetric revealed comparative advantage
SSFRs	site-specific fertilizer recommendations
STEM	science, technology, engineering and mathematics
SYRCA	scatterplot of the symmetric revealed comparative advantage
UMICs	upper-middle-income countries
UNCCD	United Nations Convention to Combat Desertification
UNCTAD	United Nations Conference on Trade and Development
UNEP	United Nations Environment Programme
UNESCO	United Nations Educational, Scientific and Cultural Organization
UNICEF	United Nations Children's Fund
UWI	University of the West Indies
VWT	virtual water trade
WASH	water, sanitation, and hygiene
WHO	World Health Organization
WRI	World Resources Institute
WWF	World Wildlife Fund for Nature

All dollars are US dollars unless otherwise indicated.

CHAPTER 1

Introduction: Economic Choices for a Finite Planet

What is a livable planet?

The world is undergoing a new transition—one that could redefine the relationship between economic growth, human well-being, and the environment. Throughout history, major transformations, such as the Industrial and Green Revolutions, have dramatically improved human living standards. But this has often been at a steep cost. Pollution surged, natural ecosystems degraded, and the stability of Earth's systems was compromised. Today, the convergence of unprecedented wealth, abundant data, and the transformative power of the digital revolution offers a unique opportunity to chart a different path—one that enhances prosperity while simultaneously minimizing environmental harm.

For the first time, progress does not have to come at the expense of the planet. Economic growth has long been linked to rising pollution and environmental destruction, but new data enable better policy making, and technologies make it possible to decouple the two. Advances in sustainable agriculture, material sciences, and the digital economy offer solutions that past generations did not have, making sustainable development a possibility. This transition is not inevitable—it requires deliberate action and structural change—but it presents an unprecedented opportunity to improve the status of natural assets to advance human well-being.

The World Bank's vision of a world free of poverty on a livable planet reflects the growing recognition that economic development cannot be sustained without safeguarding the ecosystems that support it. The interactions of land, air, and water provide not only the natural foundations for prosperity and growth, but also a wide range of essential services—purifying air and water, supporting food production, and buffering shocks. These ecosystems constitute a form of natural capital, whose productivity depends not only on their quantity, but also their integrity (Elmqvist et al. 2003; Folke et al. 2004; Walker et al. 2023). Yet these systems have long been taken for granted. Today, the degradation of land, air, and water has reached a point where a commitment to tackle poverty and boost shared prosperity is unviable without an equal commitment to manage and restore the natural resources on which economies and life depend. Without this balance, sustaining improvements in well-being will be challenging.

A *livable planet* can be defined as one that supports environmental health, together with investments in human and physical capital, to improve lives, livelihoods, and living standards for all. All life and all economic activities depend on the balance of the planet's natural systems comprising land, air, and water. In the past, these critical resources were deemed abundant and inexhaustible. Now they have reached a level of scarcity and degradation that hinders progress. Air pollution alone is responsible for around 7 million premature deaths every year, surpassing the mortality rate of all forms of violence combined (WHO 2023). Unsafe drinking water, inadequate sanitation, and poor hygiene further contribute to around 1.4 million deaths each year, predominantly in low-income regions.[1] Beyond these immediate health impacts, environmental degradation—such as the desiccation of rivers and deforestation for short-term economic gains—undermines long-term productive capacity, leading to reduced agricultural yields, water scarcity, and biodiversity loss.

Humanity's growing footprint

Humanity relies on the global ecosystem, known as the biosphere, for essential services. The biosphere—the thin, 12-mile layer of critical natural assets—supports all life and economic activity on the planet, including the critical ecosystem functions that regulate rainfall, maintain healthy oceans, filter freshwater supplies, and provide the minerals and materials modern society needs (Folke et al. 2021).

Humans are not just passive beneficiaries of the biosphere; rather, they are the dominant force in its transformation. For a better understanding of the scale of humanity's impact, it is crucial to take stock of the different components within the natural environment as benchmarks for evaluating recent and future trends. Setting clear numerical benchmarks is important to prevent the *shifting baselines syndrome*, where people slowly adjust to and accept decline and degradation as normal when it may be dangerous (Pauly 1995; Soga and Gaston 2018). Data-driven methods provide an objective way to quantify and contextualize the magnitude of changes on the planet.

A useful benchmark for assessing changes in the natural world is the weight of living organisms, referred to as biomass. This metric provides a common yardstick for assessing planetary-scale changes across different species and between living and nonliving entities. By assessing humanity's impact on different components of the biosphere, the metric offers a benchmark for evaluating recent and future trends across time and geographical scales.

Since the Industrial Revolution, there have been profound shifts in the footprint of human activity, as measured by biomass. The global distribution of the biomass of mammals is key to understanding the status of, and trends in, the biosphere. Of all the mammals on Earth, wild animals constitute a mere 5 percent of the Earth's total mammalian biomass. In stark contrast, the remaining 95 percent constitutes humans (35 percent) and domesticated animals, such as cattle and ruminants for human consumption (60 percent) (figure 1.1). Only 150 years ago, wild animals constituted 50 percent of the Earth's mammalian biomass.

The rapid acceleration of biodiversity loss parallels the rise of industrialization. Cumulative species extinctions have surged dramatically since the onset of the Industrial Revolution, coinciding with the exponential rise in gross domestic product per capita (figure 1.2). Although this is just a simple correlation, the pattern underscores important trends.

The global mass of human-made artefacts now exceeds that of all living biomass. *Anthropogenic mass,* which is the weight of all human-made objects, has risen dramatically in the past century (figure 1.3). Notably, the global mass of produced plastic is greater than the overall mass of all terrestrial and marine animals combined, and buildings and infrastructure (composed of concrete, aggregates, bricks, and asphalt) outweigh trees and shrubs. Around 80 percent of anthropogenic mass is in the form of concrete and aggregates used in construction. In contrast, the weight of all living organisms has declined slightly due in large part to deforestation (Hansen et al. 2013). These comparisons are a quantitative symbol demarcating the dominant impacts on Earth, with implications for biogeochemical cycles and the planet's overall health.

FIGURE 1.1 **The indelible footprint of humans**

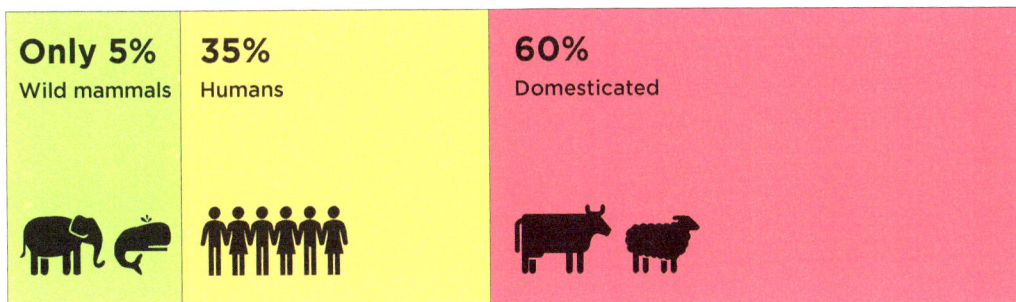

Source: Original calculations based on data from Greenspoon et al. 2023.

FIGURE 1.2 Cumulative species extinctions since 1500

Cumulative extinctions (number of species)   GDP per capita (2011 $)

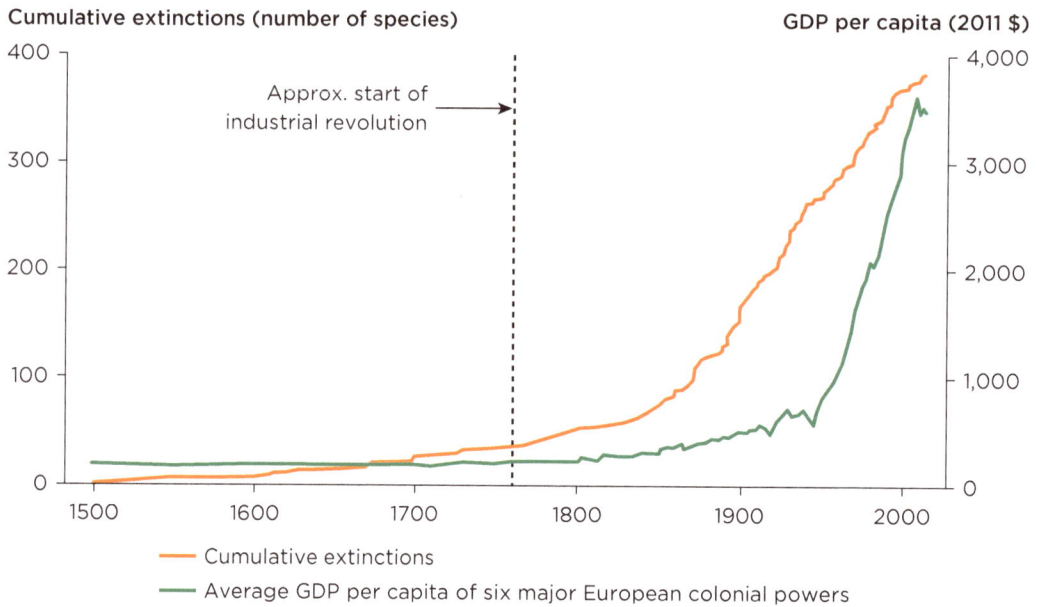

Approx. start of industrial revolution →

—— Cumulative extinctions
—— Average GDP per capita of six major European colonial powers

Source: Li, Dann, and Kompil, 2025.
Note: GDP = gross domestic product.

FIGURE 1.3 Weight of human-made objects versus all living organisms

Dry weight (teratonnes)

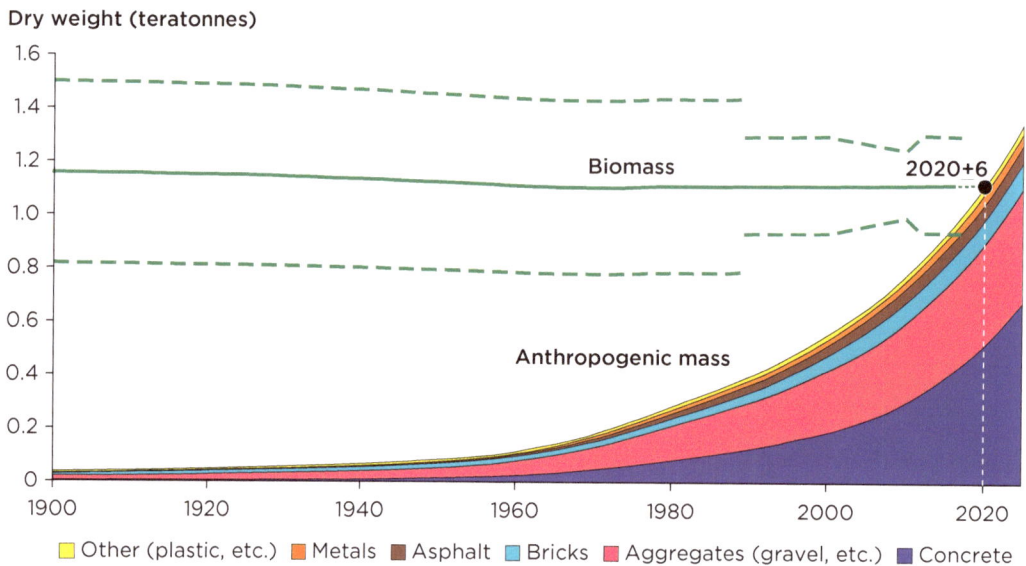

Biomass

2020+6

Anthropogenic mass

☐ Other (plastic, etc.) ■ Metals ■ Asphalt ■ Bricks ■ Aggregates (gravel, etc.) ■ Concrete

Source: Elhacham et al. 2020.
Note: The solid green line shows the total dry weight of all living organisms (global biomass).
The dashed green lines equal ±1 standard deviation. Anthropogenic mass is plotted as an area chart,
where the heights of the colored areas represent the mass of the corresponding category
accumulated until that year. The year 2020 ± 6 marks the time at which anthropogenic mass
exceeded biomass (Elhacham et al. 2020).

Humanity's footprint has grown so large that today, 92 percent of the world's population is exposed to either air pollution, or heightened water risks,[2] or land degradation (map 1.1). Approximately 78 percent of the population is exposed to at least two of these environmental pressures, and 20 percent is exposed to all three. Map 1.1 shows significant regional divides as well. For instance, in South Asia, 92 percent of the population is exposed to air pollution *and* water risks, with the remaining 8 percent exposed to all three stressors. Likewise, in Sub-Saharan Africa, 30 percent of the population is exposed to air pollution *and* water risks, and 66 percent is exposed to all three stressors. The analysis demonstrates that environmental problems often co-occur,

MAP 1.1 Ninety-two percent of the world is exposed to either water risks, air pollution, or land degradation

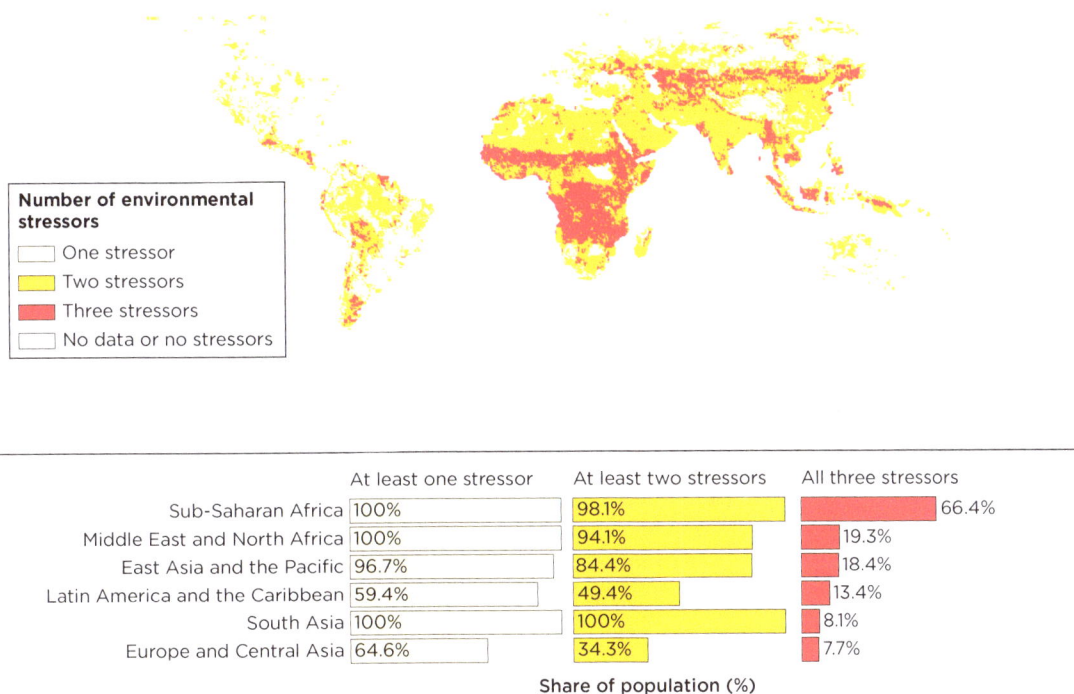

Number of environmental stressors
- One stressor
- Two stressors
- Three stressors
- No data or no stressors

	At least one stressor	At least two stressors	All three stressors
Sub-Saharan Africa	100%	98.1%	66.4%
Middle East and North Africa	100%	94.1%	19.3%
East Asia and the Pacific	96.7%	84.4%	18.4%
Latin America and the Caribbean	59.4%	49.4%	13.4%
South Asia	100%	100%	8.1%
Europe and Central Asia	64.6%	34.3%	7.7%

Share of population (%)

Sources: Original calculations for this report. Water risks, World Resources Institute 2023; air pollution, van Donkelaar et al. 2016; land degradation, Bai et al. 2025.
Note: The map shows land areas exposed to water risks, air pollution, and land degradation. Yellow areas are exposed to one of these stressors, orange areas are exposed to two, and red areas are exposed to all three. Water risks refer to the World Resources Institute's water risk index (which includes physical quantity, physical quality, and regulatory and reputational risks) at values above medium-high risk. Air pollution refers to levels of fine particulate matter—particles with a diameter of 2.5 micrometers or less—exceeding 15 micrograms per cubic meter of air, the World Health Organization's intermediate threshold for unsafe air. Land degradation is defined as areas experiencing a long-term (1980–2022) decline in ecosystem function. These are areas where net primary productivity is declining due to factors other than changes in temperature and rainfall. The factors can include soil erosion, nutrient depletion, loss of biodiversity, deforestation, desertification, or reduced agricultural productivity. This indicator aligns with the Food and Agriculture Organization and International Soil Reference and Information Centre approach (Bai et al. 2008). EAP = East Asia and the Pacific; ECA = Europe and Central Asia; LAC = Latin America and the Caribbean; MENA = Middle East and North Africa; SAR = South Asia; SSA = Sub-Saharan Africa.

adding to the risks faced by local populations and to the challenges in designing effective policies. Addressing one environmental stressor in isolation may provide limited benefits if other overlapping risks—such as poor air quality, unsafe water, and degraded land—persist.

These environmental challenges disproportionately impact developing countries, slowing down progress toward ending poverty and raising living standards. In low-income countries, an unacceptable 79 percent of the population is impacted by all three risks, while less than 1 percent of the inhabitants are free from all three environmental stressors (figure 1.4). In contrast, in high-income countries, 43 percent of the population is not exposed to any stressors, and only 1 percent is impacted by all three. This demonstrates that addressing these environmental challenges is not merely something that would be desirable—it represents smart development.

Reinforcing the point that environmental stressors are particularly burdensome on the poor, figure 1.5 shows the top 10 countries in terms of the share of the population that is both poor and exposed to one of the stressors. In many low-income countries, particularly in Sub-Saharan Africa, large shares of the population are both poor and exposed to stressors. This forcefully demonstrates that the old paradigm that pollution is a necessary evil that comes with industrialization, and diminishes in post-industrialized societies, is no longer true. Many of the countries that are most affected by environmental degradation have not yet industrialized. They face the double burden of poverty and a degraded environment, without the benefits of industrialization for improving living standards.

FIGURE 1.4 Water risks, air pollution, and land degradation disproportionately impact developing country inhabitants

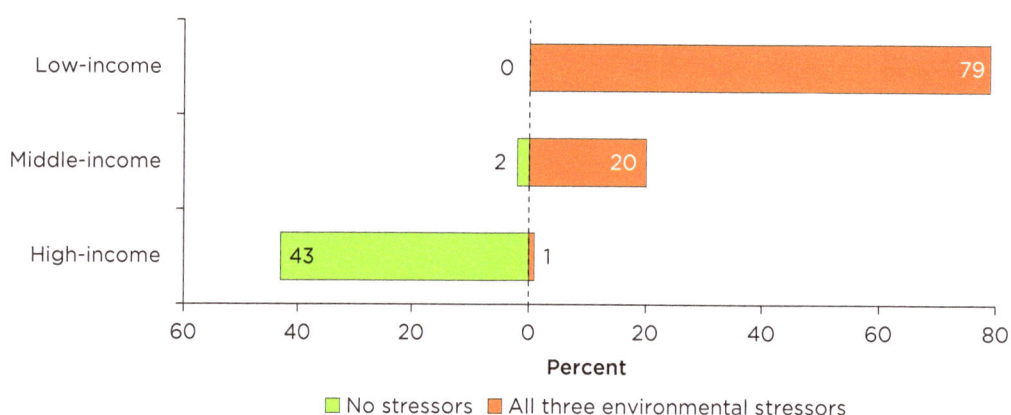

Sources: Water risks, World Resources Institute 2023; air pollution, van Donkelaar et al. 2016; land degradation, Bai et al. 2025.
Note: The figure shows the share of inhabitants in each income group exposed to water risks, air pollution, or land degradation. Definitions of the risks are given in the notes in map 1.1.

FIGURE 1.5 **Poverty and environmental pressures often co-occur: The 10 most-affected countries**

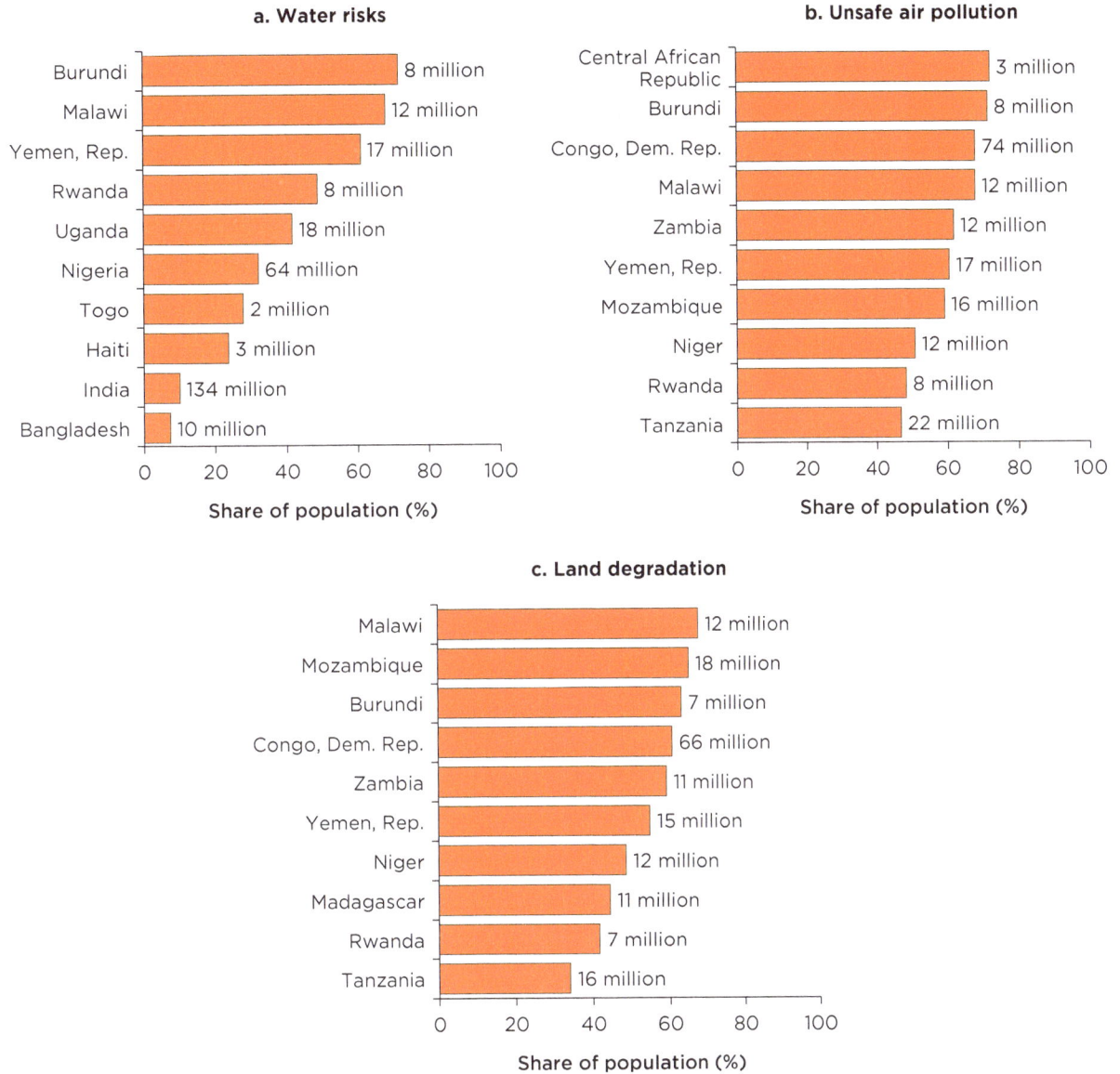

a. Water risks

Country	Share of population (%)	Value
Burundi		8 million
Malawi		12 million
Yemen, Rep.		17 million
Rwanda		8 million
Uganda		18 million
Nigeria		64 million
Togo		2 million
Haiti		3 million
India		134 million
Bangladesh		10 million

Share of population (%)

b. Unsafe air pollution

Country	Share of population (%)	Value
Central African Republic		3 million
Burundi		8 million
Congo, Dem. Rep.		74 million
Malawi		12 million
Zambia		12 million
Yemen, Rep.		17 million
Mozambique		16 million
Niger		12 million
Rwanda		8 million
Tanzania		22 million

Share of population (%)

c. Land degradation

Country	Share of population (%)	Value
Malawi		12 million
Mozambique		18 million
Burundi		7 million
Congo, Dem. Rep.		66 million
Zambia		11 million
Yemen, Rep.		15 million
Niger		12 million
Madagascar		11 million
Rwanda		7 million
Tanzania		16 million

Share of population (%)

Sources: Original calculations for this report. Water risks, World Resources Institute 2023; air pollution, van Donkelaar et al. 2016; land degradation, Bai et al. 2025.
Note: The figure shows the top 10 countries with the highest percentages of their population experiencing both poverty and exposure to environmental stressors. Extreme poverty is defined as living on less than $2.15 per day, adjusted for 2017 purchasing power parity. Definitions of the risks are given in the notes in map 1.1.

The science of biosphere limits

Recent scientific advances show that human activity is encroaching on critical thresholds that define the stability of the Earth's systems. Although there are many ways to assess these environmental limits, several major frameworks broadly align in identifying key areas of concern (box 1.1). The Planetary Boundaries framework defines nine Earth system processes that regulate planetary health and the safe operating limits, beyond which disruptions may occur (Richardson et al. 2023; Rockström et al. 2009) (figure 1.6). Assessments indicate that the Earth has already crossed six of the nine boundaries: biogeochemical flows (nitrogen and phosphorus cycles), freshwater use, land system change, biodiversity loss, climate change, and novel entities (chemical pollution), increasing the risk of destabilizing planetary systems (Richardson et al. 2023; Rockström et al. 2009).

BOX 1.1
Most major frameworks converge on key planetary limits

A growing body of research underscores that humanity is operating beyond critical environmental limits, and multiple frameworks have emerged to assess these risks. Although their methodologies vary, they share a common recognition that degradation of the Earth's life-support systems threatens both economic stability and human well-being.

The Planetary Boundaries framework defines nine Earth system processes that regulate planetary health. The framework establishes thresholds for these processes and defines a "safe operating space" for humanity—an environmental boundary within which human societies can thrive without triggering large-scale ecological disruptions (Richardson et al. 2023; Rockström et al. 2009).

The Doughnut Economics model integrates ecological ceilings with social foundations to create a safe and just operating space for humanity. This framework highlights the need to balance economic development with environmental constraints, ensuring that human societies thrive without exceeding planetary limits (Raworth 2012).

The Planetary Health framework emphasizes the direct links between environmental decline and human well-being. It focuses on how biodiversity loss, pollution, and climate change exacerbate health risks, particularly for vulnerable populations (Whitmee et al. 2015).

Natural capital accounting frameworks quantify environmental degradation as a loss of economic wealth. The Dasgupta Review on the economics of biodiversity and the World Bank's *Changing Wealth of Nations* report assess how ecosystem depletion translates into financial and societal risks, reinforcing the need for sustainable resource management (Dasgupta 2021; World Bank 2024a).

Despite their differences, these frameworks broadly converge on the same urgent planetary challenges. Freshwater scarcity, nitrogen and chemical pollution, land use change, biodiversity loss, air pollution, and climate change are consistently identified as critical threats. The findings of the frameworks underscore the need for policies that safeguard environmental stability while also supporting economic and social well-being.

FIGURE 1.6 Humanity has crossed six of the nine planetary boundaries

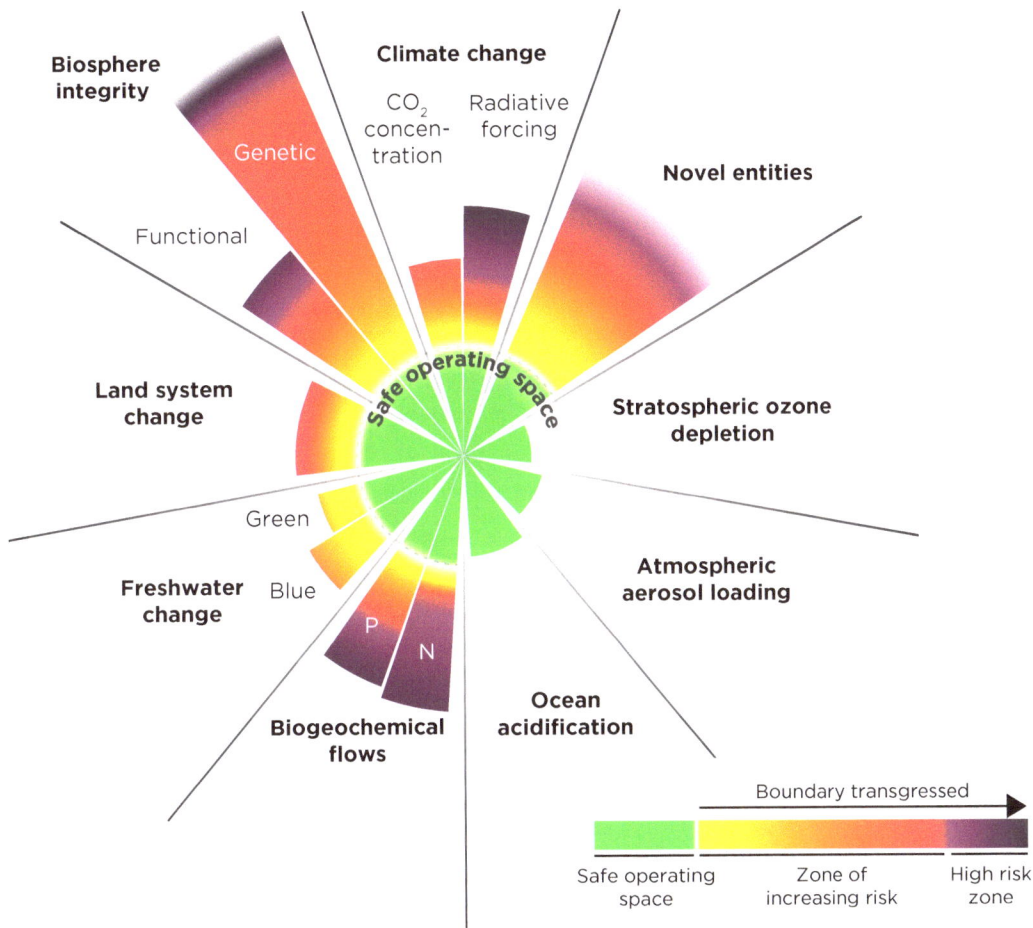

Source: Azote for Stockholm Resilience Centre, based on analysis in Richardson et al. 2023.
Note: CO_2 = carbon dioxide; N = nitrogen; P = phosphorus.

While the Planetary Boundaries framework has been influential in shaping environmental policy and sustainability discussions, it is not without criticism. Some argue that the boundaries, while useful conceptually, lack precise tipping points due to scientific uncertainties and regional variations (Montoya, Donohue, and Pimm 2018). Others highlight that the framework does not fully account for social dimensions, such as inequalities in resource consumption or the disproportionate vulnerabilities of low-income countries (Biermann and Kim 2020). In addition, some planetary processes—such as interactions between multiple boundaries—remain poorly understood, raising questions about how transgressing one boundary may exacerbate risks in others. Despite these limitations, the framework remains a powerful synthesizing tool for describing global environmental risks and guiding policies aimed at keeping the planet within livable limits.

Critical foundations: Revisiting the economics of a livable planet

If planetary boundaries are finite, then achieving lasting prosperity necessitates that economic activities operate within these limits. Traditional economic approaches treat pollution, resource depletion, and ecosystem degradation as externalities or secondary effects of market activity that can be addressed after economic transactions take place. The typical policy response is to internalize these costs through command-and-control approaches such as emissions standards, environmental taxes that penalize resource degradation, or hybrid approaches that combine these two mechanisms, such as emissions or water trading systems (Nordhaus 1994). When carefully designed, such approaches have proven to be effective and impactful. They are most appropriate where there is certainty about the extent of damage and where impacts are local and do not cross boundaries that generate systemic risks that erode the productive capacity of the economy. As the scale and complexity of environmental challenges have grown, it has become evident that there is a need to account for the deeper interactions between nature and the economy.

Embeddedness and externalities

Recognizing that economic production and human well-being depend on maintaining ecological stability implies that the economy is embedded within nature and not separate from it (box 1.2). As noted in the Dasgupta Review, commissioned by the United Kingdom Treasury, all economic production relies on nature's inputs—clean air, water, fertile soils, and a stable climate—and as such, the economy should be viewed as being embedded within nature (Dasgupta 2021).

BOX 1.2

Externalities and embeddedness

The classical environmental economics literature provides invaluable insights to assess the role of natural resources on growth and well-being. In the usual formulations, natural resources impact well-being through two channels: directly for consumption and production, such as fish stocks and timber that may be consumed or used as inputs; and indirectly through externalities, such as pollution and soil erosion that influence the productivity of some other factor of production, such as labor or

(continued)

BOX 1.2
Externalities and embeddedness *(continued)*

land. This relationship can be modeled in a simple production function where natural capital (N) and some other input (say) human capital (H), determine output (y):[a]

$$y = f(N(H), H(N))\qquad\text{(B1.2.1)}$$

The function allows for interdependence, with natural capital affecting the productivity of labor ($H(N)$) and vice versa. For instance, air pollution (a decline in N) is an externality that typically lowers labor productivity ($H(N)$).

In this formulation, the decline in natural capital will not impact growth (much) if natural capital and human capital are perfect (or close) substitutes. In this case, natural capital can be depleted and replaced with human capital, allowing for "weak sustainability." There will be uninterrupted progress even when N is depleted so long as it is replaced by another form of capital. This approach offers helpful guidance when dealing with environmental impacts that are relatively small, or in cases where there are close substitutes for the environmental resource that is being degraded. For example, if a river is dammed and diverted but groundwater is available to farmers, this approach provides useful insights for assessing trade-offs and appropriate policies. However, if groundwater is itself being depleted without reinvestment in other forms of capital, or if its depletion undermines ecological functions—such as sustaining river baseflows—the conditions for weak sustainability may not hold.

When the environmental impacts are of a greater scale, there is a need to acknowledge that the economy dwells in nature and there may be irreplaceable services for which there are no close substitutes. Drawing on developments in renewable resource economics, the Dasgupta Review shows that a seemingly innocuous technical extension of the production function in equation (B1.2.1), which captures embeddedness, has far-reaching implications (Dasgupta 2021). To see why, let S represent the biosphere (or some aspect of it) upon which the economy depends. Then a simple way of incorporating "embeddedness" is to include S as follows:

$$y = AS^{\eta} f(N(H), H(N))\ (\eta > 0)\qquad\text{(B1.2.2)}$$

This formulation implies that as the biosphere (S) depletes, so does economic output (y). In a dynamic formulation, any gains from exploiting the environment will eventually be reversed. Technological progress (A) can delay this outcome but cannot fully disconnect the economy from the biosphere.[b] When S is damaged, there is a need for both passive and active investments in ecosystems to restore S and hence potential output. This formulation of the basic economic problem is more closely aligned with the insights of the science literature on planetary boundaries.

a. It is assumed that the usual regularity conditions apply for an interior solution. Resource dynamics are suppressed for expositional purposes.

b. Formally, as $S \to 0$, then $y \to 0$.

Imperfect substitutability

Economies' dependence on natural capital would be of limited significance if there were close substitutes for the life-sustaining services provided by nature. Economic models often assume that if one resource becomes scarce, a substitute will emerge. But unlike many human-made goods, natural systems typically lack viable substitutes. As a result, the economic costs associated with the depletion of natural capital escalates as ecological services become scarcer. Clean air, fresh water, fertile soils, and other services provided by nature cannot be easily replaced. Although a face mask or filter may alleviate some of the harmful impacts of air pollution, there is no known substitute for breathable air. Similarly, forests and wetlands provide myriad services that no engineered system can fully replicate at scale. With weak substitution, depletion of resources and threats of critical scarcity or tipping points in the future would translate into greater consideration of the depletion of natural capital. Equally important is the need to understand the often-invisible complementarities between natural capital and other assets—such as how intact ecosystems enhance agricultural productivity, reduce disaster risks, or improve human health. These complementarities are frequently overlooked in economic models, yet they are critical in practice. The way in which this is to be achieved remains among the greatest theoretical, empirical, and practical challenges in economics. Because environmental services are irreplaceable, their degradation poses systemic and sometimes existential risks, often disproportionately affecting the poor (box 1.3).[3]

BOX 1.3
Poverty, prosperity, and the planet

The World Bank's 2024 *Poverty, Prosperity, and Planet Report* highlights troubling trends in poverty, inequality, and climate change (World Bank 2024b). Global poverty reduction has stalled, making the 2020s a lost decade for progress. Although extreme poverty fell from 38 to 8.5 percent between 1990 and 2024, recent setbacks—due to economic stagnation, conflict, and climate shocks—have slowed gains. Nearly 700 million people live on less than $2.15/day, and 3.5 billion fall below $6.85/day. Without major policy changes, lifting people out of poverty could take decades. Furthermore, inequality remains widespread. Although fewer countries have extreme income gaps, 1.7 billion people—mainly in Sub-Saharan Africa and Latin America—live in high-inequality economies. Achieving a global prosperity standard of $25/day would require a fivefold increase in average income. Post-COVID-19, rising income gaps and slow growth have made upward mobility more difficult.

Climate risks are rising, hitting the poorest hardest. One in five people face high risks from extreme weather, especially in South Asia and Sub-Saharan Africa. Poor countries, with weak infrastructure and services, are highly vulnerable. Undernourishment is rising, and global emissions—now trapping 50 percent more heat than in 1990—worsen climate threats.

The world faces a polycrisis of economic strain, debt, and climate risks. High debt in poor countries diverts funds from development, weakening resilience. Without international cooperation—especially as richer countries cause 80 percent of emissions—progress on poverty and climate goals will falter.

(continued)

Reboot Development

BOX 1.3
Poverty, prosperity, and the planet *(continued)*

Balancing poverty reduction with climate action is possible. The report shows that lifting people above extreme poverty would only slightly raise emissions. However, raising more than 3 billion people above $6.85/day would increase emissions more significantly. Synergistic policies—like investing in renewables, energy efficiency, and climate-smart agriculture—can reduce the trade-offs. But these require better incentives, training, and support to scale. Targeted aid, education, and market access are crucial for success.

Source: World Bank (2024b).

Uncertainty and safe limits

Humankind does not fully understand where the limits of ecological stability lie, but it is clear that exceeding them could have catastrophic consequences. Traditional cost-benefit analysis assumes that risks can be quantified, and policy decisions should be based on expected costs and benefits. However, environmental risks often involve deep uncertainty, where probability distributions are unknown and damages could be extreme (Stern, Stiglitz, and Taylor 2022; Weitzman 2009). The concept of planetary boundaries highlights that beyond certain ecological thresholds, feedback loops may trigger irreversible and potentially catastrophic changes (Rockström et al. 2009). The difficulty is that there is uncertainty about where exactly these thresholds lie. As a result, there should be science-determined safe limits to avoid high-risk scenarios (Sureth et al. 2023).

Irreversibility and the value of options

Environmental damages can be irreversible, and failing to act can lead to permanent loss. Economics has long recognized the value of preserving options in the face of irreversibility, often referred to as *option value* (Arrow and Fisher 1974). This is especially relevant in environmental policy, where crossing certain ecological thresholds can trigger irreversible change. For example, biodiversity loss is not just about species extinction; rather, it represents the degradation of ecosystems that provide essential services, such as pollination, water purification, and climate regulation. By the time the full consequences of these losses become apparent, they may be impossible to reverse. Lock-ins and path dependence represent a milder form of irreversibility, where systems become entrenched in specific pathways due to historical decisions and reinforcing mechanisms, yet there remains potential for change, albeit at a high cost. Just as in financial market analyses, including option values in environmental decision-making provides a way of formalizing and incorporating the risks of decisions that are irreversible, and the incorporation of such risks will often lead toward precautionary approaches.

Taken together, these insights call for a shift in the way environmental consequences are assessed and addressed. Rather than treating nature as a passive backdrop to economic activity, policy makers must recognize its central role in the economy, when this is relevant and material. This recognition requires moving beyond marginal adjustments and embracing changes that align economic incentives with ecological limits, to decouple economic prosperity from natural capital degradation (refer to spotlight 1).

To address environmental challenges, countries will need to weigh the costs and benefits and recognize trade-offs (box 1.4). Not all environmental problems may warrant intervention, at least when evaluated through strict economic cost-benefit assessments. A challenge in assessing the trade-offs is that environmental benefits are often nonmonetizable, unmeasured, and in many cases still unknown or poorly understood. This complexity implies that a straightforward comparison of costs against a limited accounting of benefits can lead to misleading conclusions. To make appropriate choices, recognizing the inherent uncertainties in measuring environmental impacts is vital. In such cases, it may be prudent to operate within the safe zone or "guardrails" determined by science to manage risks effectively when data are incomplete or ambiguous.

BOX 1.4
Trade-offs between economic growth and ecological systems

New data and techniques are providing much more detailed evidence of the magnitude and form of environmental trade-offs that often occur. Examples from this report include the following:

- *The costs of deforestation-induced changes in rainfall and soil moisture.* New estimates presented in chapter 2 suggest that the depletion of freshwater resources due to deforestation may cost more than $14 billion in annual gross domestic product losses due to reductions in rainfall, and nearly $380 billion annually due to reductions in soil moisture. Adding this to the additional costs of deforestation, in terms of ecosystem services, may alter the balance between the costs and benefits of deforestation.

- *The multifaceted costs of nitrogen.* Estimates presented in chapter 4 indicate that the cost of nitrogen pollution and its resulting impacts on ecosystems and human health range from $300 billion to $3,400 billion (depending on the region), while nitrogen pollution abatement would cost around $10 billion. Although the chasm between the benefits and financial costs of abatement demonstrates the economic viability of such solutions, the costs come in the form of externalities and public bads, so the high benefit-cost ratios do little to incentivize change.

- *The high health and economic costs of air pollution.* Estimates discussed in chapter 5 suggest that air pollution costs around $8 trillion per year in health and productivity. Recent trials from

(continued)

BOX 1.4
Trade-offs between economic growth and ecological systems *(continued)*

India show that interventions costing less than $4 million per year provided annual benefits exceeding $100 million, and that international cooperation on air pollution management in South Asia can dramatically lower pollution abatement costs from $2.6 billion to $278 million per microgram of fine particulate matter per cubic meter.

- *The job creation opportunities of the green transition.* The shift to a greener economy is not just an environmental imperative; it can also be a significant driver of job creation. Estimates presented in chapter 9 suggest that investing in the forestry, renewable energy, sustainable agriculture, and circular economy sectors generates more jobs per dollar spent than investing in more polluting industries. But while green jobs offer economic opportunities, they also require workforce transitions that must be actively managed. Without proper planning, workers in high-emissions industries could face job displacement, reinforcing the need for reskilling programs, labor protections, and social policies that ensure an equitable transition.

Even when the benefits of environmental action far outweigh the costs, implementation is often hindered by the mismatch between private costs and public benefits, as the above examples illustrate. Many environmental policies and interventions generate broad societal gains but require upfront investment from industry, business, or government. With the costs often concentrated among a few actors and the benefits widely shared, there is little financial incentive for private entities to act voluntarily. Overcoming this barrier requires well-designed policies—such as market-based mechanisms, regulations, or subsidies—to realign incentives and ensure that those who bear the costs of action are rewarded for the broader benefits they help create.

Cost-benefit analysis also struggles with the complexity of ecological systems, which are often nonlinear. For instance, the services of a forest ecosystem are greater than the sum of the trees (or land) that comprise that forest—a phenomenon termed super-additivity. Suppose that part of the forest is to be converted for a development project. Assessing the costs and benefits of losing a small part of the forest fails to account for how this change impacts future cost-benefit analyses of further land conversion. For instance, if the first forest conversion reduced ecological values, it would render some other project—that generated lower benefits than the first—worth pursuing. This has two far-reaching implications. First, it implies that conversion of the remaining forests would be justified by cost-benefit analysis at a lower threshold of development benefits. Second, if the entire forest were converted to farmland, the lost ecological values may exceed the development benefits of conversion. In this case, the simple cost-benefit intuition breaks down due to time inconsistency.[4]

The structure of this report and the way forward

The report is divided into three parts: the economic stakes of planetary health; the environmental impact of two defining twenty-first century trends, cities and commerce; and finally, possible solutions. In part 1, chapters 2 to 5 explore the trends of critical natural capital, focusing on topics that cover new science and economics, emphasizing interlinkages between different forms of natural capital. In part 2, chapters 6 and 7 examine the environmental impacts of two forces of the twentieth and twenty-first centuries: increasing urbanization and interconnectivity through trade. In part 3, chapters 8 and 9 discuss crosscutting policies, green jobs, and other solutions for achieving a sustainable future.

Environmental risks are no longer distant concerns; they are unfolding in real time, and their consequences are already impacting economies, ecosystems, and human well-being. The evidence presented in this report underscores the urgent need for bold policy action, highlighting that many solutions are not just necessary, but also highly cost-effective. Addressing air pollution, restoring degraded land, and improving water management have been shown to generate economic benefits that far exceed their costs, reinforcing the idea that sustainability and prosperity can go hand in hand. But overcoming barriers to action requires rethinking traditional economic models that treat environmental degradation as an externality, and instead embedding planetary health at the core of decision-making.

A new paradigm is needed to align economic development with environmental sustainability, rather than treat them as opposing forces. This requires environmental considerations to be integrated into financial systems, regulatory frameworks, and technological innovation to ensure long-term resilience. Those designing markets, policies, and incentives must ensure that they reflect the true value of natural resources and the true cost of environmental degradation, encouraging sustainable investments and innovation. This report explores how targeted policy interventions, governance mechanisms, and market-based solutions can drive this transition, ensuring that future generations inherit a world that is both economically prosperous and ecologically stable.

Notes

1. https://www.who.int/data/gho/data/themes/topics/water-sanitation-and-hygiene-burden-of-disease.
2. Water risks, determined by the World Resources Institute's (2023) water risk index, include risks to physical quantity, risks to physical quality, and regulatory and reputational risks. The last category correlates highly with risks to physical quality (correlation coefficient of ~0.7 when correlated at the grid cell level).
3. This reflects the logic of *strong sustainability*, which recognizes that natural capital and other forms of capital are not easily interchangeable. When asset classes are highly complementary, such as ecological systems and human well-being, the benefits of hedging through investment in unrelated assets (for example, education or infrastructure) diminish. In contrast, *weak sustainability* assumes a greater degree of substitutability across assets, allowing losses in

one area to be offset by gains in another. The strong sustainability framework is more applicable where natural systems provide irreplaceable services and where crossing environmental thresholds can result in irreversible damage.

4. A numerical example may help fix ideas. Consider the case of two contiguous parcels of forested land, P1 and P2. Assume that the ecological services these provide are super-additive. This may occur if the benefits generated by one parcel of forested land depend on the existence of other forest patches. Hence, converting parcel P1 to farmland reduces the environmental values flowing from the remaining parcel P2.

    Let $b(P1,P2) = \$100$ be the ecological benefits of the *combined forest* when both parcels of land are forested. Let $b(P1(P2)) = \$20$ be the benefits from plot P1, when P2 has *not* been converted to some other use. Let $b(P2(-P1)) = \$10$ be the benefits from plot P2 when P1 *has been converted* to some other use.

    Suppose there is a proposal to convert P1 to farmland, yielding profits of $D1 = \$40$. By conventional cost-benefit analysis, the conversion is warranted as $D1 = \$40 > b(P1(P2)) = \$20$. Subsequently, a second proposal emerges to convert plot P2 to farmland. This proposal generates profits of $D2 = \$30$. This conversion will also be deemed to be acceptable as $D2 = \$30 > b(P2(-P1)) = \$10$. But note that such sequential cost-benefit comparisons have led to a net loss to society, since by super-additivity, $b(P1,P2) = \$100 > D1 + D2 = \$70$. There is an interdependence between parcels, which means that the simple cost-benefit intuition breaks down. Since spatial interdependencies are widespread in natural systems such as rivers and rainforests, it must imply that choices based on piecemeal comparisons have led to inappropriate decisions. Martin and Pindyck (2015) provide analogous reasoning.

References

Arrow, K. J., and A. C. Fisher. 1974. "Environmental Preservation, Uncertainty, and Irreversibility." *Quarterly Journal of Economics* 88 (2): 312–19.

Bai, Z. G., D. Dent, L. Olsson, and M. E. Schaepman. 2008. "Proxy global assessment of land degradation." *Soil Use and Management* 24(3): 223–234.

Bai, Z., J.D. Russ, K. F. Mayr, and D. Dent. 2025. "How is Gaia doing? Trends in global land degradation and improvement." *Ambio*, 1–37.

Biermann, F., and R. E. Kim. 2020. "The Boundaries of the Planetary Boundary Framework: A Critical Appraisal of Approaches to Define a 'Safe Operating Space' for Humanity." *Annual Review of Environment and Resources* 45: 497–521.

Dasgupta, S. P. 2021. *The Economics of Biodiversity: The Dasgupta Review*. Cambridge, UK: Cambridge University Press.

Elhacham, E., L. Ben-Uri, J. Grozovski, Y. M. Bar-On, and R. Milo. 2020. "Global Human-Made Mass Exceeds All Living Biomass." *Nature* 588 (7838): 442–44.

Elmqvist, T., C. Folke, M. Nyström, G. Peterson, J. Bengtsson, B. Walker, and J. Norberg. 2003. "Response diversity, ecosystem change, and resilience." *Frontiers in Ecology and the Environment* 1(9): 488–494.

Folke, C., S. Carpenter, B. Walker, M. Scheffer, T. Elmqvist, L. Gunderson, and C. S. Holling. 2004. "Regime shifts, resilience, and biodiversity in ecosystem management." *Annual Review of Ecology, Evolution, and Systematics* 35(1): 557–581.

Greenspoon, L., E. Krieger, R. Sender, et al. 2023. "The Global Biomass of Wild Mammals." *Proceedings of the National Academy of Sciences of the United States of America* 120 (10): e2204892120.

Hansen, M. C., P. V. Potapov, R. Moore, et al. 2013. "High-Resolution Global Maps of 21st-Century Forest Cover Change." *Science* 342 (6160): 850–53.

Li, Y., C. Dann, and M. Kompil. 2025. *Transport Infrastructure and the Nature-Growth Trade-Off: A Comparative Perspective.* Available at SSRN: http://dx.doi.org/10.2139/ssrn.5311766.

Martin, I. W. R., and R. S. Pindyck. 2015. "Averting Catastrophes: The Strange Economics of Scylla and Charybdis." *American Economic Review* 105 (10): 2947–85.

Montoya, J. M., I. Donohue, and S. L. Pimm. 2018. "Planetary Boundaries for Biodiversity: Implausible Science, Pernicious Policies." *Trends in Ecology and Evolution* 33 (2): 71–73.

Nordhaus, W. 1994. *Managing the Global Commons: The Economics of Climate Change.* Cambridge, MA: MIT Press.

Pauly, D. 1995. "Anecdotes and the Shifting Baseline Syndrome of Fisheries." *Trends in Ecology & Evolution* 10 (10): 430.

Raworth, K. 2012. *A Safe and Just Space for Humanity: Can We Live Within the Doughnut?* Oxford, UK: Oxfam.

Richardson, K., W. Steffen, W. Lucht, et al. 2023. "Earth beyond Six of Nine Planetary Boundaries." *Science Advances* 9 (37): eadh2458.

Rockström, J., W. Steffen, K. Noone, et al. 2009. "A Safe Operating Space for Humanity." *Nature* 461 (7263): 472–5. doi:10.1038/461472a.

Soga, M., and K. J. Gaston. 2018. "Shifting Baseline Syndrome: Causes, Consequences, and Implications." *Frontiers in Ecology and the Environment* 16 (4): 222–30.

Stern, N., J. Stiglitz, and C. Taylor. 2022. "The Economics of Immense Risk, Urgent Action and Radical Change: Towards New Approaches to the Economics of Climate Change." *Journal of Economic Methodology* 29 (3): 181–216.

Sureth, M., M. Kalkuhl, O. Edenhofer, and J. Rockström. 2023. "A Welfare Economic Approach to Planetary Boundaries." *Jahrbücher für Nationalökonomie und Statistik* 243 (5): 477–542.

van Donkelaar, A., R. V. Martin, M. Brauer, et al. 2016. "Global Estimates of Fine Particulate Matter Using a Combined Geophysical-Statistical Method with Information from Satellites, Models, and Monitors." *Environmental Science & Technology* 50: 3762–72. doi:10.1021/acs.est.5b05833.

Walker, B., A. S. Crépin, M. Nyström, J. M. Anderies, J. Andersson, E., T. Elmqvist, C. Queiroz. 2023. "Response diversity as a sustainability strategy." *Nature Sustainability* 6(6): 621–629.

Weitzman, M. L. 2009. "On Modeling and Interpreting the Economics of Catastrophic Climate Change." *Review of Economics and Statistics* 91 (1): 1–19.

Whitmee, S., A. Haines, C. Beyrer, et al. 2015. "Safeguarding Human Health in the Anthropocene Epoch: Report of the Rockefeller Foundation–Lancet Commission on Planetary Health." *The Lancet* 386 (10007): 1973–2028.

WHO (World Health Organization). 2023. *World Health Statistics 2023: Monitoring Health for the SDGs.* Geneva: WHO.

Wollburg, P., S. Hallegatte, and D. G. Mahler. 2023. "Ending Extreme Poverty Has a Negligible Impact on Global Greenhouse Gas Emissions." *Nature* 623 (7989): 982–86.

World Bank. 2024a. *The Changing Wealth of Nations 2024: Revisiting the Measurement of Comprehensive Wealth.* Washington, DC: World Bank. http://documents.worldbank.org/curated/en/099100824155021548/P17844617dfe6e0241ad25120b1320904c2.

World Bank. 2024b. *Poverty, Prosperity, and Planet Report 2024: Pathways Out of the Polycrisis.* Washington, DC: World Bank.

Decoupling Environmental Degradation from Economic Growth

Introduction

Is it possible to grow an economy without degrading its natural resources? Views on this fundamental question vary widely. Optimists note that some countries have slashed pollution, improving public health and the environment, while simultaneously growing their economies. Such success is most often observed in high-income countries (HICs) with stringent environmental regulations and the resources to monitor and enforce compliance. A dissenting view points to evidence that such outcomes are not universal, do not happen fast enough, and are not an assured consequence of economic progress.

This spotlight casts light on this debate by examining the data on the links between economic growth and environmental degradation across available indicators for air, land, and water. It uses standard decomposition methods—mathematical techniques—to break down changes in resource use to identify the underlying drivers of change, which can inform policies.

The links between economic growth and environmental degradation are often described in terms of *decoupling*. This refers to whether economic growth has been accompanied by corresponding increases in environmental degradation. There are three types of decoupling (figure S1.1). *Absolute decoupling* occurs when economic activity increases and negative environmental impacts decline or stay constant. This is the most desirable outcome if there is a decline in the negative effects, and it happens fast enough. *Relative decoupling* occurs when environmental damage continues to increase, but at a slower rate than economic growth. As a result, the environmental damage per unit of output decreases, and this is sometimes viewed as a mark of progress in achieving environmental goals. *Negative decoupling* is the most concerning situation, where environmental degradation accelerates at a faster pace than economic growth, and it may indicate that economic activities exacerbate environmental degradation.

The decoupling framework

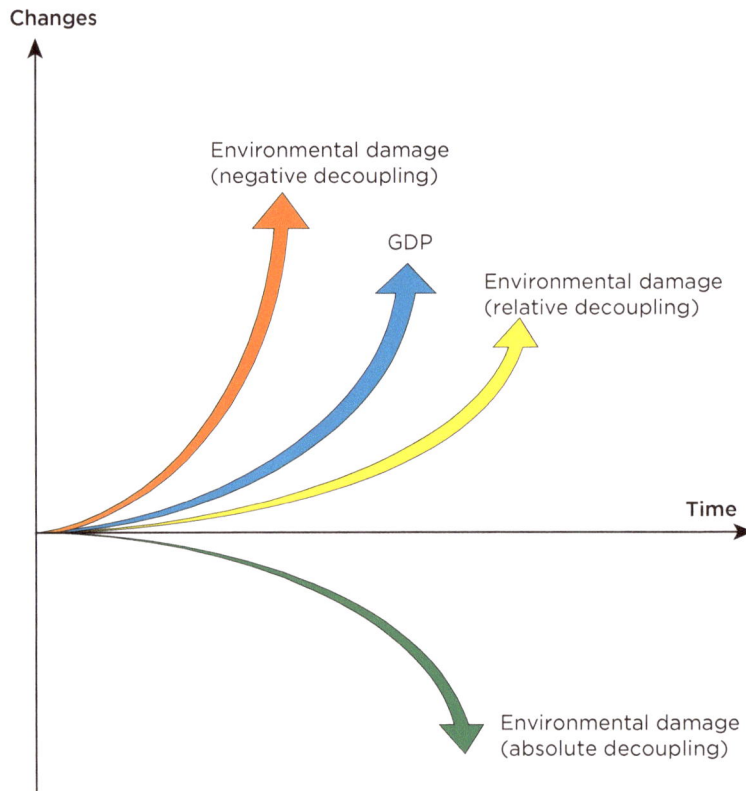

Source: Original figure for this report.
Note: GDP = gross domestic product.

For environmental damage to decline, absolute decoupling is necessary, but achieving absolute decoupling alone may not be enough to ensure sustainability. If there are thresholds beyond which environmental damage is irreversible, then sustainability calls for reducing environmental damage before irreversible tipping points are transgressed. In this case, it is important to consider both the speed of absolute decoupling and the cumulative damage that has already occurred. Box S1.1 identifies the five conditions necessary for decoupling to be effective.

BOX S1.1
The Five-S decoupling framework

For decoupling to sustain a livable planet, it needs to be achieved at sufficient *size*, across a wide *scope* of environmental harms, at large geographic *scale*, at the right *speed*, and over a long time *span* (Parrique et al. 2019). To address global environmental harms, this report proposes the Five-S decoupling framework—representing size, scope, scale, speed, and span—to assess whether economic growth is environmentally harmful, benign, or beneficial.

(continued)

BOX S1.1

The Five-S decoupling framework *(continued)*

1. *Size.* Decoupling must be absolute to reduce adverse impacts. With relative decoupling, environmental harm grows, albeit more slowly than gross domestic product. This implies that dangerous planetary boundaries or safe thresholds will eventually be transgressed. Relative decoupling could be viewed as an interim sign of progress, but it could also represent a business-as-usual outcome.

2. *Scope.* Decoupling needs to occur across the range of environmental harms. This is necessary because the Earth's systems are interconnected. For example, the loss of keystone predators cascades through an ecosystem, resulting in changes in herbivory (or grass-eating), which in turn alter vegetation, hydrological systems, and rainfall.

3. *Scale.* Global environmental harms require global decoupling. If only some countries or regions decouple while others do not, it can result in no net reduction of environmental pressures. This is a particular concern when dealing with offshoring of pollution to low- and middle-income countries, or in the case of global public goods, such as rainforests, oceans, and climate.

4. *Speed.* Decoupling needs to happen fast enough to avoid exceeding ecological limits. For example, although 11 high-income countries achieved absolute decoupling of greenhouse gas emissions between 2013 and 2019, it has been argued that at the current speed it would still take more than 220 years for these countries to reduce their emissions by 95 percent (Vogel and Hickel 2023).

5. *Span.* Decoupling needs to persist over sustained periods of time. There are various instances where decoupling has occurred over limited periods, followed by recoupling. For example, global material use decoupled after the 1970s but has since recoupled.

For effective decoupling, countries must avoid falling into a cycle that fails to meet the conditions of the Five-S decoupling framework.

Source: Grinsted and Menzies (2024).

Relative decoupling does not automatically lead to environmental improvements; instead, it merely signifies that environmental degradation is not escalating at the same pace as economic activity. Relative decoupling is widespread and can occur *without* deliberate policies to address environmental damages in rapidly growing and diversified economies. For example, if the cleaner sectors of an economy grow faster than the dirtier ones, overall economic growth may exceed the increase in pollution generated by the dirty sector—resulting in relative decoupling. This outcome reflects a mechanistic relationship in which the elasticity of emissions with respect to growth is less than one,[1] and it may have little to do with environmental policies. Relative decoupling may also occur in sectors with large "rebound" effects. For example, if improvements in energy efficiency make energy bills cheaper, it could spur increased consumption, partially negating the anticipated energy savings.

TABLE S1.1 Decoupling trends, by country income group, 2000–2023

	HICs	UMICs	LMICs	LICs	Global
PM$_{2.5}$ emissions	A	R**	R	R	A*
GHG emissions	A	R	R	R	R
Agricultural water	A	R	R	R	A*
Agricultural land	A	R	R	R	R
Metals	R	R	R	N	R
Non-metallic minerals	A	R	R	N	R

Sources: Original calculations based on data from the Electronic Data Gathering, Analysis, and Retrieval database; World Development Indicators; Annual United Nations Environment Programme International Resource Panel 2024; Inter-Sectoral Impact Model Intercomparison Project 2a and 2b.
Note: A = absolute decoupling; R = relative decoupling, N = negative decoupling. Data availability is as follows: PM$_{2.5}$ emissions, 221 countries; GHG emissions, 210 countries; agricultural water, 196 countries; agricultural land, 198 countries; metals and minerals, 171 countries. The data for PM$_{2.5}$ emissions and water span 2000–22; the data for the other indicators span 2000–23.
GHG = greenhouse gas; HICs = high-income countries; LICs = low-income countries; LMICs = lower-middle-income countries; PM$_{2.5}$ = fine particulate matter; UMICs = upper-middle-income countries.
* The identified decoupling has occurred in the past few years; for example, decoupling of PM$_{2.5}$ reflects China's recent success in bringing down pollution.
** The scale of decoupling is narrow; for example, the observed decoupling of PM$_{2.5}$ in UMICs reflects China's efforts.

Most environmental harms are undergoing relative decoupling, with some showing absolute decoupling in HICs and others showing negative decoupling in low-income countries (LICs). Table S1.1 provides a snapshot of the observed decoupling trends using the available pressure indicators for fine particulate matter (PM$_{2.5}$) or ambient air pollution, greenhouse gas (GHG) emissions, land use, water abstraction, and metal and nonmetal mineral extraction (as proxies for nonrenewable resource use).

Table S1.1 suggests three broad decoupling trends:

1. Across most of the indicators, there is relative decoupling at the global level. In some cases, this may reflect actual progress in reducing the growth of environmental damage; but in others, it is likely a consequence of the mechanical relationship between the distribution of growth across clean and dirty sectors. The scale of decoupling can also be somewhat restricted. For example, the decoupling of PM$_{2.5}$ in upper-middle-income countries (UMICs) reflects China's notable success in its "war on pollution."

2. Absolute decoupling is rare and typically encountered in HICs. This is perhaps unsurprising as fixing environmental problems requires complex policies and regulations, the institutional capacity to enforce them, and the resources to discover and adopt new and cleaner technologies. Thus, it is easier for wealthier economies to transition to greener and cleaner growth paths.[2] The implication is that economic growth is not necessarily incompatible with reducing environmental impacts and

may even help countries green their economies. The challenge is to ensure that the prevailing policies and incentives are effective enough to enable the world to decouple before critical thresholds are transgressed.

3. There are notable differences in decoupling trends between global and intergenerational externalities (such as GHGs or forest loss[3]) on the one hand, and local externalities (such as $PM_{2.5}$) on the other. The former create familiar incentives for countries to free-ride on the policy efforts of other countries or future generations, in the absence of global commitments. In contrast, when the damage is local and immediate, corrective actions have more pressing political salience, rendering policy attention more urgent.

Concerns about material use are somewhat different and relate to the potential exhaustion of the nonrenewable resources that are necessary for economic growth (box S1.2). Since 2000, the fastest growth in demand has been for non-metallic minerals, such as sand, gravel, limestone, and clay. These are core inputs for cement production, which has increased at almost twice the global rate of gross domestic product growth (figure S1.2). This fast growth reflects the increasing material intensity of modern economies, particularly in infrastructure expansion and urbanization. At the same time, there has been negative decoupling of both metal and nonmetal resource use, notably in LICs. These economies are highly dependent on natural resource extraction, often due to familiar "resource curse" challenges.[4] Of the 26 LICs, 17 are affected by conflict or fragility (Mawejje 2025), which makes economic diversification more challenging by deepening their reliance on "point resources" (Ross 2004).

BOX S1.2
Resource depletion: Is there reason for concern?

Since the 1970s, oil prices have spiked at least once every decade. The first spike, in 1973, was a result of the Organization of the Petroleum Exporting Countries' oil embargo; in 1979–80, it was the Iranian revolution; and in 1990, 2011, and 2022, the Gulf War, Arab Spring, and invasion of Ukraine, respectively. Each episode has been accompanied by headlines arguing that the world had reached "peak oil," which would lead to inevitable shortages and chaos.

New oil reserves have been discovered after each price shock, and the predicted shortages have not ensued. Yet, the past may not be an accurate guide to the future and there are legitimate fears that the demand for resources—especially for nonrenewable resources—will eventually outstrip finite supplies. It is therefore paradoxical that there is no known case of the world having exhausted an economically valuable mineral (nonrenewable natural resource), while the extinction, overharvesting, and exhaustion of common-property renewable resources, such as clean air and water, are a widespread problem.

The reasons for this paradox are well known and widely documented. As they are privately owned and marketed, the demand for and extraction of minerals and subsoil assets are guided

(continued)

Resource depletion: Is there reason for concern? *(continued)*

by prices. When prices rise to signal scarcity, consumption of the commodity declines and there is greater investment in exploration and the search for substitutes, which serve to lower depletion rates.

No such signals limit and guide the use of clean air, clean water, and other common-property renewable resources, even when the impacts of their exhaustion can be growth-constraining or life-threatening. A combination of factors, often associated with open access, the lack of property rights, and the absence of price signals, makes renewable resources especially vulnerable to overuse and depletion.

Although imperfect property rights can lead to resource depletion, it would be highly misleading to assume that privatization is the policy panacea that will ensure sustainability. A significant literature explains why establishing property rights cannot completely and solely resolve problems of overextraction. For example, if a natural resource—such as old-growth forests or a population of whales—grows at a slow rate, it would be profitable for a private investor to liquidate that resource and invest the proceeds in a higher-yielding asset, making extirpation the more profitable strategy (Clark 1973).

FIGURE S1.2 Global mineral use has surged over the past 30 years, indexed to 1990

Mineral use (indexed to 1990)

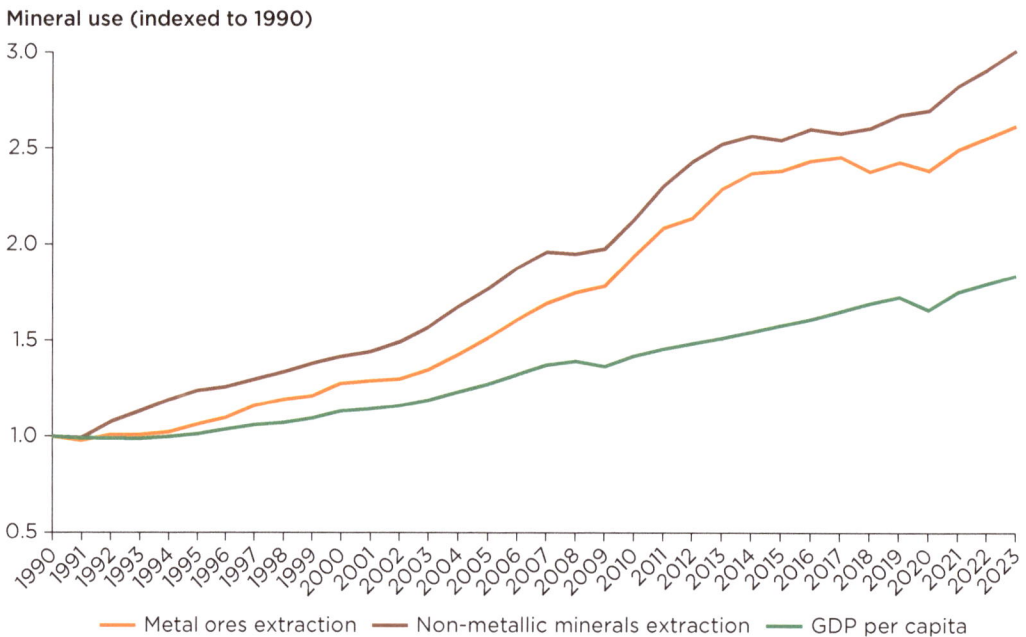

Metal ores extraction · *Non-metallic minerals extraction* · *GDP per capita*

Sources: Original calculations based on data from the United Nations Environment Programme's International Resource Panel; World Bank (GDP).
Note: GDP per capita is in constant 2015 US$. GDP = gross domestic product.

Understanding the factors that drive decoupling can guide the policies necessary for achieving absolute decoupling at a global scale. For a better understanding of these factors, the rest of this spotlight presents the results of a decomposition of the drivers of renewable resource use. The analysis uses formal decomposition methods to assess how the decoupling of renewable resources has been achieved in certain economies.

Decomposition of trends in air, land, and water

The decomposition framework illustrated in figure S1.3 can be used to describe how changes in economic activity impact natural assets through (1) the scale of economic activity, (2) the composition of economic activity, and (3) the efficiency of production. As countries expand, they use more resources and pollute more, so environmental impacts emerge from the scale effect. The composition of economic activity is also important, as some activities cast a wider environmental footprint than do others. For example, consumer purchases of animal proteins will generate different and wider patterns of land use and emissions than equivalent expenditure on plant-based proteins. Finally, cleaner (dirtier) production techniques that result in greater (lesser) efficiency can mitigate (aggravate) the impacts of greater scale. For example, pollution abatement activities or green technologies can help reverse adverse environmental effects.

The combination of the three factors—*scale, composition,* and *efficiency* effects—determines whether environmental damage declines or increases with economic activity. When the composition and efficiency effects partly offset the scale effects, pollution per unit of output will decline, even in cases where total pollution increases with growth. Because these channels fully determine environmental impacts, identifying the direction and magnitude of each allows countries to target their policies more effectively. This spotlight presents decompositions for air pollution; agricultural land use, which is the major driver of deforestation and biodiversity loss globally;

FIGURE S1.3 The decomposition framework

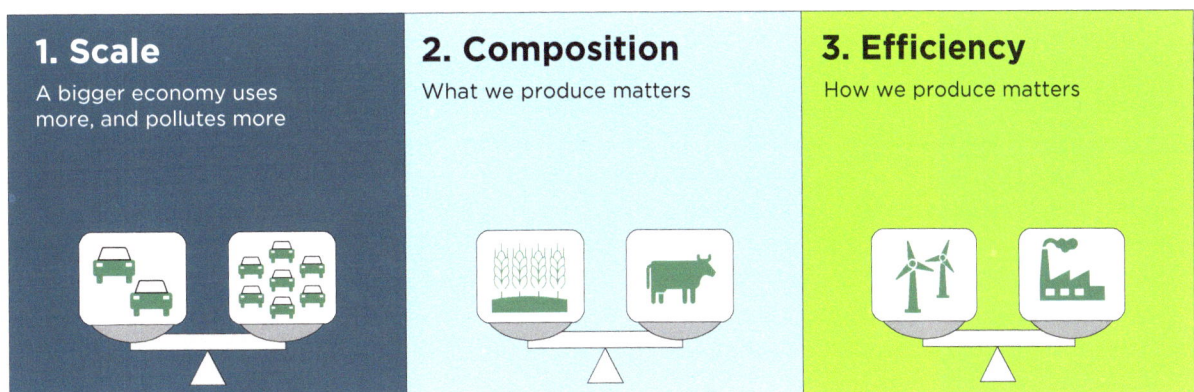

1. Scale	2. Composition	3. Efficiency
A bigger economy uses more, and pollutes more	What we produce matters	How we produce matters

Source: Original figure for this report.

and water abstraction for agricultural and nonagricultural uses. It does not cover metals and nonmetals as there are insufficient data on the uses of minerals to provide informative decompositions. Box S1.3 provides details of the decomposition method, data used, and time period of the analysis.

BOX S1.3
The decomposition method

The Logarithmic Mean Divisia Index (LMDI) decomposition method is used to analyze changes in environmental pressures. Total emissions or damage (E) can be decomposed into several factors, including the scale of economic activity (Q), efficiency of economic activity (I), and composition of economic activity (S), such that:

$$E = \sum_i Q I_i S_i \qquad (S1.3.1)$$

where $I_i = \dfrac{E_i}{Q_i}$ and $S_i = \dfrac{Q_i}{Q}$ for i = 1, ... n are sectors of the economy.

The LMDI method uses the logarithmic mean to decompose the change in emissions between two periods (0 and T). The change in emissions can be expressed as: $\Delta E = E^T - E^O$. The LMDI decomposition formula for each factor $K = Q, S, I$ is given by:

$$\Delta E_k = \sum_i \frac{E_i^T - E_i^O}{\ln E_i^T - \ln E_i^O} \ln \frac{K^T}{K^O} \qquad (S1.3.2)$$

An advantage of the LMDI method is that it yields a perfect decomposition in that the sum of the contributions from all factors equals the total change in emissions. In addition, there is consistency in aggregation since the method can be consistently applied to different levels of aggregation (for example, sectoral, national, and so on). The LMDI method does not account for the interactions between these forces. For instance, an increase in production due to higher efficiency (lower marginal cost) would be captured in the scale effect, although it is driven by increased efficiency. There are more complicated decomposition formulas available that account for interactions between scale, composition, and efficiency effects, but they require more data than are available.

Data sources for the decomposition analysis vary by resource. The greenhouse gas (GHG) and fine particulate matter ($PM_{2.5}$) analyses use the latest Global Trade Analysis Project core database (version 11), which includes national input-output tables, as introduced by Aguiar et al. (2023), for 2004–17. For GHG emissions, the fossil fuel combustion carbon dioxide (CO_2) emissions are from Aguiar et al. (2023), and the non-CO_2 GHGs and industrial process CO_2 emissions are from Chepeliev (2024a). $PM_{2.5}$ air pollutant emissions data are from Chepeliev (2024b). For the land decomposition analysis, crop production data from Portmann, Siebert, and Döll (2010) are aligned with cropland data from the Food and Agriculture Organization of the United Nations. For agricultural water, irrigation requirements from Portmann, Siebert, and Döll (2010) are matched with AQUASTAT water withdrawal data (FAO 2024). The analysis comparing agricultural water with nonagricultural water uses data from the global water sector simulation protocol under the Inter-Sectoral Impact Model Intercomparison Project 2a and 2b, which are publicly available at https://www.isimip.org/outputdata/.

GHGs and PM$_{2.5}$

Between 2004 and 2017, HICs experienced a notable reduction in both GHG and PM$_{2.5}$ emissions, contrasting sharply with trends observed in much of the rest of the world (figure S1.4). In HICs, the substantial *efficiency* improvements achieved over these years were sufficient to offset the scale effects associated with economic growth. As a result, these countries managed to decouple their GHG and PM$_{2.5}$ emissions from economic growth, indicating a remarkable shift toward more sustainable practices. Conversely, UMICs witnessed rising emissions linked to rapid economic growth and increasing demographic pressures. Although efficiency gains did occur, they were not enough to counterbalance the scale effects of heightened economic activity.

Agricultural land

Agricultural land use surged in many countries between 2004 and 2017, and was particularly pronounced in lower-middle-income countries (LMICs), which saw an increase in land use of 92.1 million hectares.[5] Efficiency effects (through increased yields) have offset much of the expansion that has resulted from scale effects (around 69 percent). In HICs, efficiency gains have neutralized the scale effects, leading to a slight decrease in cropland (figure S1.5, panel a). In LICs, however, the efficiency gains have been negligible and crop yields have remained stubbornly low. For example, the cereal yield in the average LIC is now about half that of India (Ritchie 2024). The country-level decomposition in box S1.4 shows that most countries with the largest growth in agricultural land use have seen increases in land efficiency, but these are simply not large enough to offset the scale effects.

FIGURE S1.4 Change in GHG and PM$_{2.5}$ emissions, by driver, 2004–17

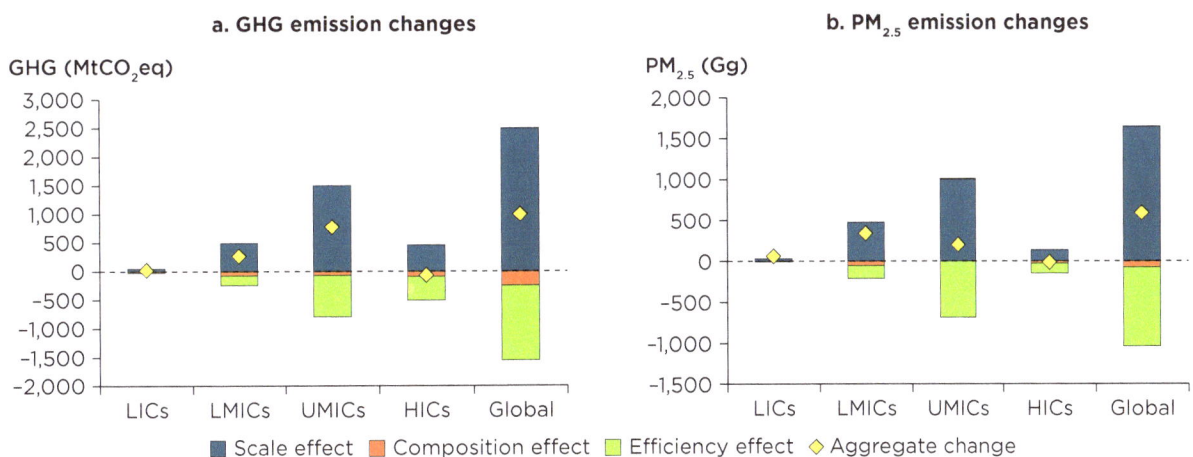

a. GHG emission changes

b. PM$_{2.5}$ emission changes

Legend: ■ Scale effect ■ Composition effect ■ Efficiency effect ◇ Aggregate change

Source: Original calculations based on data from GTAP databases.
Note: Composition is the distribution of economic activity between GTAP classified sectors of agriculture, industry, and services. *Efficiency* is defined as the intensity of emissions per unit of value added. CO_2 = carbon dioxide; Gg = gigagrams; GHG = greenhouse gas; GTAP = Global Trade Analysis Project; HICs = high-income countries; LICs = low-income countries; LMICs = lower-middle-income countries; MtCO$_2$eq = million tons of carbon dioxide equivalent; PM$_{2.5}$ = fine particulate matter; UMICs = upper-middle-income countries.

FIGURE S1.5 Change in land use and total water use, by driver, 2004–17

a. Land use changes

Land (hectares, millions)

b. Total water use changes

Water (m³, billions)

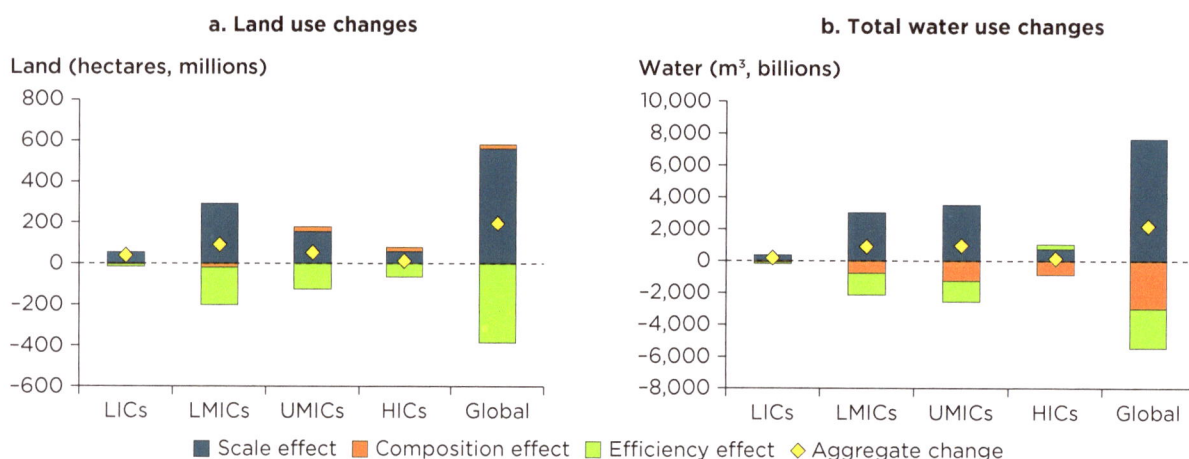

■ Scale effect ■ Composition effect ■ Efficiency effect ◇ Aggregate change

Sources: Original calculations based on land use data from GTAP databases; water use data from the Inter-Sectoral Impact Model Intercomparison Project 2a and 2b; sectoral gross domestic product data from the World Development Indicators.

Note: Composition effect includes shifts in cultivated area of the following crops: paddy rice, wheat, other cereal grains, vegetables, fruits, nuts, oilseeds, sugar crops, plant-based fibers, and other. In panel a, efficiency is defined as output per unit of land used for the cultivation of the crop under consideration. In panel b, efficiency refers to the output produced per unit of water withdrawn for both agricultural and non-agricultural activities. GTAP = Global Trade Analysis Project; HICs = high-income countries; LICs = low-income countries; LMICs = lower-middle-income countries; UMICs = upper-middle-income countries.

Water use

Between 2004 and 2017, total water use—including blue and green agricultural water, as well as industrial and household water—increased by 11.6 percent in the 169 countries included in this analysis (figure S1.5, panel b). *Blue water* refers to freshwater resources in rivers, lakes, reservoirs, and aquifers. *Green water* is the moisture stored in the soil's unsaturated zone, which is directly used by plants for growth as a part of a rainfed agricultural system. Nearly 85 percent of the increase in blue and green water use since 2004 has come from UMICs and LMICs. In all the country income groups except LICs, the composition effect, which measures the shift of water use between agricultural and nonagricultural uses, led to a reduction in water use. This was likely due to the structural transformation that most countries were undergoing. Extraction rates in agriculture, by far the largest user of water, have remained relatively constant in all income groups except LMICs. UMICs are shifting toward more water-intensive crops (composition effect); however, the efficiency effect has been nearly enough to offset both the composition effect and the scale effect. The country-level decomposition in box S1.4 shows that in the five countries with the highest increases in water use over 2004–17, water use efficiency was not increasing. Indeed, in some cases it was decreasing.

Lessons from the countries with the highest increases in land and water use

It is instructive to explore land and water use patterns in the countries that have had the most substantial increases in consumption. Scale effects were the key driver of increases in both land and water use. For land use (figure SB1.4.1), improvements in efficiency (yields) have partly offset the scale effects, but these gains have not been sufficient to generate net declines in land use.

FIGURE SB1.4.1 Changes in land and water use in the five countries with the most substantial increases, 2004–17

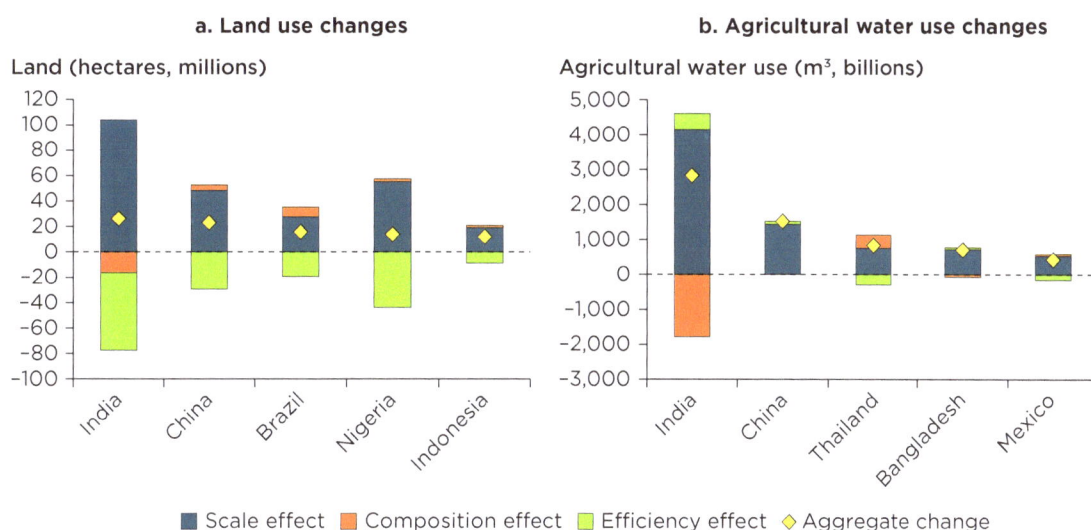

a. Land use changes

Land (hectares, millions)

b. Agricultural water use changes

Agricultural water use (m³, billions)

■ Scale effect ■ Composition effect ■ Efficiency effect ◇ Aggregate change

Source: Original calculations based on data described in the notes in figure S1.5.

Decomposition patterns across the countries with the largest increases in water use suggest a different pattern. In contrast to land use, there is generally less evidence of improvements in water efficiency to offset the scale effects. Rather, water productivity has declined, often reflecting a shift toward increased irrigation.

India, which has had the largest increases in both land and water use, illustrates the environmental challenges and trade-offs that many nations face. Figure BS1.4.1, panel a, shows that although India has had the largest expansion of agricultural land due to scale effects, around 60 percent of this has been offset by notable gains in agricultural yields. However, figure BS1.4.1, panel b, shows that this has come through increased water use. Perhaps as a result, India's water-use efficiency has deteriorated, suggesting that the rise in agricultural productivity is largely due to greater water consumption (Zaveri and Lobell 2019). A similar pattern is observed in China, which experienced the second largest increases in land area and water use in the study period. These examples highlight the need for policy approaches that recognize the intertwined nature of the drivers of environmental degradation. Where policy decisions are taken in sectoral isolation, there are risks of misalignment, unplanned trade-offs, and unintended consequences.

Conclusion

Environmental stewardship is often framed as a zero-sum game that is at odds with the economy, but this may be a misleading simplification. The economy fundamentally relies on natural resources, making environmental protection vital for long-term economic health. HICs have shown that it is possible to decouple growth from environmental harm, largely through regulations and cleaner technologies, although these require enforcement and investment. Successes like improved air quality in HICs show that absolute decoupling is achievable and compatible with prosperity—especially when transition speed is aligned with environmental thresholds.

However, such win-win outcomes are not assured. Environmental harms that are distant, diffuse, or lack alternatives (like biodiversity loss or resource use) continue to rise with income, suggesting that affluence alone does not guarantee decoupling. Even when decoupling occurs, it must be broad, fast, and sustained to avoid irreversible damage. For instance, recent global $PM_{2.5}$ declines have been largely due to China, but this could reverse if other countries continue polluting.

Decomposition analyses show that decoupling has mostly been driven by efficiency and productivity gains, not by deeper structural changes that are much harder to achieve. Efficiency gains face limits defined by the natural laws of physics. Once efficiency plateaus, reducing environmental harm will require shifts in consumption and production.

Finally, policies must reflect the interconnected nature of environmental and economic systems (box S1.5). Gains in one area (for example, water use for agriculture or reducing sulfur dioxide emissions) can harm others (for example, water scarcity or climate). Understanding these trade-offs and integrating environmental measures into economic indicators are essential for sustainable progress.

BOX S1.5
Economic growth and the rising threat to biodiversity

The global rate of species extinction is now tens to hundreds of times higher than the average over the past 10 million years. While economic growth remains a cornerstone for improving well-being and eradicating extreme poverty, drivers of biodiversity loss are closely linked to economic activities, raising concerns that continued growth could accelerate biodiversity depletion. The intersection of economic growth and biodiversity loss presents both synergies and trade-offs that must be better understood to achieve sustainable development goals.

For a better understanding of the synergies and trade-offs, Piaggio and Siikamäki (2024) estimate the contribution of economic growth to biodiversity threats. Using a difference-in-differences approach to analyze data from 179 countries between 1993 and 2019, they found that a 1 percent

(continued)

BOX S1.5

Economic growth and the rising threat to biodiversity *(continued)*

increase in gross domestic product per capita correlates with a 0.022 percent decrease in the International Union for Conservation of Nature's Red List Index, indicating a heightened biodiversity threat. This result challenges the environmental Kuznets curve hypothesis, which posits that economic growth initially leads to environmental degradation but eventually results in environmental improvements as income levels rise. Indeed, the findings suggest that economic growth has a persistent negative impact on biodiversity across various development stages.

Higher population densities increase extinction risks, and expanding protected areas has only a modest positive effect. For instance, increasing protected areas to 30 percent of a country's territory would offset just 0.4 percent of biodiversity threats by 2050 under Shared Socioeconomic Pathway 2. Projections have indicated a rise in species extinction by 2050 and 2100 if current trends continue. Piaggio and Siikamäki (2024) emphasize the need for integrated policies combining economic growth with biodiversity preservation. A shift toward development models that align economic progress with ecological protection is essential for future biodiversity.

Source: Piaggio and Siikamäki (2024).

Notes

1. *Elasticity of emissions with respect to output* defines the percentage change in emissions that results from a percentage change in output, summarizing whether emissions will rise faster (or slower) than output.

2. In addition, a large proportion of HICs' historical contribution to environmental degradation is already embedded in their built environment—for example, emissions-intensive materials like iron, steel, and cement—allowing them to reduce additional harms (saturation effect).

3. Typically a consequence of agricultural land expansion, forest loss closely tracks the spread of cropland and pasture.

4. The resource curse, also known as the paradox of plenty, refers to the phenomenon where countries rich in natural resources, such as oil or minerals, tend to experience less economic growth, less democracy, and worse development outcomes than countries with fewer natural resources. This paradox can arise due to factors like corruption, mismanagement, and overreliance on resource exports, which can hinder broader economic development.

5. Compared to increases of 52.9 million hectares in UMICs and 39.3 million hectares in LICs.

References

Aguiar, A., M. Chepeliev, E. Corong, and D. van der Mensbrugghe. 2023. "The Global Trade Analysis Project (GTAP) Data Base: Version 11." *Journal of Global Economic Analysis* 7 (2): 1–37. doi:10.21642/JGEA.070201AF.

Chepeliev, M. 2024a. "Chapter 13B: Complementary Greenhouse Gas Emissions Database." Global Trade Analysis Project (GTAP), Center for Global Trade Analysis, Purdue University, West Lafayette, IN. https://www.gtap.agecon.purdue.edu/resources/res_display.asp?RecordID=7131.

Chepeliev, M. 2024b. "Chapter 13C: Air Pollutant Emissions Database." Global Trade Analysis Project (GTAP), Center for Global Trade Analysis, Purdue University, West Lafayette, IN. https://www.gtap.agecon.purdue.edu/resources/res_display.asp?RecordID=7132.

Clark, C. W. 1973. "Profit Maximization and the Extinction of Animal Species." *Journal of Political Economy* 81 (4): 950–61.

FAO (Food and Agriculture Organization of the United Nations). 2024. *FAO AQUASTAT*. Rome: FAO. https://data.apps.fao.org/aquastat/?lang=en.

Grinsted, P., and N. Menzies. 2024. "Decoupling Economic Growth from Environmental Harms: A Conceptual Framework and Overview of Empirical Evidence." Background paper for this report. World Bank, Washington, DC.

Mawejje, J. 2025. *Fiscal Vulnerabilities in Low-Income Countries: Evolution, Drivers, and Policies*. Washington, DC: World Bank.

Parrique, T., J. Barth, F. Briens, et al. 2019. "Decoupling Debunked: Evidence and Arguments against Green Growth as a Sole Strategy for Sustainability." European Environmental Bureau, Brussels, Belgium.

Piaggio, M., and J. Siikamäki. 2024. "Growing Tensions: Economic Growth and the Rising Threat to Biodiversity." Background paper for this report. World Bank, Washington, DC.

Portmann, F. T., S. Siebert, and P. Döll. 2010. "MIRCA2000—Global Monthly Irrigated and Rainfed Crop Areas around the Year 2000: A New High-Resolution Data Set for Agricultural and Hydrological Modeling." *Global Biogeochemical Cycles* 24 (1): GB1011.

Ritchie, H. 2024. *Not the End of the World: How We Can Be the First Generation to Build a Sustainable Planet*. Random House.

Ross, M. L. 2004. "How Do Natural Resources Influence Civil War? Evidence from Thirteen Cases." *International Organization* 58 (1): 35–67.

Vogel, J., and J. Hickel. 2023. "Is Green Growth Happening? An Empirical Analysis of Achieved versus Paris-Compliant CO_2–GDP Decoupling in High-Income Countries." *The Lancet Planetary Health* 7 (9): e759–e769.

Zaveri, E. B., and D. Lobell. 2019. "The Role of Irrigation in Changing Wheat Yields and Heat Sensitivity in India." *Nature Communications* 10: 4144. doi:10.1038/s41467-019-12183-9.

Part 1

The Economic Stakes

Part 1 comprises four chapters that investigate the economic contribution of critical natural capital—land, air, and water—and the risks of addressing them in isolation. It presents new empirical insights on green water (soil moisture and rainfall recycling), biodiversity, nitrogen fertilizer overuse, and air pollution, showing how changes in one system can cascade through others and the economy. Each chapter highlights these interconnections and offers evidence-based solutions.

CHAPTER 2

Forests and the New Economics of Green Water: Moisture in Motion

"Water begets water, soil is the womb, vegetation is the midwife."
—*Millán Millán*, Spanish meteorologist (1941–2024)

Key messages

- Green water is the most abundant source of freshwater on the planet. It is the moisture held in the upper unsaturated soil layer (*stock*), which partly returns to the air through evaporation and transpiration (*flow*). Despite its dominance in the water cycle, its economic contribution is often overlooked.

- New research presented in this report finds that forests nurture and preserve green water stocks and flows. They sustain soil moisture and govern regional rainfall patterns through terrestrial moisture recycling.

- Without upwind forests and the rainfall they generate, total economic losses could reach $11.4 billion in South America, $2.5 billion in Southeast Asia, and $0.4 billion in Africa. These losses encompass declines in energy generation, agricultural productivity, and overall economic growth.

- Without upstream forests to sustain the soil moisture that is crucial for agriculture, economic growth would decline, amounting to an economic loss of about $379 billion, across the developing world. This is equivalent to about 8 percent of global agricultural gross domestic product (GDP).

- In addition, forests act as natural buffers against dry rainfall episodes (drought), by preserving soil moisture. This mitigates half of the potential economic growth losses, shielding up to $140.4 billion in GDP in developing countries.

- Natural, intact, and diverse forests with higher biodiversity provide greater drought-buffering benefits than plantations.

(continued)

- Decision makers must recognize the connections between water, soil, forests, and the atmosphere, which collectively influence rainfall and moisture availability. Ignoring these links will lead to policy missteps that have material economic consequences.

The language of water

In 1979, Brazilian scientist Eneas Salati had an idea that led to a groundbreaking discovery about the origin of rain. He used the fact that not all water is the same—its hydrogen and oxygen atoms can vary in atomic mass, forming isotopes. As water moves through the hydrological cycle—evaporating, condensing, and precipitating—these isotopes change, leaving behind a trail of signatures in molecules of water.[1] With this insight, Salati tracked water's journey through different sources. What he found was striking. Analyzing the isotopes from samples of rainwater collected from the Atlantic coast to the Peruvian border, Salati discovered that nearly half the rain in the Amazon rainforest came from trees. As air moved from the Atlantic across the Amazon basin to the west, rain clouds recycled moisture from the forests five to six times. His discovery using isotopic methods marked the first direct measurement of "terrestrial moisture recycling," indicating that water stored in soil and released by trees is essential for creating rain. Until then, observational evidence had been scarce, and the mainstream belief was that ocean evaporation was the dominant contributor to rain.

Since Salati's discovery, it has become clear that forests play a crucial role in stabilizing the hydrological cycle and governing of local, regional, and continental rainfall patterns. Forests take up water from their roots and release it into the atmosphere as vapor via transpiration, the arboreal equivalent of sweating. This vapor adds to moisture in the air, generating rain.[2] Since a single tree can transpire hundreds of liters of water a day and vapor can travel long distances, forests play a crucial role in supplying rain both locally and at continental scales.

Recent moisture-tracking research has estimated that land-based evapotranspiration, driven largely by forests, accounts for 45 percent of all rainfall received on land (De Petrillo et al. 2025), making these green water flows as important as ocean evaporation for sustaining rainfall. This chapter shows that natural, intact, and diverse forests with higher biodiversity are much more efficient at regulating water and climate than younger (monoculture) tree plantations. A clear implication is that the loss of forests is altering the hydrological cycle at all scales, from local to global, shifting and increasing the uncertainty of annual precipitation levels.

Beyond adding moisture to the air, forests also sustain and renew moisture in the soil. The dense canopy of trees provides a natural umbrella that traps rainwater, slowing the pace of rain and allowing it to enter the soil. Forest roots act as natural sponges, creating pathways for water to move underground, storing rainfall locally, adding to soil moisture, and recharging groundwater. The moisture stored in the soil root zone—called green water *stocks*—can act as a natural reservoir, buffering against drought and extreme weather.

Green water—moisture held in the upper unsaturated layer of the soil and vegetation—accounts for about 65 percent of freshwater received from rainfall (around 70,000 cubic kilometers per year). The remaining 35 percent (40,000 cubic kilometers) is *blue water*, which is held in rivers, lakes, aquifers, glaciers, and ice. Reflecting its prevalence, green water also provides 75 percent of the water consumed in food production, while blue water accounts for the remaining 25 percent. Overall, a stable supply of green water in soil is crucial to sustaining land-based ecosystems and the water cycle (refer to box 2.1). Yet, current water policies tend to deal with blue water and often overlook green water.

Despite their dominance in the water cycle, the economic impacts of green water dynamics are largely unknown to policy makers and, as a result, are neglected in decision-making. This oversight is concerning, given recent scientific assessments that have suggested that the world has surpassed the safe planetary boundary[3] for freshwater, due to climate impacts and overuse of both green and blue water resources (Richardson et al. 2023; Wang-Erlandsson et al. 2022). Understanding these dynamics is particularly urgent, given the intertwined challenges of deforestation and water deficits, which compound each other and threaten global resilience.

BOX 2.1
Stocks and flows in the water cycle

Freshwater originates from precipitation, which, on reaching land, can be categorized into two main types: blue and green water. Rainfall that infiltrates into the soil to create soil moisture or evaporates directly from the land surface—whether from canopy cover, soil, or standing water—is known as *green water*. The remainder flows as surface runoff (*blue water*), feeding rivers, wetlands, and other aquatic ecosystems. Blue water sustains aquatic ecosystems and is available to humans as an extractable resource. Green water, which is available to plants, supports terrestrial ecosystems and rainfed agriculture.

The water cycle consists of interconnected blue and green water stocks and flows. Blue water stocks are stored in lakes, behind dams, below the water table in aquifers, and in ice, glaciers, and snow. This type of water flows as runoff into rivers and recharges subsurface water tables and groundwater. Green water stocks are the moisture in the soil root zone and the water held in plants; green water flows are vapor released as transpiration and evaporation. These blue and green water systems are closely linked. For example, river water (a blue flow) can be extracted from a reservoir (a blue stock) and used for irrigation, creating soil moisture (a green stock) that subsequently turns into evaporation and transpiration (green flows).

Source: Global Commission on the Economics of Water (2024).

Deforestation is a key driver of land degradation and hydrological disruption. Since 2000, the world has lost more than 10 percent of its tree cover, with the tropics accounting for 1.48 million square kilometers of deforestation—an area larger than France, Germany, and Spain combined (Balboni et al. 2023). While forests have been recognized for their roles as hosts of biodiversity and other ecosystem services, their role in preserving green water is also vital. As such, their loss amplifies water scarcity challenges. For instance, along the West African coast, unchecked deforestation has disrupted rainfall recycling, contributing to desertification in the Sahel and deepening poverty and fragility across the region.

Water deficits are becoming the new normal across the world. With rising human populations and growing prosperity, water demand is growing exponentially, while damaging human activities degrade and diminish water supplies. The result is a water deficit. Around 60 percent of the world's population lives with periodic water stress for at least part of the year, and close to 85 percent of people affected by droughts live in developing countries. The number of extreme droughts has increased by 233 percent over the past 60 years in low-income and lower-middle-income countries (Damania et al. 2017; Mekonnen and Hoekstra 2016; Zaveri, Damania, and Engle 2023).

This chapter focuses on the seldom recognized impacts of green water on the economy. Presenting results from new research, the chapter explores the economywide impacts of green water via aboveground moisture recycling and belowground storage of soil moisture. To date, economic deliberations have overlooked these mechanisms. The analysis highlights the connections that emerge from forests in their role as conduits of green water flows and stores, and the importance of conserving these links for building resilience to climate shocks. It finds that the neglected hydrological services and economic benefits provided by forests at multiple spatial scales—from local to regional—are significant, suggesting that policy makers have routinely underestimated the contribution of forests when addressing competing demands for land.

The moisture above: The economic impacts of green water flows

Trees bring economic benefits through atmospheric moisture or green water flows. The research presented in this section is based on a background paper for this report by Araujo and Hector (2024), who employed an econometric methodology to map the effect of deforestation in the Amazon, Congo, and Southeast Asian rainforests on rainfall in their respective continents (box 2.2). The model estimates the amount of rainfall that is derived from forests in these three ecosystems, at a 0.25-degree gridded resolution.

BOX 2.2
Econometric modeling of green water flows

Araujo and Hector (2024) conducted an empirical assessment of deforestation's impact on rainfall, combining advanced statistical techniques with satellite and atmospheric data. They used various data sets, including ERA (2017), to model atmospheric trajectories of the paths of air parcels as they travel through the atmosphere. In doing so, they identified the origins of rainfall at specific locations. Forest exposure along these trajectories was quantified using high-resolution land cover data from Hansen et al.'s (2013) forest data set. This enabled measurement of the extent to which air masses pass over forested areas before precipitating.

Next, the study used a fixed-effects regression to estimate the relationship between upwind forest cover and rainfall. By controlling for location, month, and year, as well as the length of the backwind trajectory, the study estimated the plausibly causal impact of exposure to forests along air parcel trajectories on precipitation. The key identification assumption was that these trajectories, at altitudes of 2,000–8,000 meters, are exogenous to annual-level deforestation. This assumption was based on atmospheric circulation being largely driven by temperature gradients and the Earth's rotation (the Coriolis effect), making it unlikely that local deforestation events would directly influence the paths of atmospheric trajectories at scale.

The econometric method offers several strengths, particularly its empirical grounding in real-world data and the ability to infer directly the economic and policy implications of changes in forest cover. Its robustness lies in its capacity to control for confounding factors and isolate the causal effects of deforestation on rainfall. However, this method has limitations—such as assuming a uniform impact of forest cover across distances—which may oversimplify complex moisture recycling processes.

Among alternative modeling approaches, the most notable is the simulation-based uTrack model (Tuinenburg and Staal 2020), which offers a detailed representation of the mechanisms that transport moisture through the atmosphere. However, these models do not involve statistical inference and thus depend entirely on model parameterization. One important caveat when comparing the empirical methodology used by Araujo and Hector (2024) with models like uTrack is that, as in Spracklen, Arnold, and Taylor (2012), the empirical model assumes a homogeneous effect of deforestation with respect to distance, since the main explanatory variable is a nonweighted sum of forest pixels along trajectories of atmospheric transport.

Source: Araujo and Hector (2024).

After estimating the amount of rainfall derived from forests, a counterfactual simulation quantifies the impacts of historical deforestation on rainfall. This simulation is a scenario where no deforestation occurred between 2001 and 2020. An equivalent way of thinking of it is that all deforested areas during that period were completely reforested. The average annual change in rainfall from deforestation over that period is shown in map 2.1. Annex 2A presents a second simulation scenario, looking at potential future deforestation.

MAP 2.1 Estimated rainfall losses due to recent deforestation 2001–20

a. South America

b. Africa

c. Southeast Asia

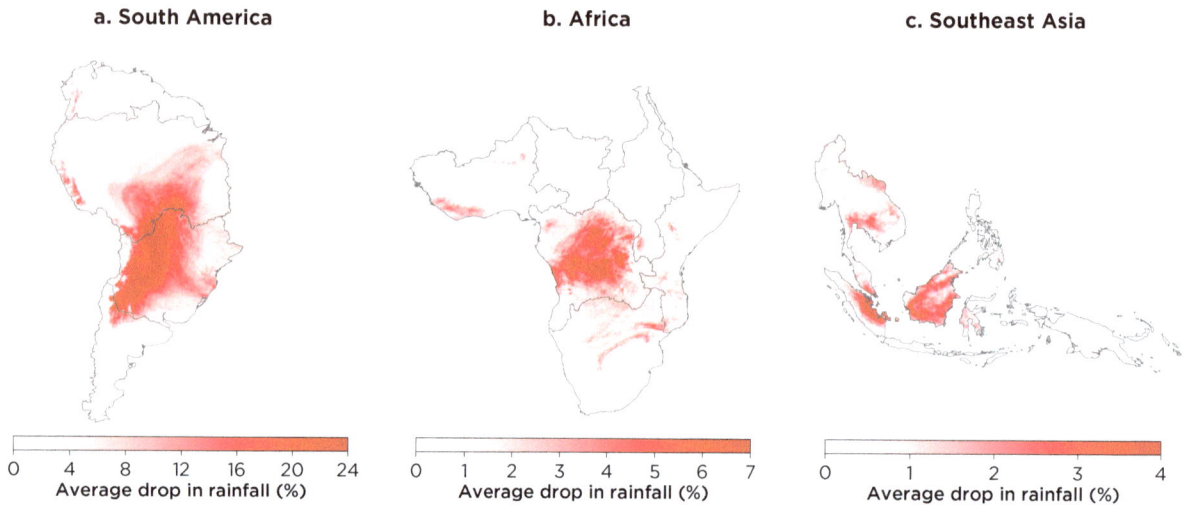

| 0 | 4 | 8 | 12 | 16 | 20 | 24 |
Average drop in rainfall (%)

| 0 | 1 | 2 | 3 | 4 | 5 | 6 | 7 |
Average drop in rainfall (%)

| 0 | 1 | 2 | 3 | 4 |
Average drop in rainfall (%)

Source: Araujo and Hector 2024.

Note: Panels a, b, and c show the results of counterfactual analysis that compares the deforestation that occurred between 2001 and 2020 with a scenario where no deforestation occurred. The effects, which show the drop in rainfall that occurred over that period due to deforestation, are shown as a percentage of the historical annual average rainfall. To improve visualization, values are capped at the 99th percentile for Africa and Southeast Asia, meaning that the top 1 percent of the most extreme rainfall reductions are omitted. For South America, values are capped at the 95th percentile.

The counterfactual scenario reveals a general decrease in rainfall due to tropical deforestation, with South America facing the most significant decline. Deforestation in the Amazon reduces rainfall not only within the forest but also up to 2,000 kilometers to the south. This is primarily driven by the region's unique wind circulation pattern, which carries rainfall to the continent's inner landmass with water. In Africa, winds are more concentrated along the equator, which keeps the impacts of deforestation on rainfall more localized to the regions where deforestation occurs. In Southeast Asia, the widespread presence of ocean water reduces the significance of forest-derived moisture in the region's overall water balance, reducing the impact of deforestation on total rainfall.

After estimating the impacts of the deforestation scenarios on rainfall, three further simulations estimate the economic impacts of reduced rainfall (described in box 2.3). These simulations explore impacts on energy potential through hydropower generation, agricultural productivity, and gross domestic product.

BOX 2.3

Three simulations to examine the economic impacts of reduced terrestrial moisture recycling due to deforestation between 2001 and 2020

Simulation 1: Estimating the impacts of reduced rainfall on energy generation

Reductions in rainfall will have subsequent impacts on river flow, which can reduce hydropower production. The simulation uses river discharge data from the Global Flood Awareness System (Harrigan et al. 2020) to estimate empirically the relationship between upstream rainfall and river flow. It then applies the physical principles of hydropower production, where production depends on the mass and velocity of water flowing through turbines. Finally, it uses the elasticity of power generation with respect to river flow to approximate the percentage change in potential energy output caused by rainfall reductions.

Simulation 2: Estimating the impacts of reduced rainfall on agricultural yields

Changes in rainfall will have significant impacts on agricultural productivity, which is highly sensitive to rainfall patterns. To estimate an econometric agricultural production function, the simulation proxies agricultural productivity by net primary productivity, a remotely sensed vegetation measure, and uses land cover data from the European Space Agency (2017) to isolate regions where at least 30 percent of land is devoted to cropland. This estimate of the relationship between rainfall and agricultural productivity is used to estimate the total impact of reductions in rainfall due to deforestation.

Simulation 3: Estimating the impacts of reduced rainfall on gross domestic product (GDP)

The literature offers recent estimates of the relationship between rainfall and GDP impacts. Damania, Desbureaux, and Zaveri (2020) provide global and developing country estimates, and Kotz, Levermann, and Wenz (2022) provide similar estimates for developing countries. The elasticities provided by these papers, combined with local GDP data provided by Chen et al. (2022), enable simulation of the loss of GDP caused by rainfall reductions from deforestation.

Source: Original calculations based on Araujo and Hector (2024).

Of the three regions examined, the Amazon is the most economically sensitive to deforestation, losing the most in terms of energy productivity, agricultural productivity, and GDP growth each year due to reduced rainfall (table 2.1). The simulations show that deforestation between 2001 and 2020 is estimated to be responsible for decreasing potential energy production by 17.8 gigawatts, or about 8 percent of current levels. The most impacted basins are the Rio de la Plata and Amazon River basins, which account for 45 and 32 percent of the region's installed capacity, respectively. Agricultural productivity in the region declines by approximately 0.6 percent per year due to reductions in rainfall caused by historical deforestation. Finally, GDP losses range

TABLE 2.1 Summary of simulated annual economic losses due to decreased rainfall caused by deforestation, 2001–20

Rainforest region	Energy potential (gigawatts lost)	Agricultural productivity (yield decline, %)	Economywide GDP (billions of $ lost)
Amazon	17.8	0.6	2.9–11.4
Congo	1.6	0.1	0.11–0.41
Southeast Asia	1.5	0.1	0.34–2.54

Source: Original calculations based on Araujo and Hector 2024.
Note: GDP = gross domestic product.

between $2.9 billion and $11.4 billion per year due to deforestation between 2001 and 2020, and between $268 million and $1.02 billion per year in the future deforestation scenario (table 2A.1 in annex 2A). The large range of GDP losses stems from different estimates in the literature on the relationship between rainfall and GDP.

Losses due to changes in rainfall from deforestation in the Congo and Southeast Asia rainforest regions are significantly lower than those in the Amazon, due to both lower impacts on rainfall and lower levels of rainfall-dependent economic activity. The simulations show that deforestation between 2001 and 2020 caused reductions of 1.6 and 1.5 gigawatts of energy potential in Africa and Southeast Asia, respectively, or about 1.7 and 1.2 percent of current production levels. Agricultural productivity losses are estimated at 0.1 percent per year in both regions. Annual GDP impacts from past deforestation range from $110 million to $410 million for Africa, and from $340 million to $2.54 billion for Southeast Asia.

Although the average economic impacts over entire regions might appear modest, some countries experience significant economic impacts due to rainfall losses from deforestation. Figure 2.1 shows the top 10 countries in terms of losses in the growth rate of GDP (panel a) and translates them into total annual GDP losses by multiplying the growth losses by 2019 GDP (panel b). In terms of loss rates, eight of the top 10 impacted countries are in South America, with Paraguay estimated to lose up to 0.83 of a percentage point of GDP growth each year. When growth rates are translated into dollars, Brazil has the highest losses, at nearly $7 billion per year.

The results are similar to recent assessments. The Global Commission on the Economics of Water (2024) assessments confirmed that deforestation significantly reduces rainfall, with serious implications for regional hydrology and economic activity. While the Global Commission uses the uTrack model—which simulates terrestrial moisture recycling with high spatial and temporal resolution—this report employs an econometric approach that statistically infers causal relationships from observed data. The uTrack model examines an extreme scenario of eliminating all terrestrial moisture

FIGURE 2.1 Top 10 countries with GDP losses due to historical deforestation-induced rainfall reductions, 2000–20

a. Mean annual GDP growth rate loss

Change (percentage points)

b. Annual GDP loss

Loss ($, millions)

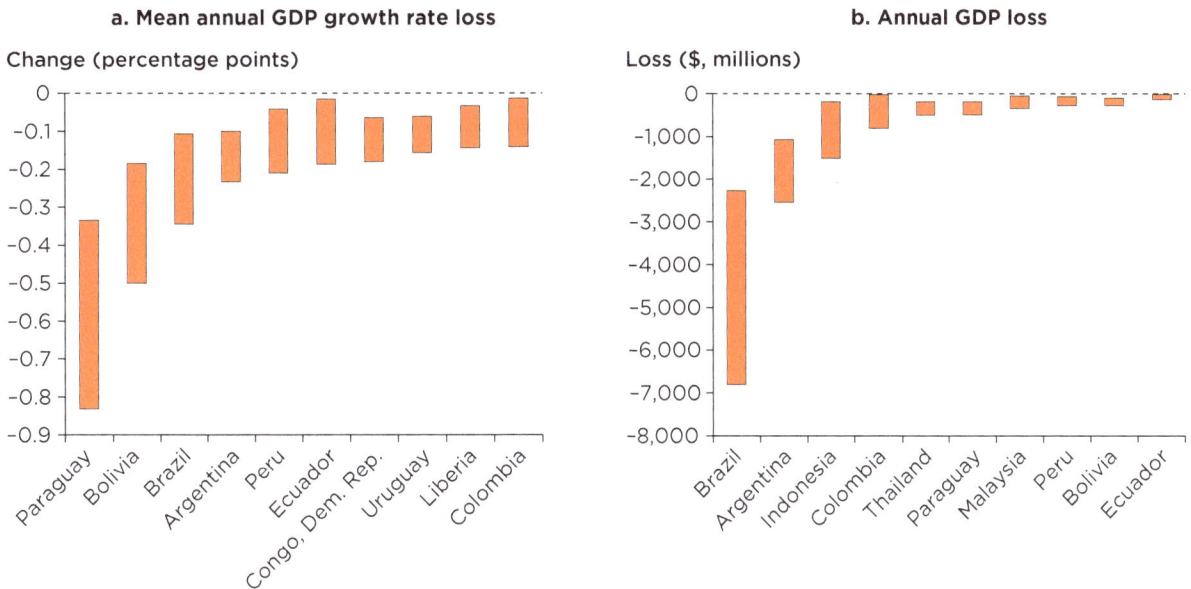

Sources: Original calculations based on data from Chen et al. 2022 (gridded GDP) and Araujo and Hector 2024 (deforestation-induced rainfall losses).
Note: Losses are estimated at the 0.5-degree grid cell level and aggregated by country. GDP = gross domestic product.

recycling, whereas this report focuses on actual forest loss in three key regions. Despite these methodological differences, both studies produce similar estimates of rainfall loss—20–25 percent reductions in affected regions—highlighting the robustness of the results. Together, the findings underscore the critical importance of forest preservation for climate resilience and economic stability, and the need for integrated forest and water management policies.

The moisture underground: The economic impacts of green water stocks

Important moisture processes also play out below ground. Forests influence water availability *downwind* (as described in the previous section) as well as *downstream*. Although it is widely accepted that the loss of forests leads to reduced rainfall in downwind areas, the effects of forest loss on soil moisture and downstream water flow are more contested. On the one hand, forests absorb moisture and hence reduce surface runoff. Some studies have shown that trees decrease the flows available in nearby rivers (Zhang and Wei 2021), particularly for monoculture plantations, compared with older and mixed native forests (Ellison et al. 2017; Krishnaswamy et al. 2013). On the other hand, the forest canopy intercepts and slows runoff and tree roots move moisture from wetter to dryer soil layers. In addition, soil porosity determines moisture storage capacity (Hall 2003; Manoli et al. 2014). These factors impact baseflows, which maintain soil moisture levels during dry periods, while runoff is more rapid and depends on current rainfall. Native trees improve baseflow by facilitating greater infiltration and retention of water in soil. Such impacts on soil moisture, rather than surface runoff alone, are often more relevant for long-term water security.

The impacts of forest loss on soil moisture (green water stocks) are also complex. On the one hand, forests lower soil moisture levels due to increased rates of evapotranspiration, where the combination of evaporation from the soil and transpiration from leaves leads to significant moisture loss, especially in warm, dry climates. On the other hand, forests can enhance soil moisture retention by improving surface infiltration (partly through the addition of organic matter), increasing hydraulic conductivity through the presence of macropores and preferential flow paths. As a result, case studies have yielded inconsistent findings, making it difficult to determine whether the overall effects of forests on soil moisture are predominantly positive or negative on a global or regional scale.

Newly available global-scale data sets provide an opportunity to clarify the impacts of forests on soil moisture at a wider spatial scale. New analysis has suggested a strong positive correlation between forests and soil moisture. Figure 2.2, panel a, shows that on average, areas with a higher share of upstream forest cover exhibit higher soil moisture levels (box 2.4). This result provides direct empirical evidence at scale of the role of forest cover in conserving green water stocks. One clear implication is that retaining upstream forests plays a role akin to water storage infrastructure in securing water supplies.

Since soil moisture is a crucial determinant of plant growth, its loss could have significant economic implications. Green water accounts for 75 percent of the water withdrawn in food production. Hence, fluctuations in soil moisture directly impact agricultural productivity. This implies that reductions in soil moisture due to forest loss could also have cascading effects on agriculture and the broader economy. Econometric estimates can unravel and quantify the magnitude of these impacts.

The economic significance of upstream forests for downstream agriculture is substantial and remains largely unacknowledged. Figure 2.2, panel b, and table 2.2 illustrate the impacts across regions. Eliminating upstream forest cover would lead to annual losses in GDP growth of 0.1–0.7 of a percentage point, on average. These declines constitute a significant growth slump in the impacted regions. The projected loss of upstream forests would result in a GDP decline of around $379 billion in developing nations, or 7.8 percent of global agricultural GDP. Thus, the water provisioning services of forests are more economically important than has previously been recognized.

FIGURE 2.2 **Links between upstream forest cover, soil moisture, and economic growth**

a. Association between upstream forest cover and soil moisture

Average soil moisture (meter)

Share of upstream forest cover

○ Bin means ── Lowess fit

b. Average change in GDP growth from removing all upstream forest cover in developing countries, by region

Mean annual GDP growth rate loss (percentage points)

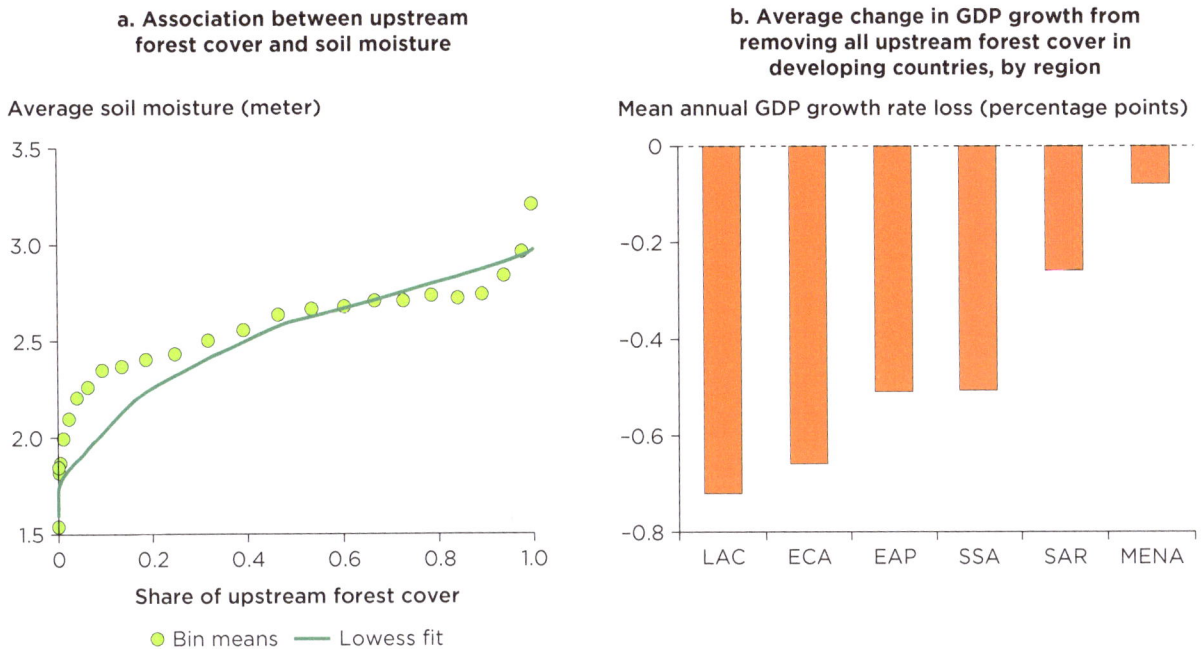

Source: Original calculations based on the data described in box 2.4.
Note: Panel a plots the output from a binned correlation using grid-level data on upstream forest cover share and average soil moisture. Each dot represents an equally sized bin of observations (grouped over the x-axis). Within these bins, the average of the x- and y-variables is visualized in a scatterplot. Average losses in GDP growth in panel b are calculated using estimates of the impact of soil moisture on economic growth (refer to box 2.4). EAP = East Asia and the Pacific; ECA = Europe and Central Asia; GDP = gross domestic product; LAC = Latin America and the Caribbean; MENA = Middle East and North Africa; p.p. = percentage points; SAR = South Asia; SSA= Sub-Saharan Africa.

BOX 2.4
Econometric modeling of green water stocks

Disaggregated data sets and statistical approaches were used to investigate the role of forests in maintaining green water stocks and their consequent impact on economic growth. For this analysis, land areas were divided into grid cells measuring 0.5 degree on each side (approximately 56 × 56 kilometers at the equator), covering 1992 to 2014. Soil depth–weighted averages of root-zone soil moisture (RZSM) were calculated using a novel global soil moisture data set, which has been shown to provide accurate measurements at various depths (Wang et al. 2021). Past local and upstream forest cover shares were measured using satellite data from the European Space Agency, with 1992 as the baseline. To identify upstream forested areas, HydroBASINS level 12 watershed data—comprising more than a million watersheds—were used to measure the proportion of forest cover in upstream basins for each grid cell, extending up to 0.25 degree around the grid cell. Additional data sources included historical weather records from Matsuura and Willmott (2018) and annual grid-level estimates of gross domestic product (GDP) from Kummu, Taka, and Guillaume (2018).

(continued)

BOX 2.4
Econometric modeling of green water stocks *(continued)*

At a global scale, grid cells with a higher share of upstream forest cover experienced higher average cumulative RZSM. To illustrate the cross-sectional correlation between upstream forest cover share and soil moisture, the sample was divided into equally sized groups based on the distribution of upstream forest cover. Mean values of upstream forest cover share and average soil moisture—controlling for other factors such as average precipitation and temperature—were then calculated within each group and visualized in a scatterplot (figure 2.2, panel a). The findings show that on average, a 10-percentage-point increase in upstream forest cover share is associated with a 4.4-meter increase in RZSM.

The impact of soil moisture on economic growth was analyzed using a fixed-effects panel regression. Control variables—including rainfall, temperature, fixed effects, and time trends—were included to isolate the impact of soil moisture from other confounding factors. These controls accounted for baseline differences in economic growth and other variations over time. Standard errors were clustered at the administrative 1 level (one level below the country level, such as a state in the United States) to account for spatial and serial correlation. On average, an additional 1 meter of RZSM increased GDP growth rates by up to 2.8 percentage points.

A counterfactual exercise was conducted to estimate potential growth losses in the absence of upstream forest cover. In this scenario, reductions in soil moisture were simulated under the hypothetical condition of zero upstream forest cover. These estimates were then combined with elasticity measures linking soil moisture to economic growth. The resulting growth losses were averaged across grid cells in various regions of the developing world (figure 2.2, panel b) and converted into dollar amounts using 2014 GDP levels (table 2.2). Together, the results illustrate the economic consequences of soil moisture reductions caused by upstream deforestation.

TABLE 2.2 Summary of mean annual economic losses from removing all upstream forest cover, 1992–2014

Region	Economywide GDP growth loss (percentage points)	Economywide GDP loss ($, billions)
Latin America and the Caribbean	0.72	74.4
Europe and Central Asia	0.65	76.2
East Asia and the Pacific	0.51	150
Sub-Saharan Africa	0.50	27
South Asia	0.26	47
Middle East and North Africa	0.07	4.1

Source: Original calculations based on the data described in box 2.4.
Note: Column 1 shows the growth losses in figure 2.2, panel b. Column 2 translates these losses into dollar amounts by multiplying by GDP in 2014 (refer to box 2.4). GDP = gross domestic product.

The benefits of forests and green stocks extend to times of extreme weather and especially drought. This is particularly important in developing countries, where the incidence of dry shocks has been steadily increasing over time, as shown in figure 2.3 (Zaveri, Damania, and Engle 2023). The vast literature on adaptation strategies has focused on issues such as resilient infrastructure or the role of economic diversification in spreading risks, drought-resistant crop management techniques or early warning systems, and insurance schemes (Adger, Arnell, and Tompkins 2005; Kirschke, Staszak, and Vliet 2021; Lobell, Deines, and Tommaso 2020). The literature has not explored the role of forests in preserving soil moisture in ways that could shield economic growth (box 2.5).

FIGURE 2.3 **Incidence of extreme droughts in low-income and lower-middle-income countries, 1950s to 2014**

Average number of months

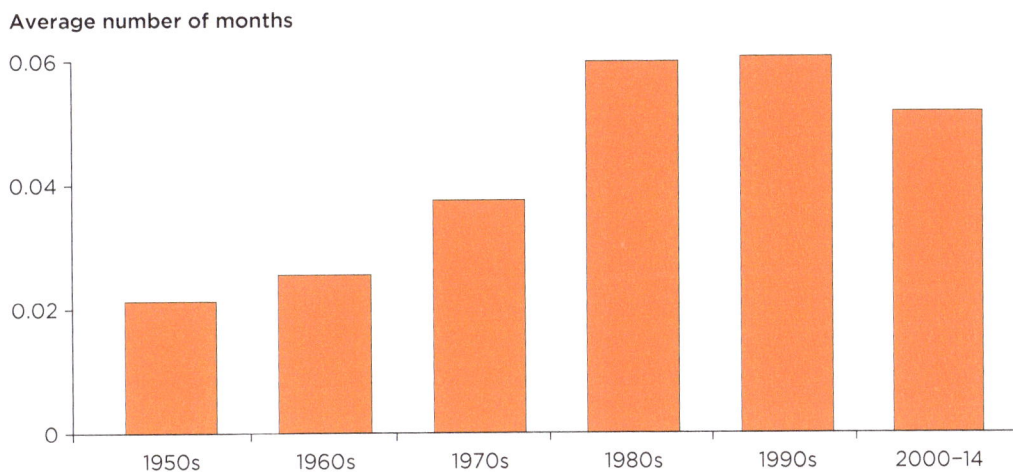

Sources: Original calculations based on data from Matsuura and Willmott 2018 and Zaveri, Damania, and Engle 2023.
Note: Extreme droughts are defined here as shocks that are at least two standard deviations below the long-term mean.

BOX 2.5
Buffering impacts of green water stocks and upstream forest cover during droughts

The analysis described in box 2.4 was extended to examine the buffering effects of green water stocks and upstream forests on economic growth during drought periods. A dry rainfall shock—defined as a year in which rainfall is below the long-term annual mean by at least one standard deviation—is found to have a statistically significant negative impact on economic growth (Zaveri, Damania, and Engle 2023). However, green water stocks play a crucial role in mitigating these effects.

(continued)

BOX 2.5
Buffering impacts of green water stocks and upstream forest cover during droughts
(continued)

To assess whether upstream forest cover helps buffer the adverse economic impacts of droughts, separate regressions were run for high- and low-forested areas (figure 2.4). High (low) forest cover areas are defined relative to the country median, with "high" areas having greater upstream forest cover than the median grid cell and "low" areas having less. A counterfactual exercise was run to estimate the growth losses that would have occurred if high-forested areas had instead been low-forested areas under rainfall reductions of 10 to 30 percent.

Forestry and biodiversity metrics in upstream areas help assess how forest characteristics influence soil moisture retention and drought buffering (figure 2.5). Forestry metrics include (1) forest intactness via the Forest Fragmentation Index (Ma et al. 2023), which captures patch density, edge effects, and isolation; (2) forest management types (Lesiv et al. 2022), distinguishing naturally regenerating forests with limited human impact and plantations with long or short rotation cycles or oil palm; and (3) tree species richness, based on species per hectare (Liang et al. 2022). Biodiversity metrics align with the Kunming-Montreal Global Biodiversity Framework and include the biodiversity habitat index (Harwood et al. 2022), which estimates biodiversity persistence based on habitat area, connectivity, and integrity.

To examine how built water storage complements the buffering effects of upstream forests in figure 2.6, additional irrigation metrics were used. Irrigation shares at the grid level, with 1990 as the baseline, were derived from Mehta et al. (2024), who spatially disaggregated subnational irrigation data at the global scale from Food and Agriculture Organization statistics, the Food and Agriculture Organization's Global Information System on Water and Agriculture, and the statistical office of the European Union. A grid was classified as irrigated if it had any irrigation during the baseline period.

Upstream forests act as critical buffers against dry shocks through their influence on soil moisture. By enabling soil to absorb, retain, and transfer moisture, upstream forests increase the moisture memory in soil. This sustains vegetation during periods of limited rainfall, enabling plants to maintain physiological processes. The estimates suggest that forests neutralize about half of the potential economic growth losses caused by rainfall deficits (figure 2.4).

The drought-neutralizing benefits are significant. To get a sense of the magnitude, it is instructive to consider a counterfactual scenario where: (1) rainfall is assumed to decline by an arbitrary 10–30 percent, and (2) tree density degrades in all regions. Simulations suggest that there would be hypothetical losses of $39.9 billion to $140.4 billion across the developing world, equivalent to 1 to 3.6 percent of global agricultural GDP.

FIGURE 2.4 **Buffering effects of upstream forest cover on economic growth during a dry rainfall shock**

Change in per capita GDP growth (p.p.)

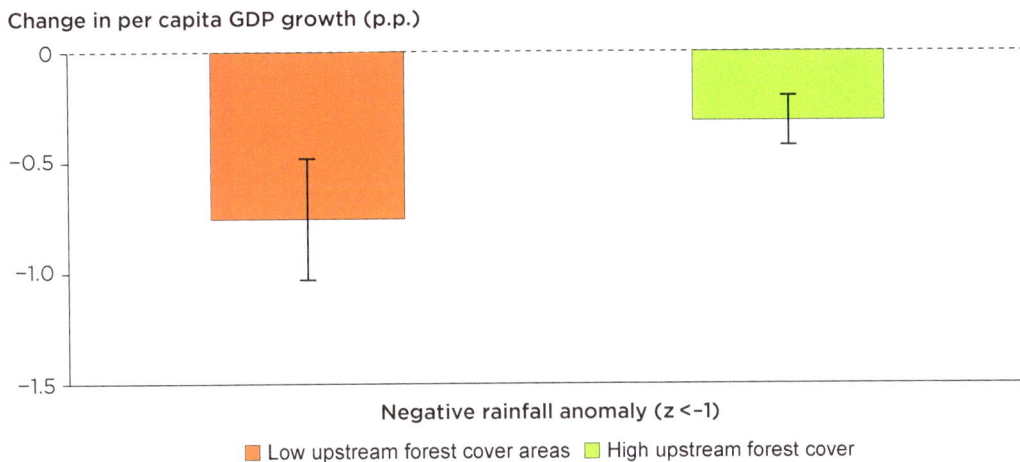

Negative rainfall anomaly (z <−1)

■ Low upstream forest cover areas ■ High upstream forest cover

Sources: Original calculations based on the data described in boxes 2.4 and 2.5.
Note: High upstream forested areas denote places upstream from the grid cell where the forest area is more than the 50th percentile of the forest distribution among all grid cells in each country. GDP = gross domestic product; p.p. = percentage points.

Forests are not all identical and the buffering impacts vary by forest type. Natural, intact, and diverse forests are known to be more resilient, harbor greater biodiversity, and provide greater water provisioning benefits relative to managed plantations. In part, this is due to niche effects, where diverse plant species have different root structures and water uptake patterns, which enhance water infiltration (Hua et al. 2022; Yu et al. 2019). Biodiversity, through its impact on vegetation density and composition, also influences hydrological conditions—a phenomenon termed a trophic cascade, which is explored in more detail in chapter 3 of this report. With recently available data, it is possible to test whether moisture retention and the buffering effects of forests vary systematically with forest type and the presence (or absence) of biodiversity[4].

Natural forests, with their diverse species composition, are more effective at retaining soil moisture and buffering drought impacts than plantation monoculture or degraded forests. As illustrated in figure 2.5, panel a, regions with extensive upstream natural forests experience considerably lower economic growth losses during droughts compared to areas with managed plantations, more forest fragmentation, or low tree species richness. Furthermore, a comparison of natural forests with varying levels of biodiversity shows that areas with a lower biodiversity habitat index suffer greater growth losses due to weaker drought buffering effects (figure 2.5, panel b). The literature on trophic cascades has suggested the mechanisms through which changes in the abundance of species cascade through the food web and modify hydrological regimes (refer to chapter 3 for a detailed discussion of these links).

FIGURE 2.5
Buffering effects of different types of forest management on economic growth during a dry rainfall shock

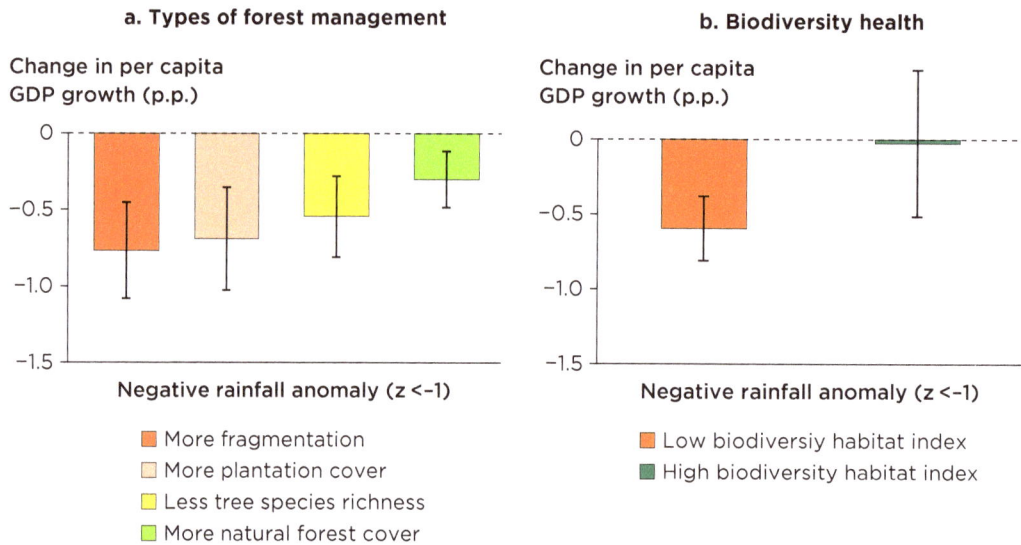

a. Types of forest management

Change in per capita GDP growth (p.p.)

Negative rainfall anomaly (z <−1)

■ More fragmentation
■ More plantation cover
■ Less tree species richness
■ More natural forest cover

b. Biodiversity health

Change in per capita GDP growth (p.p.)

Negative rainfall anomaly (z <−1)

■ Low biodiversiy habitat index
■ High biodiversity habitat index

Sources: Original calculations based on the data described in box 2.5.
Note: For all management types, *high* denotes grid cells where the upstream extent is more than the 50th percentile of the distribution among all grid cells. GDP = gross domestic product; p.p. = percentage points.

The findings highlight that ecological imbalances negatively impact both the economy and the natural resources that underpin progress. Therefore, understanding the role of nature in enabling economic progress is critical. Collectively, these findings suggest that current land use policies permitting forest and biodiversity loss may generate far greater economic costs than previously recognized.

Built water storage infrastructure is the typical remedy to address dry rainfall episodes (BenYishay et al. 2024). Figure 2.6 demonstrates that the buffering effect of upstream forests holds whether or not an area is rainfed or irrigated highlighting that healthy landscapes provide complementary benefits even in irrigated areas. Forests also serve as a "green safety-net": in areas with high forest cover, drought-related out-migration is negligible (Zaveri et al., 2021). Globally, compensating for the buffering effect lost from a 10-percentage-point decline in forest cover would cost an estimated $0.8–3 trillion in irrigation infrastructure, making forests a far more cost-effective way to buffer incomes while also complementing gray infrastructure. The policy implication suggests the need to consider the role of natural infrastructure, which remains largely unseen, under-researched, and under-recognized. Strengthening landscapes through sustainable agricultural practices that emphasize soil restoration, water retention, and ecosystem health can significantly boost both productivity and resilience in agriculture (refer to spotlight 2, on regenerative agriculture).

FIGURE 2.6 **Buffering effects of upstream forest cover on economic growth during a dry rainfall shock in areas with varying levels of irrigation**

Change in per capita GDP growth (p.p.)

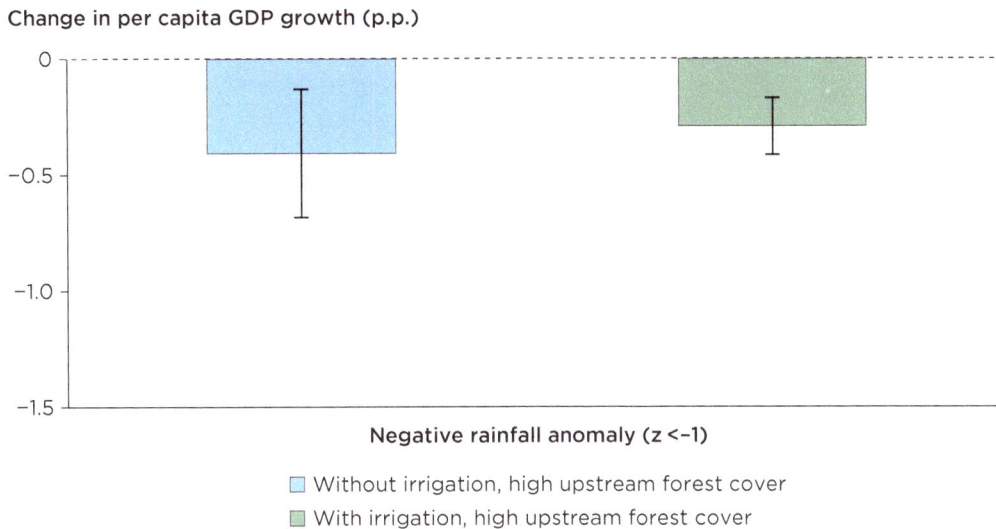

Negative rainfall anomaly (z <–1)

☐ Without irrigation, high upstream forest cover
☐ With irrigation, high upstream forest cover

Source: Original calculations based on the data described in box 2.5.
Note: GDP = gross domestic product; p.p. = percentage points.

Stewardship of water resources

From their canopies to their roots, forests are integral to water supplies, acting as upwind sources of rainfall and upstream sources of soil moisture. However, ongoing changes to forests—from deforestation to intensified afforestation—are reshaping water availability across spatial scales. These shifts highlight the urgent need for stewardship of landscapes that enhance rainfall, improve soil health, and increase water retention. New analysis in this chapter reveals that such efforts can significantly boost agricultural productivity, support hydropower generation, and stabilize overall economic growth, with especially strong protective effects during extreme weather events. Notably, these growth benefits are most pronounced in developing countries, many of which lack the fiscal capacity to invest in costly infrastructure, positioning landscape-based solutions as a vital and cost-effective alternative.

Although forest and landscape rehabilitation can restore water-related ecosystem services, the land-water connection is often misunderstood. Tree planting efforts frequently involve single-species plantations for commercial use, reducing biodiversity and disrupting evapotranspiration, with adverse effects on water availability (Ellison et al. 2017; Manna 2024). Replacing diverse forests with plantations or managing forests for short-term timber gains weakens their ability to regulate the water cycle and climate (Gies 2023; Makarieva et al. 2023). Planting trees in nonforested ecosystems, like savannas or peatlands, can further harm biodiversity and groundwater (Fleischman et al. 2020). Yet, tree-planting initiatives have been widespread, and the global area of planted forests increased from 170 million to 293 million hectares between 1990 and 2020 (Manna 2024).

Native, intact, and mixed forests regulate water and climate far more effectively than monoculture plantations, supporting long-term ecosystem health. Native species, adapted to local hydrology, help sustain soil moisture over time. A mature native forest transpires more water than younger tree plantations and contains diverse undergrowth, fertile soil, and decomposing wood, which create a spongy, moist environment, reducing soil evaporation losses and increasing infiltration capacity, which simultaneously improves underground water storage (Creed et al. 2019; Ellison et al. 2017; Gies 2023).[5] Clear-cutting desiccates the system by stripping away the canopy, reducing shade, and disrupting the moisture-retaining capacity of the soil and vegetation. Unsurprisingly, projects designed in natural or seminatural ecosystems—such as restoring riparian zones or maintaining intact forests—tend to outperform those that require significant intervention or artificial creation, such as in afforestation (Chausson et al. 2020). Natural regeneration, already adapted to local conditions, also typically requires less oversight and is more cost-effective than tree planting.

The future of water security may well be rooted in the stewardship and management of landscapes. While this chapter has laid important groundwork, much more remains to be done. In particular, more location-specific research is needed to understand how different types of land cover influence green water dynamics over time, which can in turn guide restoration and management efforts. Similarly, helping developing countries invest in nature-based solutions will require addressing deeper structural barriers, including governance, poverty, and competing land pressures. Closing these knowledge and financing gaps is essential for designing restoration strategies that truly integrate water, biodiversity, and economic goals—themes that are explored further in chapter 3.

Annex 2A: Simulation of the effects of future deforestation on terrestrial moisture recycling

This annex presents a second simulation assessing the economic impacts of reduced terrestrial moisture recycling from potential complete deforestation in the hotspots identified by Pacheco et al. (2021), shown in map 2A.1.

MAP 2A.1 Estimated rainfall losses due to future deforestation

a. South America

b. Africa

c. Southeast Asia

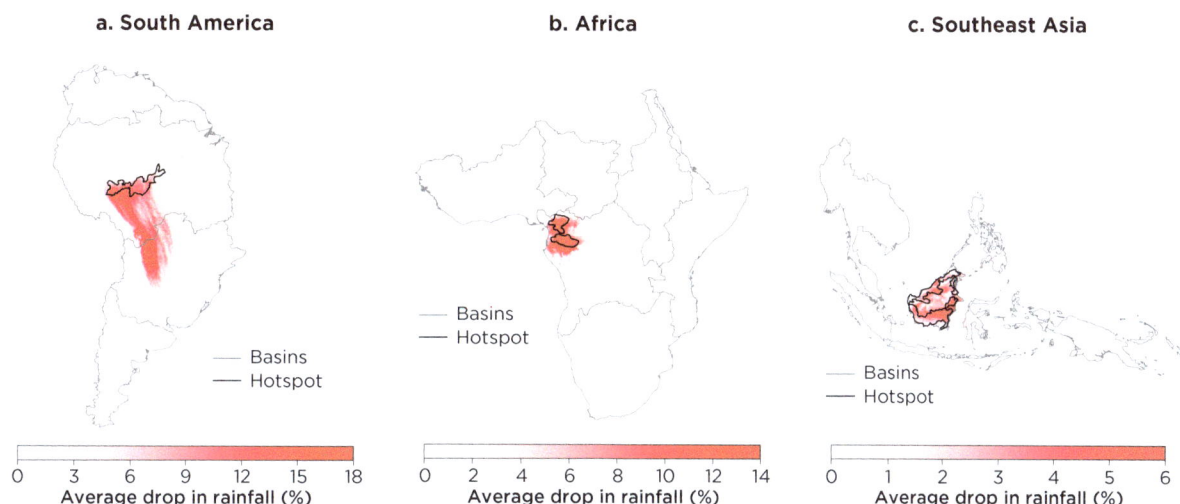

Legend (South America): Basins, Hotspot

Scale: 0 — 3 — 6 — 9 — 12 — 15 — 18 Average drop in rainfall (%)

Legend (Africa): Basins, Hotspot

Scale: 0 — 2 — 4 — 6 — 8 — 10 — 12 — 14 Average drop in rainfall (%)

Legend (Southeast Asia): Basins, Hotspot

Scale: 0 — 1 — 2 — 3 — 4 — 5 — 6 Average drop in rainfall (%)

Sources: Araujo and Hector 2024, with hotspots identified by Pacheco et al. 2021.
Note: The maps show the results of the second counterfactual analysis, which modeled a scenario where hotspots of deforestation are completely deforested. The effects are shown as a percentage of the historical annual average rainfall. To improve visualization, values are capped at the 99th percentile for Africa and Southeast Asia, meaning the top 1 percent of the most extreme rainfall reductions are omitted. For South America, they are capped at the 95th percentile.

In this future deforestation scenario, there are two distinct patterns. Deforestation in Africa and Southeast Asia would cause a reduction in rainfall in nearby regions, although this would be notably smaller in Southeast Asia than Africa. The greatest impact is in South America, where areas far from the deforestation hotspot would face substantial rainfall reductions. As in the previous deforestation scenario, this happens due to wind circulation patterns.

Simulated economic impacts from the future deforestation scenario are presented in table 2A.1. The impacts in this scenario are lower than those in the previous deforestation scenario, largely because the future scenario simulates the loss of forest in relatively small, isolated areas.

Compared with the previous deforestation scenario, there is a notable shift, with countries in Africa and Southeast Asia being the most impacted. In terms of gross domestic product growth, Equatorial Guinea and Gabon become the most sensitive. In terms of total annual GDP losses, Indonesia is the most at risk, with annual losses estimated to reach $730 million.

TABLE 2A.1 Summary of simulated annual economic losses due to losses in rainfall caused by future deforestation hotspots

Rainforest region	Energy potential (gigawatts lost)	Agricultural productivity (yield decline, %)	Economywide GDP (millions of $ lost)
Amazon	1.2	0.04	268–1,029
Congo	0.8	0.01	99–567
Southeast Asia	1.0	0.02	28–284

Source: Original calculations based on Araujo and Hector 2024.
Note: GDP = gross domestic product.

Notes

1. Refer to https://climatewaterproject.substack.com/p/measuring-the-small-water-cycle-the.
2. This phenomenon is part of the "small water cycle." Refer to https://climatewaterproject.substack .com/p/map-of-the-small-water-cycle.
3. *Planetary boundary* refers to the environmental limits within which humanity can safely operate.
4. The estimates can be interpreted as approximating how the causal effect of drought differs in locations with varying upstream forest characteristics, rather than a causal effect of changing forest quality.
5. The experience of the Loess Plateau illustrates the evolving understanding of land-water linkages. Early phases of China's "Grain for Green" program focused on rapidly stabilizing degraded slopes using monoculture plantations of fast-growing, often non-native species, which reduced soil erosion but also lowered soil moisture and downstream water availability (World Bank 2003). Subsequent restoration efforts shifted toward more diverse, ecologically appropriate vegetation, improving soil infiltration and stabilizing baseflows (Gong et al. 2024).

References

Adger, W. N., N. W. Arnell, and E. L. Tompkins. 2005. "Successful Adaptation to Climate Change across Scales." *Global Environmental Change* 15 (2): 77–86.

Araujo, R., and V. Hector. 2024. "Deforestation, Rainfall, and Energy: Mapping the Impacts of Deforestation on Rainfall and Hydropower Generation." Background paper for this report. World Bank, Washington, DC.

Balboni, C., A. Berman, R. Burgess, and B. A. Olken. 2023. "The Economics of Tropical Deforestation." *Annual Review of Economics* 15 (1): 723–54.

BenYishay, A., R. Sayers, K. Singh, et al. 2024. "Irrigation Strengthens Climate Resilience: Long-Term Evidence from Mali Using Satellites and Surveys." *PNAS Nexus* 3 (2): 022.

Chausson, A., B. Turner, D. Seddon, et al. 2020. "Mapping the Effectiveness of Nature-Based Solutions for Climate Change Adaptation." *Global Change Biology* 26 (11): 6134–55.

Chen, J., M. Gao, S. Cheng, et al. 2022. "Global 1km×1km Gridded Revised Real Gross Domestic Product and Electricity Consumption during 1992–2019 Based on Calibrated Nighttime Light Data." *Scientific Data* 9 (1): 202.

Creed, I. F., J. A. Jones, E. Archer, et al. 2019. "Managing Forests for Both Downstream and Downwind Water." *Frontiers in Forests and Global Change* 2: 64. doi:10.3389/ffgc.2019.00064.

Damania, R., S. Desbureaux, M. Hyland, et al. 2017. *Uncharted Waters: The New Economics of Water Scarcity and Variability*. Washington, DC: World Bank.

Damania, R., S. Desbureaux, and E. Zaveri. 2020. "Does Rainfall Matter for Economic Growth? Evidence from Global Sub-National Data (1990–2014)." *Journal of Environmental Economics and Management* 102: 102335.

De Petrillo, E., S. Fahrländer, M. Tuninetti, et al. 2025. "Reconciling Tracked Atmospheric Water Flows to Close the Global Freshwater Cycle." *Communications Earth & Environment* 6 (347).

Ellison, D., C. E. Morris, B. Locatelli, et al. 2017. "Trees, Forests and Water: Cool Insights for a Hot World." *Global Environmental Change: Human and Policy Dimensions* 43: 51–61.

ERA. 2017. *Fifth Generation of ECMWF Atmospheric Reanalyses of the Global Climate*. Brussels, Belgium: Copernicus Climate Change Service Climate Data Store, European Commission.

European Space Agency. 2017. *Land Cover CCI Product User Guide Version 2.0*. Paris: European Space Agency. https://www.esa-landcover-cci.org.

Fleischman, F., S. Basant, A. Chhatre, et al. 2020. "Pitfalls of Tree Planting Show Why We Need People-Centered Natural Climate Solutions." *Bioscience* 70 (11): 947–50.

Gies, E. 2023. "More than Carbon Sticks." *Nature Water* 1 (10): 820–23.

Global Commission on the Economics of Water. 2024. *The Economics of Water: Valuing the Hydrological Cycle as a Global Common Good*. Paris: Global Commission on the Economics of Water.

Gong, C., Q. Tan, G., Liu, and M. Xu. 2024. "Positive effects of mixed-species plantations on soil water storage across the Chinese Loess Plateau." *Forest Ecology and Management* 552, p.121571.

Hall, R. L. 2003. "Interception Loss as a Function of Rainfall and Forest Types: Stochastic Modelling for Tropical Canopies Revisited." *Journal of Hydrology* 280 (1–4): 1–12.

Hansen, M. C., P. V. Potapov, R. Moore, et al. 2013. "High-Resolution Global Maps of 21st-Century Forest Cover Change." *Science* 342 (6160): 850–53.

Harrigan, S., E. Zsoter, L. Alfieri, et al. 2020. "GloFAS-ERA5 Operational Global River Discharge Reanalysis 1979–Present." *Earth System Science Data* 12 (3): 2043–60.

Harwood, T., J. Love, M. Drielsma, C. Brandon, and S. Ferrier. 2022. "Staying Connected: Assessing the Capacity of Landscapes to Retain Biodiversity in a Changing Climate." *Landscape Ecology* 37 (12): 3123–39.

Hua, F., L. A. Bruijnzeel, P. Meli, et al. 2022. "The Biodiversity and Ecosystem Service Contributions and Trade-Offs of Forest Restoration Approaches." *Science* 376 (6595): 839–44.

Kirschke, S., A. Staszak, and M. V. Vliet. 2021. "Climate Change Adaptation Pathways for Water Infrastructure: A Review of Global Case Studies." *Journal of Environmental Management* 291: 112650. doi:10.1016/j.jenvman.2021.112650.

Kotz, M., A. Levermann, and L. Wenz. 2022. "The Effect of Rainfall Changes on Economic Production." *Nature* 601 (7892): 223–27.

Krishnaswamy, J., M. Bonell, B. Venkatesh, et al. 2013. "The Groundwater Recharge Response and Hydrologic Services of Tropical Humid Forest Ecosystems to Use and Reforestation: Support for the 'Infiltration-Evapotranspiration Trade-Off Hypothesis.'" *Journal of Hydrology* 498: 191–209.

Kummu, M., M. Taka, and J. H. Guillaume. 2018. "Gridded Global Datasets for Gross Domestic Product and Human Development Index over 1990–2015." *Scientific Data* 5 (1): 1–15.

Lesiv, M., D. Schepaschenko, M. Buchhorn, et al. 2022. "Global Forest Management Data for 2015 at a 100 m Resolution." *Scientific Data* 9: 199. doi:10.1038/s41597-022-01332-3.

Liang, J., J. G. P. Gamarra, N. Picard, et al. 2022. "Co-Limitation towards Lower Latitudes Shapes Global Forest Diversity Gradients." *Nature Ecology and Evolution* 6: 1423–37. doi:10.1038/s41559-022-01831-x.

Lobell, D. B., J. M. Deines, and S. D. Tommaso. 2020. "Changes in the Drought Sensitivity of US Maize Yields." *Nature Food* 1 (11): 729–35.

Ma, J., J. Li, W. Wu, and J. Liu. 2023. "Global Forest Fragmentation Change from 2000 to 2020." *Nature Communications* 14: 3752. doi:10.1038/s41467-023-39221-x.

Makarieva, A. M., A. V. Nefiodov, A. Rammig, and A. D. Nobre. 2023. "Re-Appraisal of the Global Climatic Role of Natural Forests for Improved Climate Projections and Policies." *Frontiers in Forests and Global Change* 6: 1150191.

Manna, D. 2024. "Why Repairing Forests Is Not Just about Planting Trees." *Nature* 633 (8028): 30–31.

Manoli, G., S. Bonetti, J. C. Domec, M. Putti, G. Katul, and M. Marani. 2014. "Tree Root Systems Competing for Soil Moisture in a 3D Soil–Plant Model." *Advances in Water Resources* 66: 32–42.

Matsuura, K., and C. J. Willmott. 2018. "Terrestrial Air Temperature and Precipitation: Monthly and Annual Time Series (1900–2017)." http://climate.geog.udel.edu/~climate/html_pages/download.html.

Mehta, P., S. Siebert, M. Kummu, et al. 2024. "Half of Twenty-First Century Global Irrigation Expansion Has Been in Water-Stressed Regions." *Nature Water* 2 (3): 254–61.

Mekonnen, M. M., and A. Y. Hoekstra. 2016. "Four Billion People Facing Severe Water Scarcity." *Science Advances* 2 (2): e1500323.

Pacheco, P., K.-K. Mo, N. Dudley, et al. 2021. *Deforestation Fronts: Drivers and Responses in a Changing World*. Gland, Switzerland: World Wildlife Fund.

Richardson, K., W. Steffen, W. Lucht, et al. 2023. "Earth beyond Six of Nine Planetary Boundaries." *Science Advances* 9 (37): eadh2458.

Spracklen, D. V., S. R. Arnold, and C. Taylor. 2012. "Observations of Increased Tropical Rainfall Preceded by Air Passage over Forests." *Nature* 489 (7415): 282–85.

Tuinenburg, O. A., and A. Staal. 2020. "Tracking the Global Flows of Atmospheric Moisture and Associated Uncertainties." *Hydrology and Earth System Sciences* 24 (5): 2419–35.

Wang, Y., J. Mao, M. Jin, et al. 2021. "Development of Observation-Based Global Multilayer Soil Moisture Products for 1970 to 2016." *Earth System Science Data* 13: 4385–4405.

Wang-Erlandsson, L., A. Tobian, R. J. Van der Ent, et al. 2022. "A Planetary Boundary for Green Water." *Nature Reviews Earth & Environment* 3 (6): 380–92.

World Bank. 2003. "China: Loess Plateau Watershed Rehabilitation Project." Washington, DC. https://documents1.worldbank.org/curated/en/142661468762366534/pdf/307770CHA0Loess0Plateau01see0also0307591.pdf.

Yu, Z., S. Liu, J. Wang, et al. 2019. "Natural Forests Exhibit Higher Carbon Sequestration and Lower Water Consumption Than Planted Forests in China." *Global Change Biology* 25 (1): 68–77.

Zaveri, E., J. Russ, A. Khan, R. Damania, E. Borgomeo, and A. Jägerskog. 2021. *Ebb and Flow, Volume 1: Water, Migration, and Development*. Washington, DC: World Bank. http://hdl.handle.net/10986/36089.

Zaveri, E., R. Damania, and N. Engle. 2023. "Droughts and Deficits: The Global Impact of Droughts on Economic Growth." Policy Research Working Paper WPS 10453, World Bank, Washington, DC.

Zhang, M., and X. Wei. 2021. "Deforestation, Forestation, and Water Supply." *Science* 371 (6533): 990–91.

The Significance of Soil: Regenerative Agriculture as an Imperative

Soil health is the foundation of a productive and resilient agricultural system. Sustainable agricultural practices that include regenerative solutions can go a long way toward rebuilding healthy soil. Although lacking a formal definition, *regenerative agriculture* prioritizes restoring and enhancing soil health by increasing organic matter and fostering a thriving soil microbiome (Dabalen, Goyal, and Song 2024; Schreefel et al. 2020; Soliman 2024). It shares similarities with other sustainable farming practices—such as conservation agriculture—but differs from conventional farming, which primarily focuses on maximizing yields, often at the expense of long-term soil health and ecosystem stability (Dabalen, Goyal, and Song 2024; box S2.1). Since regenerative practices can enhance crop resilience to drought and extreme weather, they are also a vital part of climate-proofing agriculture (Soliman 2024).

The effectiveness of regenerative approaches depends on local conditions, underscoring the need for context-specific strategies that balance sustainability with food production. Well-established techniques—such as no-till farming, mulching, pitting, contouring, and terracing—help retain soil moisture by reducing evaporation, increasing infiltration, and preventing runoff (He and Rosa 2023). Cover cropping, where non-harvested plants are grown to protect the soil, can also protect against soil erosion, enrich soil nutrients, and support the microbial community, which is essential for soil health. Similarly, crop rotation—swapping out crops, rather than growing the same monoculture in the same field for years on end—can improve soil health without sacrificing productivity (Soliman 2024; Venter, Jacobs, and Hawkins 2016). One study found that diversified cropping increased yields by 14 percent and biodiversity by 24 percent (Beillouin et al. 2021), and broader evidence has shown that such systems enhance both yields and ecosystem services 63 percent of the time (Tamburini et al. 2020).

BOX S2.1
Recipe for a livable planet: Climate and beyond

The world's food system requires urgent reform because it is damaging the planet and plays a major role in climate change. It significantly impacts the global environment and is responsible for deforestation and a third of global greenhouse gas emissions. The largest contributors are livestock and net forest conversion—accounting for 25.9 and 18.4 percent of agri-food system emissions, respectively.

A World Bank report, *Recipe for a Livable Planet: Achieving Net Zero Emissions in the Agrifood System*, outlined actions that can address these challenges (Sutton, Lotsch, and Prasann 2024). Although that report's primary focus was climate change mitigation, there were synergies with other environmental assets, such as soil moisture. Many of the recommended interventions—such as agroforestry, improved land management, and forest restoration—not only lower emissions, but also build agricultural resilience. These practices improve soil structure, enhance water-holding capacity, and reduce reliance on chemical inputs. By fostering healthier root systems and supporting soil biology that feeds plants naturally, they help in sustaining productivity and adapting to climate extremes—benefits that go far beyond carbon alone.

High-income countries (HICs) can lead the way, supporting low-income countries (LICs) and middle-income countries (MICs) by providing financial aid, transferring technology, and leading capacity-building initiatives. Other actions include adopting low-emission farming methods and technologies, offering technical assistance for forest conservation programs that generate high-integrity carbon credits, and shifting subsidies away from high-emission food sources.

If MICs transitioned to more sustainable land use, they could cost-effectively cut one-third of global agri-food emissions. MIC emissions are higher, partly due to sheer numbers (there are many more MICs than HICs or LICs). Effective mitigation strategies include conserving, managing, and restoring forests and ecosystems, especially in tropical areas, which could prevent 5 gigatonnes of carbon dioxide emissions annually in MICs alone.

LICs can achieve sustainable economic development and greener, more competitive economies by avoiding the mistakes of richer nations and embracing climate-smart opportunities. Forest restoration can help them meet climate goals and drive development, offering a net benefit of $7–$30 for every dollar invested through ecosystem services. Agroforestry, which integrates trees into croplands, provides additional benefits such as increased land productivity, livelihood opportunities, diversified diets, and enhanced ecosystem resilience.

Source: Sutton, Lotsch, and Prasann (2024).

There are also links between soil health and the nutritional content of food. Nutrient-rich soil contributes to the growth of crops that are higher in essential vitamins and minerals (Lal 2009). Healthy soils also support diverse microbial communities, which are vital for nutrient cycling and plant health. However, the imbalanced use of fertilizers has created widespread deficiency of secondary and micronutrients such as iron, manganese, sulfur, and zinc in the soil, reducing the nutritional content of food (De Groote et al. 2021). Zinc deficiency in soil, for example, is an important constraint to crop production and the most ubiquitous micronutrient deficiency in crops worldwide. Emerging research has suggested a connection between soil health and gut health, as nutrient-dense food—grown in healthy soil—may support a more balanced gut microbiome, although more evidence is needed. In Nepal's Tarai, low soil zinc levels were causally linked to child stunting, highlighting the need for soil health improvements as a public health priority (Bevis, Kim, and Guerena 2023).

Economic barriers remain a significant challenge for the widespread adoption of regenerative agriculture. Although regenerative agriculture can increase long-term farm profitability, high upfront investment costs often prevent farmers from making the transition (Gill et al. 2025). A meta-analysis found that nature-based solutions increased farmers' profits by an average of 19 percent, but the initial costs—such as implementing erosion control or reduced tillage—can be prohibitive. Erosion control measures alone raised costs by 7 percent, while reduced tillage required a 62 percent increase in upfront investment (Steward et al. 2023). For smallholder farmers in low- and middle-income countries, these financial barriers create significant liquidity constraints, limiting their ability to adopt regenerative techniques. This suggests that these transition costs should be acknowledged and treated akin to capital and investment expenditures that are often subsidized when there are beneficial externalities.

Yet despite these costs, new research has shown that the benefits of regenerative practices can be financially worthwhile over time. A meta-analysis conducted for this report (Gill and Srivastava 2024), covering 66 studies from 17 countries, assessed four regenerative practices focused on improving soil moisture: agroforestry, biochar, cover crops, and residue mulching. The findings are encouraging. On average, these practices increase gross revenues more than costs, resulting in stronger profit margins at the farm level. Residue mulching in particular emerged as the most profitable, improving soil structure and moisture retention. The analysis further showed that returns increase with each additional year of implementation, consistent with the view that nature-based practices often take time to mature and that learning-by-doing can lower costs. Still, for smallholder farmers facing liquidity constraints and uncertain payoffs, even profitable solutions may remain out of reach without targeted support.

Redirecting subsidies and offering financial incentives can drive the adoption of regenerative practices. Governments worldwide allocate $650 billion annually in agricultural subsidies that often promote high-input, yield-driven production systems that degrade soil health (Damania et al. 2023). Agriculture currently accounts for over 30 percent of global greenhouse gas emissions, generating up to $10 trillion in hidden environmental and social costs (Gill et al. 2025). Repurposing a portion of the subsidies toward sustainable land management practices could encourage farmers to adopt regenerative practices. Conservation programs under the US Farm Bill and agri-environmental measures and eco-schemes in the European Union's Common Agricultural Policy are examples of initiatives that support sustainable agriculture. In Africa, there is growing policy momentum around soil health, including shifts in fertilizer strategies that emphasize not only boosting yields but also enhancing long-term soil productivity.

More targeted financial support is needed, particularly in developing countries. Temporary subsidies, credit guarantees, and matching grants can help offset transition costs, ensuring that farmers, particularly in low-income regions, have the financial flexibility to implement regenerative techniques (Gill et al. 2025). By creating the right enabling conditions, governments and development practitioners can unlock the potential of regenerative agriculture—ensuring that the "secret in our dirt" becomes the foundation for a more resilient global food system (Soliman 2024).

While regenerative agriculture strengthens resilience at the farm level, nature-based solutions (NbS) can operate at larger landscape and urban scales. Box S2.2 highlights the role of off-farm NbS in managing water, climate, and ecosystems.

BOX S2.2
Beyond the farm: Nature-based solutions for land, water, and climate resilience

Nature-based solutions (NbS) harness ecosystems to deliver resilience and development benefits. They work with, rather than against, nature by embedding natural processes into engineering designs. Common forms include green or blue infrastructure such as mangroves, wetlands, and riparian buffers. NbS often combine with gray infrastructure in hybrid systems—like seawalls paired with coastal vegetation—to reduce flood and erosion risks (Danielsen et al. 2005).

Restoring forests and natural landscapes improves soil and reduces erosion. Afforestation and assisted natural regeneration (ANR) can lower sediment runoff, enhance soil carbon, and increase green water storage. China's Conversion of Cropland to Forests Program reduced erosion and improved soil health (Jia et al. 2017; Trac et al. 2013), and ANR has shown strong results in tropical regions (Ilstedt et al. 2007; Yang et al. 2018). In contrast, monoculture plantations may cause water stress in arid areas (Cao 2008).

Wetlands and coastal ecosystems act as natural flood defenses. Coral reefs, salt marshes, and mangroves reduce wave heights by 70, 72, and over 90 percent, respectively (Danielsen et al. 2005; Narayan et al. 2016). Losing inland wetlands increases flood damage, with US studies showing losses up to $8,290 per hectare in developed areas (Taylor and Druckenmiller 2022). In cities, bioswales, green roofs, and constructed wetlands help manage stormwater (Hobbie and Grimm 2020; Kabisch et al. 2017).

NbS contribute to both climate mitigation and adaptation. Improved land management and restoration could deliver 37 percent of the emissions reductions needed by 2030 (Griscom et al. 2017). Their adaptive benefits have been well documented: 59 percent of empirical cases have shown reduced climate impacts such as erosion and flood risk, post-intervention (Chausson et al. 2020).

(continued)

Effectiveness tends to be highest in projects that restore or protect natural ecosystems, which typically outperform more artificial or heavily engineered interventions (Chausson et al. 2020). This is because natural ecosystems are better adapted to local conditions and more likely to provide services sustainably over time. In contrast, interventions that involve high levels of artificial creation or rely on monocultures are less reliable and may generate trade-offs.

Emerging approaches like ANR and rewilding offer promising, cost-effective restoration strategies. ANR supports natural regrowth by removing barriers to vegetation recovery, and can reduce restoration costs by 66 percent compared to tree planting (World Bank, forthcoming). In Brazil, restoring the Atlantic Forest through ANR could save $90.6 billion relative to tree planting (Crouzeilles et al. 2020). In China, ANR reduced surface runoff by up to 50 percent while increasing plant diversity and aboveground biomass (Yang et al. 2018). Rewilding—through species reintroduction or river and wetland restoration—has also boosted carbon storage and ecosystem function (Berzaghi et al. 2022; Sobral et al. 2017), although social acceptance and land-use trade-offs can be limiting factors (Ripple and Beschta 2012).

Despite growing momentum, scaling up NbS faces financing, governance, and design challenges. Many NbS are underused in low-income countries where returns are high but fiscal and institutional capacity is limited. Economic appraisals often fail to capture non-market co-benefits such as biodiversity or recreation (Wild, Henneberry, and Gill 2017). Governance fragmentation and weak stakeholder coordination further constrain scale (Seddon et al. 2020; Vojinovic et al. 2021).

To succeed at scale, NbS must be integrated into development planning and treated as essential infrastructure. Policy frameworks should promote hybrid gray–green solutions, expand research on co-benefits and cost-effectiveness, and foster multi-stakeholder governance. With the right support, NbS can provide durable development outcomes.

Source: Zhou and Choi (2024).

References

Beillouin, D., T. Ben-Ari, E. Malézieux, V. Seufert, and D. Makowski. 2021. "Positive but Variable Effects of Crop Diversification on Biodiversity and Ecosystem Services." *Global Change Biology* 27 (19): 4697–4710.

Berzaghi, F., R. Chami, T. Cosimano, and C. Fullenkamp. 2022. "Financing Conservation by Valuing Carbon Services Produced by Wild Animals." *Proceedings of the National Academy of Sciences* 119 (22): e2120426119. www.doi.org/10.1073/pnas.2120426119.

Bevis, L., K. Kim, and D. Guerena. 2023. "Soil Zinc Deficiency and Child Stunting: Evidence from Nepal." *Journal of Health Economics* 87: 102691.

Cao, S. 2008. "Why Large-Scale Afforestation Efforts in China Have Failed to Solve the Desertification Problem." *Environmental Science & Technology* 42 (6): 1826–31. www.doi.org/10.1021/es0870597.

Chausson, A., B. Turner, D. Seddon, et al. 2020. "Mapping the Effectiveness of Nature-Based Solutions for Climate Change Adaptation." *Global Change Biology* 26 (11): 6134–55. www.doi.org/10.1111/gcb.15310.

Crouzeilles, R., H. L. Beyer, L. M. Monteiro, et al. 2020. "Achieving Cost-Effective Landscape-Scale Forest Restoration through Targeted Natural Regeneration." *Conservation Letters* 13 (3): e12709. www.doi.org/10.1111/conl.12709.

Dabalen, A., A. Goyal, and R. Song. 2024. "Regenerative Agriculture in Practice: A Review." Policy Research Working Paper 10919, World Bank, Washington, DC.

Damania, R., E. Balseca, C. De Fontaubert, et al. 2023. *Detox Development: Repurposing Environmentally Harmful Subsidies*. Washington, DC: World Bank.

Danielsen, F., M. K. Sørensen, M. F. Olwig, et al. 2005. "The Asian Tsunami: A Protective Role for Coastal Vegetation." *Science* 310(5748): 643–43. www.doi.org/10.1126/science.1118387.

De Groote, H., M. Tessema, S. Gameda, and N. S. Gunaratna. 2021. "Soil Zinc, Serum Zinc, and the Potential for Agronomic Biofortification to Reduce Human Zinc Deficiency in Ethiopia." *Scientific Reports* 11 (1): 1–11.

Gill, J., C. Pirela, S. Zorya, A. D. Darko, M. Ahmed, and B. Srivastava. 2025. "Nature-Based Solutions: Bridging the Gap between Sustainability and Economic Viability." Agriculture and Food Blog, World Bank, February 26, 2025. https://blogs.worldbank.org/en/agfood/nature-based-solutions--bridging-the-gap-between-sustainability-.

Gill, J., and B. Srivastava. 2024. "Economics of NbS for Improving Resilience to Climate Change." Background paper for this report. World Bank, Washington, DC.

Griscom, B. W., J. Adams, P. W. Ellis, et al. 2017. "Natural Climate Solutions." *Proceedings of the National Academy of Sciences* 114 (44): 11645–50. www.doi.org/10.1073/pnas.1710465114.

He, L., and L. Rosa. 2023. "Solutions to Agricultural Green Water Scarcity under Climate Change." *PNAS Nexus* 2 (4): gad117.

Hobbie, S. E., and N. B. Grimm. 2020. "Nature-Based Approaches to Managing Climate Change Impacts in Cities." *Philosophical Transactions of the Royal Society B: Biological Sciences* 375 (1794): 20190124. www.doi.org/10.1098/rstb.2019.0124.

Ilstedt, U., A. Malmer, E. Verbeeten, and D. Murdiyarso. 2007. "The Effect of Afforestation on Water Infiltration in the Tropics: A Systematic Review and Meta-Analysis." *Forest Ecology and Management* 251 (1-2): 45–51. www.doi.org/10.1016/j.foreco.2007.06.014.

Jia, X., M. Shao, Y. Zhu, and Y. Luo. 2017. "Soil Moisture Decline Due to Afforestation across the Loess Plateau, China." *Journal of Hydrology* 546: 113–22. www.doi.org/10.1016/j.jhydrol.2017.01.011.

Kabisch, N., H. Korn, J. Stadler, and A. Bonn, eds. 2017. *Nature-Based Solutions to Climate Change Adaptation in Urban Areas: Linkages between Science, Policy and Practice*. Theory and Practice of Urban Sustainability Transitions Series. Springer Nature. https://link.springer.com/book/10.1007/978-3-319-56091-5.

Lal, R. 2009. "Soil Degradation as a Reason for Inadequate Human Nutrition." *Food Security* 1 (1): 45–57.

Narayan, S., M. W. Beck, B. G. Reguero, et al. 2016. "The Effectiveness, Costs and Coastal Protection Benefits of Natural and Nature-Based Defences." *PLoS One* 11 (5): e0154735. www.doi.org/10.1371V/journal.pone.0154735.

Ripple, W. J., and R. L. Beschta. 2012. "Trophic Cascades in Yellowstone: The First 15 Years after Wolf Reintroduction." *Biological Conservation* 145 (1): 205–13. www.doi.org/10.1016/j.biocon.2011.11.005.

Schreefel, L., R. P. Schulte, I. J. de Boer, A. P. Schrijver, and H. H. van Zanten. 2020. "Regenerative Agriculture—The Soil Is the Base." *Global Food Security* 26: 100404. doi:10.1016/j.gfs.2020.100404.

Seddon, N., A. Chausson, P. Berry, C. A. J. Girardin, A. Smith, and B. Turner. 2020. "Understanding the Value and Limits of Nature-Based Solutions to Climate Change and Other Global Challenges." *Philosophical Transactions of the Royal Society B: Biological Sciences* 375 (1794): 20190120. www.doi.org/10.1098/rstb.2019.0120.

Sobral, M., K. M. Silvius, H. Overman, L. F. B. Oliveira, T. K. Raab, and J. M. V. Fragoso. 2017. "Mammal Diversity Influences the Carbon Cycle through Trophic Interactions in the Amazon." *Nature Ecology & Evolution* 1 (11): 1670–76. www.doi.org/10.1038/s41559-017-0334-0.

Soliman, A. 2024. "How to Climate-Proof Crops: Scientists Say the Secret's in the Dirt." *Nature*, November 5. doi:10.1038/d41586-024-03480-5.

Steward, P., N. Joshi, G. Kacha, et al. 2023. "Economic Benefits and Costs of Nature-Based Solutions in Low- and Middle-Income Countries." Working Paper, Alliance of Bioversity-CIAT, Rome.

Sutton, W. R., A. Lotsch, and A. Prasann. 2024. *Recipe for a Livable Planet: Achieving Net Zero Emissions in the Agrifood System.* Agriculture and Food Series. Washington, DC: World Bank. http://hdl.handle.net/10986/41468.

Tamburini, G., R. Bommarco, T. C. Wanger, et al. 2020. "Agricultural Diversification Promotes Multiple Ecosystem Services without Compromising Yield." *Science Advances* 6 (45): eaba1715.

Taylor, C. A., and H. Druckenmiller. 2022. "Wetlands, Flooding, and the Clean Water Act." *American Economic Review* 112 (4): 1334–63. www.doi.org/10.1257/aer.20210497.

Trac, C. J., A. H. Schmidt, S. Harrell, and T. M. Hinckley. 2013. "Environmental Reviews and Case Studies: Is the Returning Farmland to Forest Program a Success? Three Case Studies from Sichuan." *Environmental Practice* 15 (3): 350–66. www.doi.org/10.1017/S1466046613000355.

Venter, Z. S., K. Jacobs, and H. J. Hawkins. 2016. "The Impact of Crop Rotation on Soil Microbial Diversity: A Meta-Analysis." *Pedobiologia* 59 (4): 215–23.

Vojinovic, Z., A. Alves, J. P. Gómez, et al. 2021. "Effectiveness of Small- and Large-Scale Nature-Based Solutions for Flood Mitigation: The Case of Ayutthaya, Thailand." *Science of The Total Environment* 789 (October): 147725. www.doi.org/10.1016/j.scitotenv.2021.147725.

Wild, T. C., J. Henneberry, and L. Gill. 2017. "Comprehending the Multiple 'Values' of Green Infrastructure—Valuing Nature-Based Solutions for Urban Water Management from Multiple Perspectives." *Environmental Research* 158: 179–87. www.doi.org/10.1016/j.envres.2017.05.043.

World Bank. Forthcoming. *Assessing Forest Restoration Potential across Low-and-Middle-Income Countries.* Washington, DC: World Bank.

Yang, Y., L. Wang, Z. Yang, et al. 2018. "Large Ecosystem Service Benefits of Assisted Natural Regeneration." *Journal of Geophysical Research: Biogeosciences* 123 (2): 676–87. www.doi.org/10.1002/2017JG004267.

Zhou, W., and E. Choi. 2024. "Nature-Based Solutions for Sustainable Development." Background paper for this report. World Bank, Washington, DC.

CHAPTER 3

Biodiversity and the Economic Value of Species Abundance

> "Like winds and sunsets, wild things were taken for granted until progress began to do away with them."
>
> —*Aldo Leopold*, American writer and scientist (1887–1948)

Key messages

- Biodiversity is in crisis, with all major indicators showing rapid declines. This matters for development because biodiversity affects ecosystems that in turn have economic impacts.

- Information on the linkages between biodiversity and the economy is limited. Nevertheless, a rapidly growing empirical literature provides invaluable insights and identifies three major pathways through which impacts are channeled:

  - The introduction or removal of a single species can have wide-reaching— and often unforeseen—economic consequences. Interactions between species are complex, but new data have been helpful in revealing impacts on the economy, which can be surprisingly significant.

  - The diversity and abundance of species are crucial determinants of the productivity and resilience of ecosystem services, affecting everything from agricultural yields, to soil moisture and water purification.

  - Ecosystems regulate temperature, rainfall, nutrient recycling, water flows, and other environmental conditions. Disruptions to these regulatory functions can have substantial economic impacts.

- Countries can use various approaches to reverse the decline of ecosystems, from payments for ecosystem services, to conservation tenders, to protected areas, and collaborative management partnerships. There is no policy panacea, although well-designed policies that are mindful of incentives have been more successful than blunt regulations.

Introduction

The Indian peacock, with its iridescent plumage and extravagant tail feathers, stands out as one of the most beautiful birds on Earth. In stark contrast, the vulture, with its unkempt appearance and scavenging habits, may never be celebrated for its beauty or grace. Yet the vulture plays a pivotal economic role, far beyond its unassuming looks. A path-breaking study by Frank and Sudarshan (2024), in the prestigious *American Economic Review*, tracked a hidden link between the collapse of vulture populations in South Asia and the resulting economic damages, estimated at a staggering $69 billion annually. Vultures feed primarily on the carcasses of dead animals and are equipped with stomach acids that neutralize harmful bacteria and toxins. Their disappearance left a void, leading to rotting carcasses that became breeding grounds and vectors for infectious diseases like rabies and anthrax, costing billions in mortality and morbidity.

As this example highlights, biodiversity loss often has unpredictable effects and is typically irreversible. Ecosystems are highly complex, with species interconnected through a web of relationships such as predation, competition, and symbiosis. The loss of a single species in a food web can cascade through an ecosystem with surprising economic consequences as the example of the loss of vultures illustrates.[1] Habitat loss and fragmentation are the main causes of biodiversity loss. The size of habitat patches and their connectivity are crucial. As smaller patches are created, the diversity within them declines. Recent assessments have suggested that the ecosystem services provided by biodiversity, such as the capacity to pollinate crops, regulate pests, and limit soil erosion, decline significantly when a habitat area falls below 20–25 percent per square kilometer (Mohamed et al. 2024). This chapter highlights recent empirical research that has unveiled some of the unseen dependencies and economic consequences.

Biodiversity is in crisis and in the midst of a "mass extinction" (McCallum 2015). Approximately 75 percent of the world's land surface has been significantly altered by humanity, 70 percent of monitored wildlife has vanished in the past 40 years, and 60 percent of global extinctions have been influenced by the introduction of invasive alien species (IPBES 2019, 2023). Of all the mammalian biomass on Earth today, wild mammals comprise a mere 5 percent—the remaining 95 percent consists of humans (35 percent) and mammals for human consumption (60 percent). Such is the impact of the anthropogenic forces that drive this decline, much of which is due to habitat loss driven by the agriculture, forestry, and mining sectors (McElwee et al. 2024).

Despite the magnitude of these shifts, understanding of the economic consequences is limited. The term biodiversity was coined but half a century ago (Wilson and Peter 1988) and is only now beginning to capture policy attention. In general, the costs of species extinction are hard to estimate as there is a scarcity of comprehensive data linking ecosystems to their economic impacts. Most available information tends to focus on marginal changes in ecosystems, which limits understanding of the consequences of large collapses. In addition, establishing causal relationships is challenging due to the lack of extensive data on species population counts.

Estimates based on accounting approaches suggest that more than half of the world's total gross domestic product is "moderately or highly dependent" on nature and its services (World Economic Forum 2020). New research by S&P has indicated that 85 percent of the world's largest companies that make up the S&P Global 1200 Index have a significant dependency on nature and biodiversity across their direct operations (S&P Global 2023). The consequences of biodiversity loss also reverberate through financial markets—albeit to a more limited extent (Cherief, Sekine, and Stagnol 2025; Giglio et al. 2024). There have also been heroic efforts to model the macroeconomic impacts of biodiversity losses (for example, Costanza et al. 1997; Johnson et al. 2021). However, in the absence of sufficient information on the way biodiversity connects with the economy, or data on the magnitude involved, and the functional form of these interactions, such efforts remain informative but incomplete. As a result, there are more unknowns than knowns in this nascent field of research.

This chapter delves into new and groundbreaking empirical research on the interaction between the economy and biodiversity. At least three pathways have been identified in recent research on how biodiversity and ecosystems exert economic impacts. First, species within a food web are interconnected, implying that alterations in species composition can trigger chain reactions—known as trophic cascades. The literature has suggested that these can have significant, although often elusive, economic consequences. These are among the most complex and least understood issues in environmental economics. Second, the diversity and abundance of species are crucial determinants of the productivity and resilience of ecosystem services, affecting everything from agricultural yields to soil moisture and water purification. Third, well-functioning ecosystems can regulate temperature, evapotranspiration, nutrient recycling, water flows, and other environmental conditions. Disruptions to these regulatory functions can have substantial economic repercussions. A better understanding of these pathways is critical to improving economic efficiency and policy effectiveness.

Species interactions

To a large degree, ecosystems are shaped and defined by the way species interact, which affects the flow of nutrients and energy (Estes et al. 2011). Species interactions can be direct or indirect. The former include predator-prey relationships, competition over resources, and provision of mutual benefits. *Trophic cascades* describe situations where population changes in one part of the food web directly or indirectly affect the abundance and behavior of species in another part of the web. These can be *top-down cascades*, such as the predatory pressure on herbivores changing the abundance of vegetation, or *bottom-up cascades*, such as a decline in plankton narrowing the food base and resulting in lower fish population levels further up the food chain. Indirect interactions can also occur when species change the environment by affecting the flow of water or impacting the microclimate, which then impacts other species.

The stabilizing role of biodiversity in an ecosystem can deteriorate well before any species becomes locally extinct. Depleted population levels of species can result in "functional extinctions" and the rapid degradation of ecosystem services (Valiente-Banuet et al. 2014). Although the thresholds for these tipping points where species experience a collapse in their ability to perform their function are hard to quantify, they are recognized as vital to understanding how biodiversity losses can affect human well-being (Díaz et al. 2006). This is particularly so when considering species that have an effect on the ecosystem that is disproportionate to their abundance (Power et al. 1996).

This section considers the economic impacts of three types of species interaction: predation, biological pest control, and damage from invasive species.

Economic implications of predation

Top-down control of herbivores can dramatically alter ecosystems, with impacts ranging from flood risk to pest control. An iconic and much debated example is the reintroduction of gray wolves into Yellowstone National Park in the United States (Beschta and Ripple 2006, 2019). Their return reduced browsing by herbivores on vegetation by up to 80 percent (Ripple and Beschta 2003, 2012), leading to cascading impacts from vegetation recovery and the return of beavers, and even altered river flows (Allen et al. 2017; Mech 2017). Beavers thrive in habitats created by more robust willow growth, which in turn depends on reduced herbivory (consumption of plant material). Analysis has shown that willow regeneration is better explained by browsing pressures than by climate shifts (Ripple and Beschta 2006).

Although debate remains over the precise causal chain in Yellowstone, the broader pattern of trophic cascades has been well-established. As presented in chapter 2 of this report, trophic cascades can have significant impacts on water regulation services, enhancing soil moisture regulation and buffering the impacts of droughts—benefits with clear economic value.

Biodiversity's contributions can often manifest in unexpected places. The recolonization of wolves in Wisconsin, United States, has been linked to a 23.7 percent drop in deer-vehicle collisions due to behavioral shifts in deer prompted by the "ecology of fear" (Raynor, Grainger, and Parker 2021). Similar effects have been observed in Quebec, Canada, where wolf presence correlates with fewer animal-related traffic accidents (Frank, Missiran, et al. 2024). Wolves and other large predators often hunt near roads or trails, pushing deer away from these high-risk areas (Ripple and Beschta 2004).

Predation also benefits aquatic environments. In coastal regions, sea otters help restore wetlands by controlling crab populations, which overgraze marsh vegetation. This predation improves plant biomass and soil stability (Dedman et al. 2024; Heithaus et al. 2014; Hughes et al. 2024).

Economic implications of biological pest control

Trophic interactions foster a balanced and diverse assemblage of species, which in turn may provide natural pest control services and promote ecosystem stability. However, despite growing evidence of the scale of beneficial services provided through natural pest control, the importance of biological pest control remains largely underappreciated until damage becomes visible.

The extent of damage resulting from the loss of biological pest control services usually becomes apparent only when the harmful effects manifest in tangible losses. In South Asia the use of the painkiller diclofenac to treat cattle proved lethal to vultures, causing their sudden collapse. As noted in the Introduction to this chapter, the collapse of the vulture population resulted in the loss of sanitation services. This culminated in diseases such as anthrax and $69.4 billion in health damages (Frank and Sudarshan 2024).

In another prominent example, the eradication of sparrows in China was a component of the "Four Pests Campaign" launched in 1958, which aimed to eliminate rats, flies, mosquitoes, and sparrows. Sparrows were targeted because they were believed to consume significant quantities of grain. However, the campaign failed to recognize the crucial ecological role that sparrows played in controlling insect populations. The large-scale reduction of sparrow populations led to an insect population boom, which exacerbated crop damage and contributed to significant agricultural losses. This ecological imbalance is considered one of the factors that worsened the Great Chinese Famine (Frank, Wang, et al. 2024).

In the absence of effective biological pest control, farmers are often compelled to increase insecticide applications, with consequences for surrounding communities. Investigating the expansion of a bat-killing wildlife disease known as white-nose syndrome, novel research documented that as local bat populations perished, farmers compensated for the decline in pest control services with higher insecticide use (31 percent), which generated negative health impacts in the form of elevated infant mortality rates (7.9 percent) (Frank 2024). Perhaps more surprisingly, the expansion of this bat disease and the resulting reduction in biological pest control also caused a decline in agricultural rental land values (Manning and Ando 2022).

Economic implications of invasive species

Introducing new species into an ecosystem with no natural population controls can lead to large fluctuations in ecosystem functioning. Such introductions are called non-native species. When they cause sharp, observable changes, they are considered invasive species. All invasive species are non-native, but many non-native species cause little or no meaningful impact.

In 2023, the Intergovernmental Science-Policy Platform on Biodiversity and Ecosystem Services conducted a thematic assessment of invasive alien species. The assessment estimated that damages have "quadrupled every decade since 1970" and cost roughly $423 billion a year (IPBES 2023). Scientists warn about the projected growth in the number of invasive species due to increased international trade, land use change, climate change, and inadequate prevention and control measures (Paini et al. 2016; Pyšek et al. 2020). Quantifying the full scope of damages from all invasive species is challenging; however, estimates for the United States suggest that the damages are more than $100 billion a year (Pimentel et al. 2000; Pimentel, Zuniga, and Morrison 2005; Shwiff et al. 2018).[2] Among other channels, invasive species spread through trade networks (Chapman et al. 2017), exacerbated by the burgeoning demand for exotic plants and wild animals as pets—some of which are illegal (Harrison, Roberts, and Hernandez-Castro 2016; Humair et al. 2015).

Productivity, resilience, and household incomes

A longstanding hypothesis in ecology is that more diverse ecosystems are more productive. A seminal experiment by Tilman, Wedin, and Knops (1996) showed that grassland plots with higher plant diversity have higher productivity. Subsequent analysis has shown that more diverse grassland plots have higher levels of stability or resilience, measured as lower volatility in productivity (Tilman, Reich, and Knops 2006). These two observed relationships are referred to as the *diversity-productivity* and *diversity-stability hypotheses*.[3]

The diversity-productivity and diversity-stability relationships can have significant beneficial economic impacts, which are gradually being investigated. One conclusion ecologists have drawn from the impacts of higher plant diversity is that agricultural systems might be able to increase their yields and reduce year-on-year variation in yields if they reduce their reliance on monoculture plots and plant more diverse plots. This follows from an observed positive correlation between crop diversity and agricultural productivity (Bareille and Dupraz 2020). Research comparing monoculture plots with more diverse plots found that, on average, high-diversity plots outperformed even the most successful monoculture plots in terms of biomass gain in the long run (Tilman et al. 2001). One potential explanation lies in different plant species having different optimal environmental conditions, allowing more diverse plots to fill gaps left open in less diverse ones (Tilman, Lehman, and Thomson 1997). Similarly, chapter 2 demonstrated that natural—and therefore more diverse— forests can buffer the effects of drought, whereas monoculture forest plantations aggravate drought impacts by absorbing more soil moisture when it is most needed for crops.

A large literature has documented the role that forests play as nature's safety net in times of crisis. For example, analysis of data from 23 countries found that rural households compensated for declining incomes due to droughts by increasing extraction from nearby forests by 30 percent (Noack, Riekhof, and Di Falco 2019). More diverse forests buffered larger shares of the negative income shock. A forest that was more diverse (by one standard deviation) could fully offset the reduction in crop income after a drought. This robust finding emerged from different geographies across developing countries.

Productivity can decline even without a change in species diversity, particularly when the distribution and abundance of key species are affected. For instance, in marine ecosystems, plankton are crucial for food webs and a decline in their populations can lead to reduced biomass in higher trophic levels (box 3.1 discusses the broader crisis in ocean ecosystems, including the impacts of pollution, ocean acidification, and overfishing). A study in Japan highlighted that increased use of neonicotinoids has led to an 83 percent decline in zooplankton biomass, resulting in significant drops in smelt and eel yields by 91 and 74 percent, respectively (Yamamuro et al. 2019). In coastal areas, mangroves play a vital role in supporting fish habitats, and research in Indonesia has revealed a fishery income–mangrove elasticity of 5.3–9.8 percent (Yamamoto 2023). The study indicated that households facing income losses from mangrove decline cut back on nonfood consumption and increased labor in other sectors, yet continued to fish despite reduced returns. Protecting mangrove resources has proved economically beneficial, with the mangrove-fishery link valued at $12,364 to $22,861 per hectare annually, compared to $7,610 for aquaculture and $9,630 for palm plantations (Yamamoto 2023).

BOX 3.1
The dismal state of the ocean

Ocean covers over 70 percent of the Earth's surface and is crucial for climate regulation, food supply, and biodiversity. Despite this, the state of the ocean in 2025 is marked by challenges like pollution, changing chemistry, and overfishing beyond economically and ecologically desirable limits.

- *Pollution.* Plastic waste has continued to rise, with 2 million metric tons entering the oceans annually via rivers (OECD 2022). Most plastic remains near coastal areas, posing threats to marine life and human health (Leslie and Depledge 2020; Street et al. 2024) (map B3.1.1). Nutrient pollution from runoff creates dead zones, depleting oxygen and harming marine life (Breitburg et al. 2018). Even in developed countries this leads to large economic losses, often exceeding $2.4 billion annually in some subregions (Boehm 2020).
- *Changing ocean chemistry.* The ocean absorbs carbon dioxide and becomes more acidic, which affects calcifying organisms like corals and mollusks (Guinotte and Fabry 2008;

(continued)

BOX 3.1
The dismal state of the ocean *(continued)*

Hoegh-Guldberg et al. 2014). This disrupts marine ecosystems and food webs. Coral reefs face bleaching from rising temperatures and acidification, with significant losses in the Great Barrier Reef (Bozec et al. 2022). Data platforms can aid reef health by monitoring environmental indicators and providing early warnings.

- *Ocean salinity.* Rising temperatures intensify the hydrological cycle, altering salinity levels and affecting coastal communities that are reliant on fisheries and agriculture (Akhter et al. 2021; Bricheno, Wolf, and Sun 2021; Dasgupta, Wheeler, and Ghosh 2022).
- *Overfishing.* Overfishing depletes fish stocks, with 38 percent harvested unsustainably (FAO 2024). Subsidies and illegal fishing worsen the depletion, although satellite technology aids enforcement. Some improvements have been noted since 2004, but challenges have persisted in balancing exploitation with conservation.

MAP B3.1.1 Global ocean plastic density (mean, 2017–18)

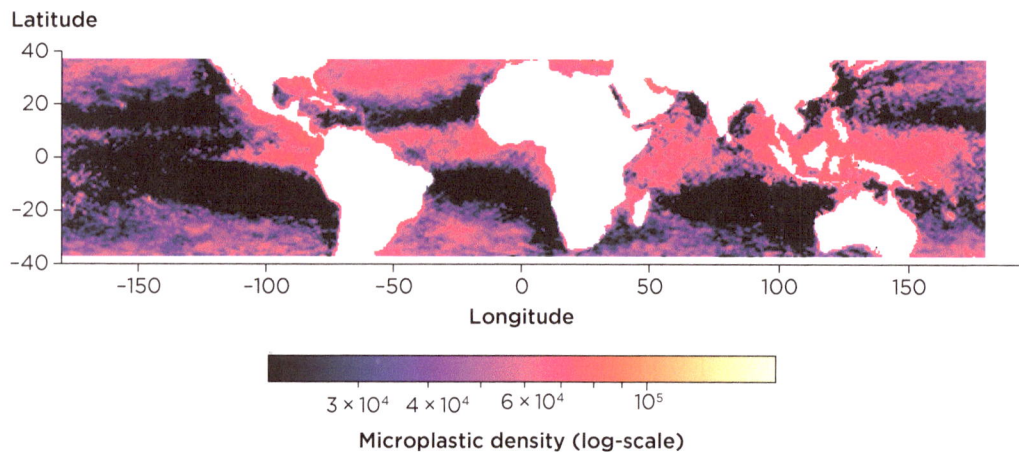

Source: Original calculation based on CYGNSS 2021.
Note: The latitude restriction in the map is due to the data source.

Marine protected areas (MPAs) remain the cornerstone of efforts to mitigate damage to ocean fisheries. MPAs aim to protect 30 percent of the ocean by 2030, offering benefits for fish stocks and ecosystems (from the 15th meeting of the Conference of the Parties to the United Nations Convention on Biological Diversity). Studies have shown that MPAs enhance fish biomass and coral reef resilience (Carassou 2013; Gill et al. 2017). Socioeconomic impacts vary, with successful outcomes linked to no-take, well-enforced, and older MPAs (Ban et al. 2019). Effective MPAs boost local economies and fisheries, although some have faced resistance due to access restrictions (Christie 2004; Weigel et al. 2015). Co-management MPAs, which often involve shared responsibilities among governments, communities, and nongovernmental organizations, have shown to balance ecological and socioeconomic goals, improving governance and compliance (Barley Kincaid, Rose, and Mahudi 2014).

Pollination is a crucial ecological process that is vital for the reproduction of many flowering plants, which in turn provide food and habitat. In addition, approximately 75 percent of the world's food crops depend on animal pollinators, making pollination essential for global food security (Klein et al. 2007). The decline in pollinator populations due to habitat loss, pesticide use, and climate change poses risks to biodiversity and agriculture. Several studies have suggested that the decline of pollinators has led to large economic losses. However, causality is difficult to establish and most work has relied on correlations between pollinator presence and crop yields.

Ambient environmental quality

Well-functioning ecosystems can regulate temperature, evapotranspiration, nutrient recycling, water flows, and other environmental conditions. Disruptions to these regulatory functions can increase air pollution levels and the likelihood of flooding and reduce productivity and land values, leading to the loss of environmental amenities (Berryman 1989; van Breugel et al. 2024). One such example is deforestation in the Amazon region, which has altered regional rainfall patterns, which, in turn, impact agricultural productivity, hydropower generation, and economic performance in affected areas (chapter 2). Linkages have been observed across multiple ecosystems, affecting freshwater, air quality, and temperature (Brockerhoff et al. 2017; D'Odorico et al. 2013; De Frenne et al. 2013; McLaughlin and Cohen 2013; Zellweger et al. 2020).

Trees provide a vast array of nontimber ecosystem services, from regulating temperatures to reducing local pollution, and recent evidence indicates that forested areas boost worker productivity by moderating ambient temperatures. For example, a study in Indonesia that randomly assigned workers to forested and deforested settings revealed a 2.84-degree Celsius difference in wet-bulb temperature[4] and an 8.22-percent decrease in productivity in the hotter, deforested areas (Masuda et al. 2021). Cognitive performance is also worse in deforested settings (Masuda et al. 2020). Another study linked the loss of tree cover due to the emerald ash borer in the United States to a 2.1 percentage point increase in the probability of a relatively hot day, above 32.2 degrees Celsius (Jones 2019). A vast literature has documented the temperature-regulating effects of tree cover in urban areas (refer to chapter 6). For example, research in Toronto, Canada, found that summer month temperatures in neighborhoods with higher tree cover are 0.8 degree Celsius lower, alleviating energy demand for cooling (Han et al. 2024).

Forests also regulate local microclimates, and their impacts extend far downwind or downstream. For example, temperature and precipitation regulation effects from the Great Plains Shelterbelt, a 1930 forestation program in the Midwestern United States, extend 200 kilometers downwind, and counties with high exposure to the shelterbelt have experienced, on average, precipitation that is 3 percent higher and summer-time maximum temperatures that are 0.9 percent lower, boosting corn production by 30 percent (Grosset-Touba, Papp, and Taylor 2024). By lowering the likelihood of crop failure, such changes in environmental conditions also slow down the process of farm consolidation. As noted in chapter 2, intact forests help mitigate drought effects and increase precipitation when winds pass over tropical rainforests, enhancing resilience against dry conditions and delivering essential rainfall for agriculture and hydropower downstream.

Intact forests improve human health by controlling disease vectors and reducing waterborne diseases. In Indonesia, a 1 percent decline in forest cover was linked to a 10 percent rise in malaria cases (Garg 2019). This association was impacted by the biodiversity of forests, with less diverse ecosystems showing stronger links to malaria due to the loss of mosquito predators. Another study found that a 1 percent decline in forest cover increased the probability of childhood diarrheal disease by 5.8 percent and stunting by 3.9 percent due to diminished water quality further downstream (Damania et al. 2024). However, in contrast, a study in China found mixed results from urban afforestation: although air quality improved, hospital visits for allergies increased due to a rise in pollen (Xing et al. 2023).

Reversing the decline of ecosystems

Ecosystem intactness is vital for maintaining natural balance and providing essential services that are crucial for human and economic well-being. Disruptions lead to negative effects on biodiversity and society. Restoring and preserving ecosystems is key to resilience and sustainability. Ecosystem restoration involves activities aimed at recovering ecosystem health and sustainability. Cost-effective forest restoration is achievable, at an estimated $1,382 per hectare, and the most affordable restoration areas are in Latin America and Sub-Saharan Africa (Herrera Garcia, Vincent, and Chang 2024). Common approaches to restoring or conserving ecosystems include the following:

Payment for ecosystem services (PES). PES schemes offer financial incentives to landowners for maintaining or enhancing ecosystem services (Mirzabaev and Wuepper 2023). However, they often face challenges like adverse selection, where payments are made to those already conserving, reducing additional impact. To improve effectiveness, PES programs could incorporate better targeting and monitoring to ensure that payments lead to genuine conservation actions.

Conservation tenders and auctions. These approaches create competition among environmental service providers, aligning private incentives with public conservation goals. By inviting bids for conservation contracts, tenders can achieve better outcomes at lower costs compared to fixed-price schemes. This method is especially effective in regions with well-defined property rights (Bardsley and Burfurd 2009; James, Lundberg, and Sills 2021).

Biodiversity credits. Biodiversity credits involve trading conservation efforts for credits, which can be purchased by entities looking to enhance their biodiversity footprint. Unlike offsets, which compensate for negative impacts, credits focus on proactive conservation. This market-based approach encourages investment in projects that positively impact biodiversity.

Sustainability-linked bonds. These bonds are designed to channel resources into conservation projects, often backed by public finance to attract private investors. They face challenges due to the nonmonetizable nature of biodiversity benefits, making them less profitable. However, they hold potential for funding large-scale restoration efforts if supported by government subsidies or guarantees (Flammer, Giroux, and Heal 2025).

Protected areas and collaborative approaches. Protected areas aim to conserve biodiversity but are often placed where they cause minimal conflict with economic interests, limiting effectiveness. Collaborative management partnerships between state and nonstate actors can leverage resources and expertise, leading to significant conservation gains, such as reduced deforestation and increased biodiversity (Denny, Englander, and Hunnicutt 2024; Desbureaux et al. 2024).

Infrastructure siting. Infrastructure projects like roads and dams can have extensive ecological impacts, often leading to deforestation and habitat loss far beyond the immediate area. Careful planning is essential to avoid ecologically sensitive areas, as the effects of infrastructure extend as far as 50 kilometers (AIIB 2023). Renewable energy projects should be located on already degraded lands to minimize disruptions. This strategic approach helps preserve biodiversity while still allowing for necessary development.

Monitoring biodiversity. Advances in remote sensing technologies, such as drones and satellite imagery, are crucial for real-time monitoring of ecosystems. They provide data to detect habitat loss and assess the effectiveness of conservation efforts. However, data gaps, especially in remote regions, highlight the need for investment in observation technologies and local capacity building (box 3.2).

BOX 3.2
Earth observation of the Amazon

Famed for its biodiversity, the Amazon faces severe deforestation threats, primarily driven by human activities such as agriculture, logging, and mining. Large-scale cattle ranching and soybean farming have led to the clearing of large areas, while illegal logging strips the forest of valuable timber. Infrastructure projects, such as road building, further fragment the ecosystem, opening it up to even more exploitation. This loss of forest disrupts critical biodiversity, endangering countless species and Indigenous communities that rely on the forest for their survival.

(continued)

To combat these challenges, digital technologies are pivotal in enhancing environmental monitoring and supporting climate adaptation. Key requirements include reliable internet connectivity, strong data governance, and active local community involvement with digital skills. Connectivity is crucial but challenging in the Amazon due to its geography. Efforts like Brazil's Infovia project and Peru's National Telecommunications Program aim to improve access by laying fiber-optic cables in riverbeds to connect remote areas (Ministério das Comunicações 2025).

Technologies such as the US National Aeronautics and Space Administration's Landsat satellites and drones provide critical real-time data on forest changes, aiding in the detection of deforestation and illegal activities. These tools enhance enforcement of environmental laws and support sustainable management by providing precise data for conservation efforts. Remote sensors can monitor deforestation and track the carbon cycle without direct interaction with the forest. Digital financial services, where accessible, can support conservation efforts and sustainable economic activities by enabling efficient financial transactions, such as in payment-for-ecosystem services programs. This supports eco-friendly farming and reduces reliance on deforestation-driven industries.

Conclusion

Empirical evidence is increasingly revealing the vital roles that species play in production functions that contribute to human welfare. However, further research is needed as the evidence remains fragmentary. Much of the robust quantifiable research has been conducted in high-income countries, primarily due to better data availability, highlighting a knowledge gap in developing countries, although measurement and data collection are improving. The World Bank Group's Global Biodiversity Species Global Grid database is one initiative addressing this gap by providing spatially disaggregated occurrence data for more than 600,000 species (box 3.3).

Limited access to up-to-date data, especially in developing regions, has hindered conservation efforts and quantitative analysis. Traditional metrics focus mainly on vertebrates, while emerging threats outpace slow data updates.

To address this gap, Dasgupta, Blankespoor, and Wheeler (2024) used species occurrence records by the Global Biodiversity Information Facility (GBIF) to reassess the spatial distribution of global biodiversity. GBIF's network includes more than 2 million species occurrences, with around 1.3 million new records added over the past two years. Most records include location data, enabling improved spatial estimates for under-mapped species and refining maps for those with existing data. The study developed an algorithm that generates species maps directly from GBIF data, updating

(continued)

BOX 3.3
Global Biodiversity Species Global Grid database *(continued)*

automatically as new occurrences are added. Maps are produced for around 600,000 species that meet the computational criteria, encompassing a broad range of species, including not only vertebrates, but also invertebrates, plants, fungi, and nonanimal/non-plant species.

For validation, the maps are compared with expert-verified datasets for mammals (Marsh et al. 2022), ants (Kass et al. 2022), and vascular plants (Borgelt et al. 2022). The maps have been found to be in close agreement with global distribution patterns, with regional differences reflecting technical variations or necessary updates to expert maps based on GBIF data. Map B3.3.1 illustrates the spatial distribution of the mountain viscacha in Latin America.

MAP B3.3.1 Mapping exercise results for *Lagidium viscacia* (mountain viscacha), showing overlapping boundaries

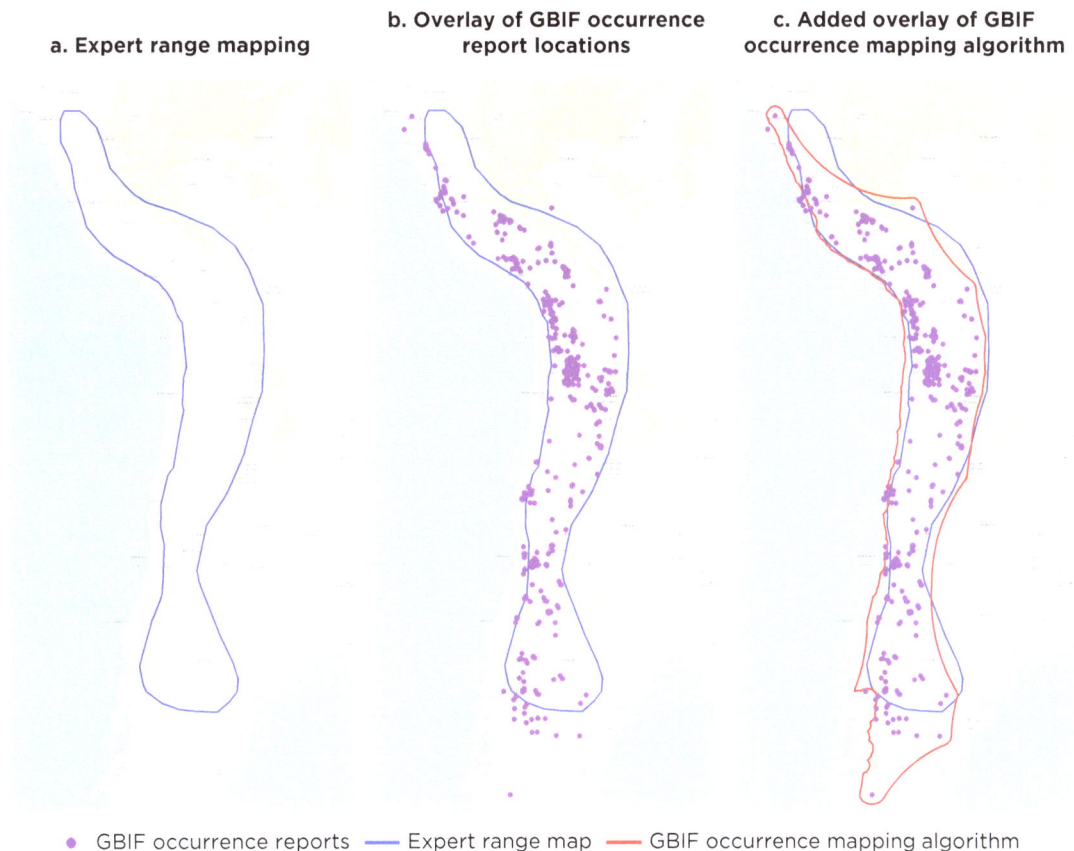

a. Expert range mapping
b. Overlay of GBIF occurrence report locations
c. Added overlay of GBIF occurrence mapping algorithm

● GBIF occurrence reports ── Expert range map ── GBIF occurrence mapping algorithm

Sources: Dasgupta, Blankespoor, and Wheeler 2024; expert range mapping (panel a) based on Burgin et al. 2020.
Note: GBIF = Global Biodiversity Information Facility.

The GBIF's expanding global species database is a valuable resource for conservation. The estimation algorithms have been designed for rapid updates, supporting the ongoing growth of open-source GBIF reports and enabling broader applications in biodiversity conservation.

Notes

1. Biodiversity is categorized into genetic, species, and ecosystem diversity. Genetic diversity refers to the variations in genes within a species, and ecosystem diversity defines the range of different habitats and ecological processes within a given area. Within these categories, alpha biodiversity refers to the diversity within a particular area or ecosystem, typically measured by the number of species in that location. In contrast, beta biodiversity measures the difference in species diversity between ecosystems. Gamma biodiversity refers to the overall diversity of species across a larger geographic area, encompassing multiple ecosystems and combining both alpha and beta diversity.
2. Refer to https://www.nifa.usda.gov/grants/programs/invasive-species-program.
3. Criticism in the literature about these two hypotheses has often focused on the lack of well-understood causal mechanisms and that grassland experiments tend to plant fewer non-native and rare species relative to their presence and abundance in real-world conditions (Dee et al. 2023).
4. This is a measure of temperature that incorporates air temperature and humidity, thus accounting for how well the human body can cool itself via perspiration.

References

AIIB (Asian Infrastructure Investment Bank). 2023. *Nature as Infrastructure*. Beijing, China: AIIB.

Akhter, S., F. Qiao, K. Wu, X. Yin, K. M. A. Chowdhury, and N. U. M. K. Chowdhury. 2021. "Seasonal and Long-Term Sea-Level Variations and Their Forcing Factors in the Northern Bay of Bengal: A Statistical Analysis of Temperature, Salinity, Wind Stress Curl, and Regional Climate Index Data." *Dynamics of Atmospheres and Oceans* 95: 101239. doi:10.1016/j.dynatmoce.2021.101239.

Allen, B. L., L. R. Allen, H. Andrén, et al. 2017. "Can We Save Large Carnivores without Losing Large Carnivore Science?" *Food Webs* 12: 64–75. doi:10.1016/j.fooweb.2017.02.008.

Ban, N. C., G. G. Gurney, N. A. Marshall, et al. 2019. "Well-Being Outcomes of Marine Protected Areas." *Nature Sustainability* 2: 524–32. doi:10.1038/s41893-019-0306-2.

Bardsley, P., and I. Burford. 2009. "Contract Design for Biodiversity Procurement." 2009 Conference (53rd), February 11-13, 2009, Cairns, Australia. No. 48047. Australian Agricultural and Resource Economics Society.

Bareille, F., and P. Dupraz. 2020. "Productive Capacity of Biodiversity: Crop Diversity and Permanent Grasslands in Northwestern France." *Environmental & Resource Economics* 77: 365–99.

Barley Kincaid, K., G. Rose, and H. Mahudi. 2014. "Fishers' Perception of a Multiple-Use Marine Protected Area: Why Communities and Gear Users Differ at Mafia Island, Tanzania." *Marine Policy* 43: 226–35. doi:10.1016/j.marpol.2013.06.005.

Berryman, A. A. 1989. "The Conceptual Foundations of Ecological Dynamics." *Bulletin of the Ecological Society of America* 70 (4): 230–36.

Beschta, R. L., and W. J. Ripple. 2006. "River Channel Dynamics Following Extirpation of Wolves in Northwestern Yellowstone National Park, USA." *Earth Surface Processes and Landforms* 31 (12): 1525–39.

Beschta, R. L., and W. J. Ripple. 2019. "Can Large Carnivores Change Streams via a Trophic Cascade?" *Ecohydrology* 12 (1): e2048.

Boehm, R. 2020. *Reviving the Dead Zone: Solutions to Benefit Both Gulf Coast Fishers and Midwest Farmers*. Cambridge, MA: Union of Concerned Scientists. https://www.ucsusa.org/resources/reviving-dead-zone.

Borgelt, J., J. Sicacha-Parada, O. Skarpaas, and F. Verones. 2022. "Native Range Estimates for Red-Listed Vascular Plants." *Scientific Data* 9: 117. doi:10.1038/s41597-022-01233-5.

Bozec, Y., K. Hock, R. A. B. Mason, et al. 2022. "Cumulative Impacts across Australia's Great Barrier Reef: A Mechanistic Evaluation." *Ecological Monographs* 92 (1): e01494. doi:10.1002/ecm.1494.

Breitburg, D., L. A. Levin, A. Oschlies, et al. 2018. "Declining Oxygen in the Global Ocean and Coastal Waters." *Science* 359 (6371): eaam7240. doi:10.1126/science.aam7240.

Bricheno, L. M., J. Wolf, and Y. Sun. 2021. "Saline Intrusion in the Ganges-Brahmaputra-Meghna Megadelta." *Estuarine, Coastal and Shelf Science* 252: 107246. doi:10.1016/j.ecss.2021.107246.

Brockerhoff, E. G., L. Barbaro, B. Castagneyrol, et al. 2017. "Forest Biodiversity, Ecosystem Functioning and the Provision of Ecosystem Services." *Biodiversity and Conservation* 26: 3005–35. doi:10.1007/s10531-017-1453-2.

Burgin, C., D. Wilson, R. Mittermeier, A. Rylands, T. Lacher, and W. Sechrest. 2020. *Illustrated Checklist of the Mammals of the World*. Barcelona, Spain: Lynx Nature Books. https://www.amazon.com/Illustrated-Checklist-Mammals-Connor-Burgin/dp/8416728364.

Carassou, L. 2013. "Does Herbivorous Fish Protection Really Improve Coral Reef Resilience? A Case Study from New Caledonia (South Pacific)." *PLoS One* 8 (4): e60564.

Chapman, D., B. V. Purse, H. E. Roy, and J. M. Bullock. 2017. "Global Trade Networks Determine the Distribution of Invasive Non-Native Species." *Global Ecology and Biogeography* 26 (8): 907–17.

Cherief, A., T. Sekine, and L. Stagnol. 2025. "A Novel Nature-Based Risk Index: Application to Acute Risks and Their Financial Materiality on Corporate Bonds." *Ecological Economics* 228: 108427.

Christie, P. 2004. "Marine Protected Areas as Biological Successes and Social Failures in Southeast Asia." *American Fisheries Society Symposium* 42: 155–64.

Costanza, R., R. d'Arge, R. de Groot, et al. 1997. "The Value of the World's Ecosystem Services and Natural Capital." *Ecological Economics* 387: 253–60. doi:10.1016/s0921-8009(98)00020-2.

CYGNSS (Cyclone Global Navigation Satellite System). 2021. *CYGNSS Level 3 Microplastic Concentration Retrievals Version 1.0*. Ver. 1.0. CA: PO.DAAC. Ann Arbor, MI: CYGNSS. doi:10.5067/CYGNS-L3M10.

D'Odorico, P., Y. He, S. Collins, S. F. J. De Wekker, V. Engel, and J. D. Fuentes. 2013. "Vegetation-Microclimate Feedbacks in Woodland-Grassland Ecotones: Vegetation-Microclimate Feedbacks." *Global Ecology and Biogeography* 22 (4): 364–79.

Damania, R., L. D. Herrera Garcia, H. Kim, et al. 2024. "Is Natural Capital a Complement to Human Capital? Evidence from 46 Countries." Policy Research Working Paper 10617, World Bank, Washington, DC. doi:10.2139/ssrn.4875797.

Dasgupta, S., B. Blankespoor, and D. J. Wheeler. 2024. "Revisiting Global Biodiversity: A Spatial Analysis of Species Occurrence Data from the Global Biodiversity Information Facility." Policy Research Working Paper 10821, World Bank, Washington, DC.

Dasgupta, S., D. Wheeler, and S. Ghosh. 2022. "Fishing in Salty Waters: Poverty, Occupational Saline Exposure, and Women's Health in the Indian Sundarban." *Journal of Management and Sustainability* 12 (1): 1.

De Frenne, P., F. Rodríguez-Sánchez, D. A. Coomes, et al. 2013. "Microclimate Moderates Plant Responses to Macroclimate Warming." *Proceedings of the National Academy of Sciences* 110 (46): 18561–565. doi:10.1073/pnas 1311190110.

Dedman, S., J. H. Moxley, Y. P. Papastamatiou, et al. 2024. "Ecological Roles and Importance of Sharks in the Anthropocene Ocean." *Science* 385 (6708): adl2362. doi:10.1126/science.adl2362.

Dee, L. E., P. J. Ferraro, C. N. Severen, et al. 2023. "Clarifying the Effect of Biodiversity on Productivity in Natural Ecosystems with Longitudinal Data and Methods for Causal Inference." *Nature Communications* 14: 2607. doi:10.1038/s41467-023-37194-5.

Denny, S., G. Englander, and P. Hunnicutt. 2024. *Private Management of African Protected Areas Improves Wildlife and Tourism Outcomes but with Security Concerns in Conflict Regions.* Washington, DC: World Bank. http://hdl.handle.net/10986/42074.

Desbureaux, S., I. Kabore, G. Vaglietti, et al. 2024. "Collaborative Management Partnerships Strongly Decreased Deforestation in the Most At-Risk Protected Areas in Africa since 2000." *Proceedings of the National Academy of Sciences* 122 (1): e2411348121. doi:10.1073/pnas.2411348121.

Díaz, S., J. Fargione, F. S. Chapin, III, and D. Tilman. 2006. "Biodiversity Loss Threatens Human Well-Being." *PLoS Biology* 4 (8): e277.

Estes, J. A., J. Terborgh, J. S. Brashares, et al. 2011. "Trophic Downgrading of Planet Earth." *Science* 333 (6040): 301–06. doi:10.1126/science.1205106.

FAO (Food and Agriculture Organization of the United Nations). 2024. *The State of World Fisheries and Aquaculture 2024—Blue Transformation in Action.* Rome: FAO.

Flammer, C., T. Giroux, and G. M. Heal. 2025. "Biodiversity Finance." *Journal of Financial Economics* 164: 103987. doi:10.1016/j.jfineco.2024.103987.

Frank, E. 2024. "The Economic Impacts of Ecosystem Disruptions: Costs from Substituting Biological Pest Control." *Science* 385: eadg0344.

Frank, E., A. Missiran, D. Parker, and J. Raynor. 2024. "Option Value of Apex Predators: Evidence from a River Discontinuity." Unpublished manuscript, University of Wisconsin–Madison.

Frank, E., and A. Sudarshan. 2024. "The Social Costs of Keystone Species Collapse: Evidence from the Decline of Vultures in India." *American Economic Review* 114 (10): 3007–40. doi:10.1257/aer.20230016.

Frank, E., S. Wang, X. Wang, Q. Wang, and Y. You. 2024. "Campaigning for Extinction: Eradication of Sparrows and the Great Famine in China." Unpublished manuscript, University of British Columbia, Vancouver, Canada.

Garg, T. 2019. "Ecosystems and Human Health: The Local Benefits of Forest Cover in Indonesia." *Journal of Environmental Economics and Management* 98: 102271.

Giglio, S., T. Kuchler, J. Stroebel, and O. Wang. 2024. "The Economics of Biodiversity Loss." NBER Working Paper No. w32678, National Bureau of Economic Research, Cambridge, MA. https://ssrn.com/abstract=4894672.

Gill, D. A., M. B. Mascia, G. N. Ahmadia, et al. 2017. "Capacity Shortfalls Hinder the Performance of Marine Protected Areas Globally." *Nature* 543: 665–69. doi:10.1038/nature21708.

Grosset-Touba, F., A. Papp, and C. A. Taylor. 2024. "Rain Follows the Forest: Land Use Policy, Climate Change, and Adaptation." Working Paper. Available at SSRN: https://ssrn.com/abstract=4333147 or http://dx.doi.org/10.2139/ssrn.4333147.

Guinotte, J. M., and V. J. Fabry. 2008. "Ocean Acidification and Its Potential Effects on Marine Ecosystems." *Annals of the New York Academy of Sciences* 1134: 320–42. doi:10.1196/annals.1439.013.

Han, L., S. Heblich, C. Timmins, and Y. Zylberberg. 2024. "Cool Cities: The Value of Urban Trees." NBER Working Paper No. 32063, National Bureau of Economic Research, Cambridge, MA.

Harrison, J. R., D. L. Roberts, and J. C. Hernandez-Castro. 2016. "Assessing the Extent and Nature of Wildlife Trade on the Dark Web." *Conservation Biology* 30 (4): 900–04.

Heithaus, M. R., T. Alcoverro, R. Arthur, et al. 2014. "Seagrasses in the Age of Sea Turtle Conservation and Shark Overfishing." *Frontiers in Marine Science* 1: 28. doi:10.3389/fmars.2014.00028.

Herrera Garcia, L. D., J. Vincent, and K. Chang. 2024. "Assessing Forest Restoration Potential across Low- and Middle-Income Countries." World Bank, Washington, DC.

Hoegh-Guldberg, O., R. Cai, E. S. Poloczanska, et al. 2014. "The Ocean." In *Climate Change 2014: Impacts, Adaptation, and Vulnerability. Part B: Regional Aspects. Contribution of Working Group II to the Fifth Assessment Report of the Intergovernmental Panel on Climate Change*, edited by K. R. Smith, A. L. Woodward, D. Campbell-Lendrum, et al.,1655–1731. Cambridge: Cambridge University Press.

Hughes, B. B., K. M. Beheshti, M. T. Tinker, et al. 2024. "Top-Predator Recovery Abates Geomorphic Decline of a Coastal Ecosystem." *Nature* 262: 111–18. doi:10.1038/s41586-023-06959-9.

Humair, F., L. Humair, F. Kuhn, and C. Kueffer. 2015. "E-Commerce Trade in Invasive Plants: E-Commerce Scanning for Invaders." *Conservation Biology* 29 (6): 1658–65.

IPBES (Intergovernmental Science-Policy Platform on Biodiversity and Ecosystem Services). 2019. *Global Assessment Report on Biodiversity and Ecosystem Services of the Intergovernmental Science-Policy Platform on Biodiversity and Ecosystem Services*, edited by E. S. Brondizio, J. Settele, S. Díaz, and H. T. Ngo. Bonn, Germany: IPBES Secretariat. doi:10.5281/zenodo.3831673.

IPBES (Intergovernmental Science-Policy Platform on Biodiversity and Ecosystem Services). 2023. *Summary for Policymakers of the Thematic Assessment Report on Invasive Alien Species and their Control of the Intergovernmental Science-Policy Platform on Biodiversity and Ecosystem Services*, edited by H. E. Roy, A. Pauchard, P. Stoett, et al. Bonn, Germany: IPBES Secretariat. doi:10.5281/zenodo.7430692.

James, N., L. Lundberg, and E. Sills. 2021. "The Implications of Learning on Bidding Behavior in a Repeated First Price Conservation Auction with Targeting." *Strategic Behavior and the Environment* 9 (1-2): 69–101. doi:10.1561/102.00000101.

Johnson, J. A., G. Ruta, U. Baldos, et al. 2021. *The Economic Case for Nature: A Global Earth-Economy Model to Assess Development Policy Pathways*. Washington, DC: World Bank. http://hdl.handle.net/10986/35882.

Jones, B. A. 2019. "Tree Shade, Temperature, and Human Health: Evidence from Invasive Species-Induced Deforestation." *Ecological Economics*, 156, 12-23.

Kass, J. M., B. Guénard, K. L. Dudley, et al. 2022. "The Global Distribution of Known and Undiscovered Ant Biodiversity." *Sciences Advances* 8: eabp9908. doi:10.1126/sciadv.abp9908.

Klein, A. M., B. E. Vaissière, J. H. Cane, et al. 2007. "Importance of Pollinators in Changing Landscapes for World Crops." *Proceedings of the Royal Society B: Biological Sciences* 274 (1608): 303–13.

Leslie, H. A., and M. H. Depledge. 2020. "Where Is the Evidence That Human Exposure to Microplastics Is Safe?" *Environment International* 142: 105807. doi:10.1016/j.envint.2020.105807.

Manning, D., and A. Ando. 2022. "Ecosystem Services and Land Rental Markets: Producer Costs of Bat Population Crashes." *Journal of the Association of Environmental and Resource Economists* 9 (6): 1235–77.

Marsh, C. J., Y. V. Sica, C. J. Burgin, et al. 2022. "Expert Range Maps of Global Mammal Distributions Harmonised to Three Taxonomic Authorities." *Journal of Biogeography* 49 (5): 979–92. doi:10.1111/jbi.14330.

Masuda, Y. J., T. Garg, I. Anggraeni, et al. 2020. "Heat Exposure from Tropical Deforestation Decreases Cognitive Performance of Rural Workers: An Experimental Study." *Environmental Research Letters* 15 (12): 124015.

Masuda, Y. J., T. Garg, I. Anggraeni, et al. 2021. "Warming from Tropical Deforestation Reduces Worker Productivity in Rural Communities." *Nature Communications* 12 (1): 1601.

McCallum, M. L. 2015. "Vertebrate Biodiversity Losses Point to a Sixth Mass Extinction." *Biodiversity and Conservation* 24 (10): 2497–2519.

McElwee, P. D., P. A. Harrison, T. L. van Huysen, et al. 2024. *IPBES Nexus Assessment: Summary for Policymakers*. Bonn, Germany: Zenodo. doi:10.5281/ZENODO.13850290.

McLaughlin, D. L., and M. J. Cohen. 2013. "Realizing Ecosystem Services: Wetland Hydrologic Function along a Gradient of Ecosystem Condition." *Ecological Applications* 23 (7): 1619–31.

Mech, L. D. 2017. "Where Can Wolves Live and How Can We Live with Them?" *Biological Conservation* 210: 310–17.

Ministério das Comunicações. 2025. *Ministério das Comunicações inicia obras de infovia sob as águas do Rio Solimões para levar internet à região amazônica*. Governo do Brasil. https://www.gov.br/mcom /pt-br/noticias/2025/janeiro/ministerio-das-comunicacoes-inicia-obras-de-infovia-sob-as-aguas -do-rio-solimoes-para-levar-internet-a-regiao-amazonica.

Mirzabaev, A., and D. Wuepper. 2023. "Economics of Ecosystem Restoration." *Annual Review of Resource Economics* 15 (1): 329–50. doi:10.1146/annurev-resource-101422-085414.

Mohamed, A., F. DeClerck, P. H. Verburg, et al. 2024. "Securing Nature's Contributions to People Requires at Least 20%–25% (Semi-) Natural Habitat in Human-Modified Landscapes." *One Earth* 7 (1): 59–71.

Noack, F., M.-C. Riekhof, and S. Di Falco. 2019. "Droughts, Biodiversity, and Rural Incomes in the Tropics." *Journal of the Association of Environmental and Resource Economists* 6 (4): 823–52.

OECD (Organisation for Economic Co-operation and Development). 2022. *Global Plastics Outlook: Economic Drivers, Environmental Impacts and Policy Options*. Paris: OECD Publishing. doi:10.1787/de747aef-en.

Paini, D. R., A. W. Sheppard, D. C. Cook, P. J. De Barro, S. P. Worner, and M. B. Thomas. 2016. "Global Threat to Agriculture from Invasive Species." *Proceedings of the National Academy of Sciences* 113 (27): 7575–79.

Pimentel, D., L. Lach, R. Zuniga, and D. Morrison. 2000. "Environmental and Economic Costs of Nonindigenous Species in the United States." *BioScience* 50 (1): 53–65.

Pimentel, D., R. Zuniga, and D. Morrison. 2005. "Update on the Environmental and Economic Costs Associated with Alien-Invasive Species in the United States." *Ecological Economics* 52 (3): 273–88.

Power, M. E., D. Tilman, J. A. Estes, et al. 1996. "Challenges in the Quest for Keystones." *BioScience* 46 (8): 609–20.

Pyšek, P., P. E. Hulme, D. Simberloff, et al. 2020. "Scientists' Warning on Invasive Alien Species." *Biological Reviews* 95 (6): 1511–34. doi:10.1111/brv.12627.

Raynor, J. L., C. A. Grainger, and D. P. Parker. 2021. "Wolves Make Roadways Safer, Generating Large Economic Returns to Predator Conservation." *Proceedings of the National Academy of Sciences* 118 (22): e2023251118.

Ripple, W. J., and R. L. Beschta. 2003. "Wolf Reintroduction, Predation Risk, and Cottonwood Recovery in Yellowstone National Park." *Forest Ecology and Management* 184 (1–3): 299–313.

Ripple, W. J., and R. L. Beschta. 2004. "Wolves and the Ecology of Fear: Can Predation Risk Structure Ecosystems?" *BioScience* 54 (8): 755–66.

Ripple, W. J., and R. L. Beschta. 2006. "Linking Wolves to Willows via Risk-Sensitive Foraging by Ungulates in the Northern Yellowstone Ecosystem." *Forest Ecology and Management* 230 (1–3): 96–106.

Ripple, W. J., and R. L. Beschta. 2012. "Trophic Cascades in Yellowstone: The First 15 Years after Wolf Reintroduction." *Biological Conservation* 145 (1): 205–13.

S&P Global. 2023. *How the World's Largest Companies Depend on Nature*. https://www.spglobal.com /esg/insights/featured/special-editorial/how-the-world-s-largest-companies-depend-on-nature -and-biodiversity#:~:text=S&P%20Global%20Sustainable1%20data%20shows,or%20highly%20 dependent%20on%20nature.

Shwiff, S. A., S. S. Shwiff, J. Holderieath, W. Haden-Chomphosy, and A. Anderson. 2018. "Economics of Invasive Species Damage and Damage Management." In *Ecology and Management of Terrestrial Vertebrate Invasive Species in the United States*, edited by W. C. Pitt, J. C. Beasley, and G. W. Witmer, 35–60. Boca Raton, FL: Taylor and Francis Group.

Street, A., R. Stringer, P. Mangesho, R. Ralston, and J. Greene. 2024. "Why Medical Products Must Not Be Excluded from the Global Plastics Treaty." *The Lancet* 404 (10464): 1708–10. doi:10.1016/s0140 -6736(24)02254-2.

Tilman, D., C. L. Lehman, and K. T. Thomson. 1997. "Plant Diversity and Ecosystem Productivity: Theoretical Considerations." *Proceedings of the National Academy of Sciences* 94 (5): 1857–61.

Tilman, D., P. B. Reich, and J. M. H. Knops. 2006. "Biodiversity and Ecosystem Stability in a Decade-Long Grassland Experiment." *Nature* 441 (7093): 629–32.

Tilman, D., P. B. Reich, J. Knops, D. Wedin, T. Mielke, and C. Lehman. 2001. "Diversity and Productivity in a Long-Term Grassland Experiment." *Science* 294 (5543): 843–45.

Tilman, D., D. Wedin, and J. Knops. 1996. "Productivity and Sustainability Influenced by Biodiversity in Grassland Ecosystems." *Nature* 379 (6567): 718–20.

Valiente-Banuet, A., M. A. Aizen, J. M. Alcántara, et al. 2014. "Beyond Species Loss: The Extinction of Ecological Interactions in a Changing World." *Functional Ecology* 29 (3): 299–307. doi:10.1111/1365 -2435.12356.

van Breugel, M., F. Bongers, N. Norden, et al. 2024. "Feedback Loops Drive Ecological Succession: Towards a Unified Conceptual Framework." *Biological Reviews* 99 (3): 928–49. doi:10.1111/brv .13051.

Weigel, J. Y., P. Morand, T. Mawongwai, J. F. Noel, and R. Tokrishna. 2015. "Assessing Economic Effects of a Marine Protected Area on Fishing Households: A Thai Case Study." *Fisheries Research* 161: 64–76.

Wilson, E. O., and F. M. Peter. 1988. "The Rise of the Global Exchange Economy and the Loss of Biological Diversity." In *Biodiversity*, edited by E. O. Wilson, chapter 23. Washington, DC: National Academies Press.

World Economic Forum. 2020. *Nature Risk Rising: Why the Crisis Engulfing Nature Matters for Business and the Economy*. Geneva: World Economic Forum.

Xing, J., Z. Hu, F. Xia, J. Xu, and E. Zou. 2023. "Urban Forests: Environmental Health Values and Risks." NBER Working Paper No. 31554, National Bureau of Economic Research, Cambridge, MA.

Yamamoto, Y. 2023. "Living under Ecosystem Degradation: Evidence from the Mangrove-Fishery Linkage in Indonesia." *Journal of Environmental Economics and Management* 118: 102788.

Yamamuro, M., T. Komuro, H. Kamiya, T. Kato, H. Hasegawa, and Y. Kameda. 2019. "Neonicotinoids Disrupt Aquatic Food Webs and Decrease Fishery Yields." *Science* 366 (6465): 620–23.

Zellweger, F., P. De Frenne, J. Lenoir, et al. 2020. "Forest Microclimate Dynamics Drive Plant Responses to Warming." *Science* 368 (6492): 772–75. doi:10.1126/science.aba6880.

CHAPTER 4

The Nitrogen Legacy: Boon or Curse?

> "Every excess causes a defect; every defect an excess.
> Every sweet hath its sour; every evil its good."
>
> —*Ralph Waldo Emerson*, American philosopher and poet (1803–1882)

Key messages

- Nitrogen is both a boon and a burden. It is essential for life and food production, yet in excess reactive form it is one of the world's largest externalities, rivaling carbon.

- Unbalanced nitrogen fertilizer use, exacerbated by subsidies, is a key culprit, diminishing crop productivity and increasing nitrogen waste. Nearly half of the fertilizer that is applied is lost, triggering a cascade of environmental impacts across soil, water, and air.

  - In soil, more nitrogen does not always mean higher yields. Excess application leads to diminishing returns and can degrade soil health.

  - In water, excess nitrogen fuels algal blooms, which can deplete oxygen in waterways and kill fish. These water quality issues also pose significant risks to human health.

  - In air, nitrous oxide is a potent greenhouse gas and major driver of ozone depletion. Nitrogen oxides and ammonia also contribute to fine particulate matter, worsening air pollution.

- Crucially, it is policy choices that drive nitrogen use inefficiencies, so better management can reduce pollution without jeopardizing agricultural productivity. Countries must urgently close efficiency gaps and pursue on- and off-farm policy options to decouple agricultural production from its harmful impacts.

The nitrogen fix

No other element on the periodic table has transformed the world as much as nitrogen. Once called *"azote,"* which is Greek for "lifeless," it is not at all lifeless. Nitrogen is a building block of life and a vital nutrient for growing crops. Yet, despite its abundance in the Earth's atmosphere, it remains largely locked away and inaccessible to plants. This paradoxical scarcity of usable nitrogen stems from its peculiar chemistry (Smil 2022). In the air, nitrogen exists as a nonreactive molecule, N_2—two nitrogen atoms bound by an incredibly strong triple bond—making it difficult to transform into a reactive form that plants can absorb. The trick is to break the bonds holding the nitrogen atoms together, freeing it to react with hydrogen and form a usable compound, such as ammonia (NH_3), a type of reactive nitrogen that serves as fertilizer for plants.

For millennia, humans relied on nature to recycle nitrogen back to the soil. The roots of legumes like peas and beans host bacteria that act as nature's chemists, extracting nitrogen from the air and converting it into a usable form. Raising livestock and spreading manure on fields was another way to replenish nitrogen levels. Then, in the nineteenth century, Germany, the United Kingdom, and other countries started enriching their soils with imported nitrogen—in the form of saltpeter from Chile and guano from Peru's Pacific islands. Nitrogen held the same geopolitical weight at that time as oil does today. But reliance on natural sources of nitrogen struggled to satisfy the growing demand for food.

In 1909, a scientific breakthrough solved the worrying shortage of nitrogen fertilizers needed to feed a growing population. Fritz Haber, a German scientist, made a groundbreaking discovery, developing an artificial way to "fix" nitrogen, or convert it into a form that was usable by plants and other organisms. With the right combination of temperature, pressure, and a catalyst, he replicated the process that peas and beans perform as they effortlessly produce ammonia from air. Carl Bosch, another German scientist, scaled this breakthrough for commercial use. Both scientists were awarded the Nobel Prize in Chemistry for their work, which was widely regarded as the greatest invention of the twentieth century. The industrial-scale transformation, famously dubbed *brot aus luft*, or bread from air, ushered in a new era for agriculture.

The Haber-Bosch process has profoundly shaped the modern world. The ability to produce fertilizers at scale, combined with large subsidies, drove a rapid surge in the use of nitrogenous fertilizers, particularly since the 1960s. Over the next two decades, ammonia production increased eightfold, reaching just over 30 million tonnes annually, as synthetic fertilizer fueled the Green Revolution. One result of the experiment is clear—it has more than doubled global rates of nitrogen fixation, enabled a 30–50 percent increase in crop yields, and supported the lives of billions who might otherwise have died prematurely or never been born (Erisman et al. 2008; Stewart et al. 2005).

However, the mass consumption of nitrogen fertilizers has come with significant costs, both on and off farms, depressing yields and causing severe health problems, including in infants, as a result of nitrogen pollution. The unique chemistry of the nitrogen cycle, which enables a single nitrogen atom to cascade through various compounds once in reactive

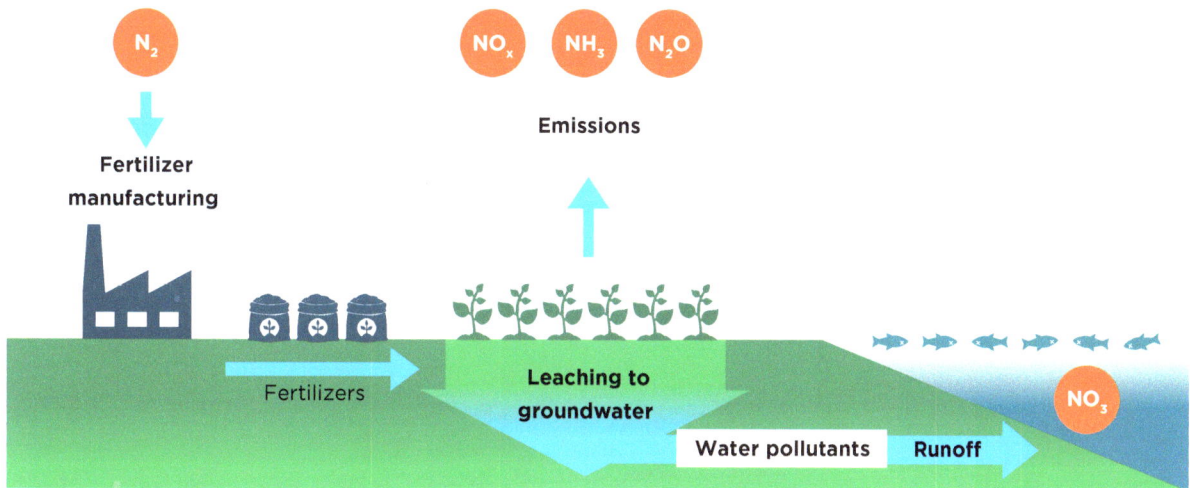

FIGURE 4.1 **The nitrogen cascade**

Source: Original figure for this report.
Note: N_2 = nitrogen; N_2O = nitrous oxide; NH_3 = ammonia; NO_3 = nitrate; NO_x = nitrogen oxides.

form (figure 4.1), means that nitrogen pollution causes much harm to the economy as it travels from soil to the air, and to water and back to the air (Kanter et al. 2020). This is why science suggests that the world may have surpassed the planetary boundary[1] for nitrogen (Richardson et al. 2023), and some have argued that it is the world's largest externality, exceeding carbon (Keeler et al. 2016). Ironically, while the twentieth century celebrated the Haber-Bosch process for feeding billions, the twenty-first century is contending with the unintended consequences. The rest of this chapter explores the unintended but significant long-term costs and outlines the solutions needed to pave the path to reform.

Diminishing returns to nitrogen use

The large-scale production of nitrogen fertilizer, coupled with subsidies that disproportionately reduce its price compared to other nutrients, has resulted in excessive and unbalanced use. Since the 1960s, nearly all the growth of nitrogen fertilizer use has been in Asia, particularly China and India (figure 4.2, panel a), coinciding with the Green Revolution in India. During this period, government policies actively supported a system of domestic price controls by way of large subsidies that distorted market prices. In many countries, fertilizer subsidies are among the largest budgetary expenditures, often emphasizing nitrogen over other fertilizers (Zaveri 2025; refer also to figure 4.2, panel b). Studies have indicated that such subsidies increase fertilizer use intensity, contributing to overuse (Cassou, Jaffee, and Ru 2018; Huang, Gulati, and Gregory 2017). Moreover, excessive reliance on nitrogen causes nutrient imbalances in many regions, as farmers apply significantly more nitrogen compared to other primary nutrients, such as potassium and phosphorous, or other secondary nutrients and micronutrients (Gautam 2015; Kurdi et al. 2020). These imbalances can ultimately degrade soil health and affect crop yields.

FIGURE 4.2 **Nitrogen fertilizer consumption and total fertilizer production, 1961–2014**

a. Consumption of nitrogen fertilizer, by region

b. Total production of fertilizer, by nutrient

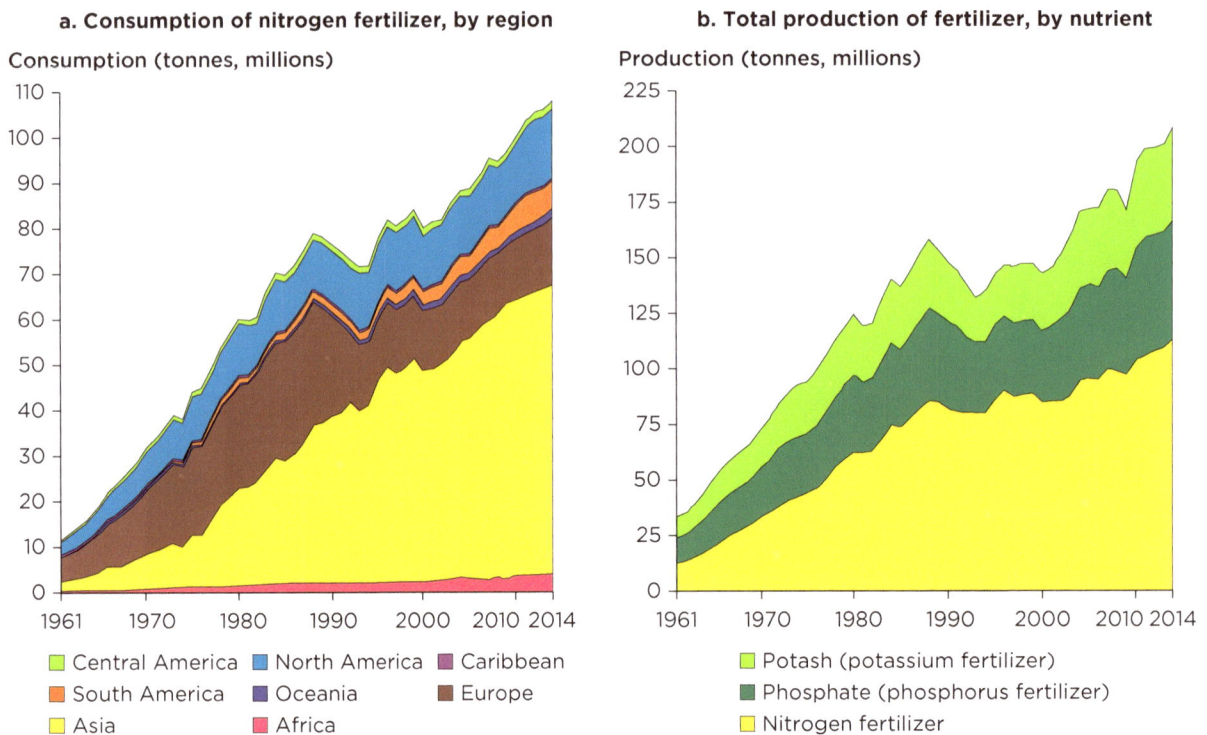

Sources: United Nations Food and Agriculture Organization; Ritchie, Roser, and Rosado 2013; adapted from ourworldindata.org.

Like much else, nitrogen fertilizer use follows the law of diminishing returns, with excess application reducing its effectiveness on yields. Indeed, there is so much inefficiency in fertilizer use that about 50 percent of global calories are produced in areas that exhibit diminishing returns. The implication is that fertilizer use can be reduced with benign impacts on food supplies (Damania et al. 2023). Although nutrients are essential for plant growth, there is a point beyond which adding more fertilizer does not boost yields. The global response curve in figure 4.3 shows that yields rise to a peak with increasing nitrogen application but then decline sharply, reflecting diminishing productivity (refer to box 4.1).

One consequence of inefficient nitrogen fertilizer use is that not all of the nitrogen applied to fields is absorbed by crops. *Nitrogen use efficiency* (NUE)—a measure of how much applied fertilizer is absorbed by harvested crops—remains stubbornly low in many regions. Studies have suggested that in India, only 32 percent of applied nitrogen fertilizer is absorbed by plants, compared with 52 percent in Europe and 68 percent in Canada and the United States (Zhang et al. 2015). A global meta-analysis showed that NUE has declined by 22 percent since 1961, suggesting that crop production is becoming increasingly less efficient at using nitrogen fertilizer (Zhang et al. 2021). Of the 161 teragrams of nitrogen applied to crops globally, only 73 teragrams are absorbed, implying that more than half is wasted (Zhang et al. 2021). While the global NUE average is 46 percent, the European Union Nitrogen Expert Panel has recommended an upper limit for NUE of 90 percent. So, if around 50 percent of nitrogen fertilizer is not used by plants where does it go?

BOX 4.1
Fertilizer feast and famine

There is vast heterogeneity in yield impacts as nitrogen usage ranges from acute deficiencies to extreme excesses across world regions.

East Asia and South Asia are at the high end of the global distribution, with much of their nitrogen usage already on the diminishing returns part of the response curve. South Asia's median nitrogen usage is near global peak response levels, while East Asia's is at the 95th percentile globally. This reflects a long history of fertilizer use in both regions.

In Sub-Saharan Africa, in contrast, median nitrogen usage is only a fraction of the global median, resulting in low yields and limited productivity. In much of the region, agricultural soil is being depleted of its nutrients due to *nitrogen mining*, whereby crops take nitrogen from the soil. In these cases, increasing nitrogen inputs can improve productivity. However, solely focusing on increasing nitrogen use without considering soil and agronomic conditions can fail to address limiting factors for plant growth that are specific to local contexts (Harou et al. 2022; Smale and Thériault 2018). Variations in soil composition and acidity can significantly affect yield responsiveness to nitrogen fertilizers in African soil, underscoring the importance of balanced, soil-specific fertilizer strategies to help farmers optimize returns.

FIGURE 4.3 **Change in global agricultural productivity due to nitrogen fertilizer**

Increase in yield productivity (%)

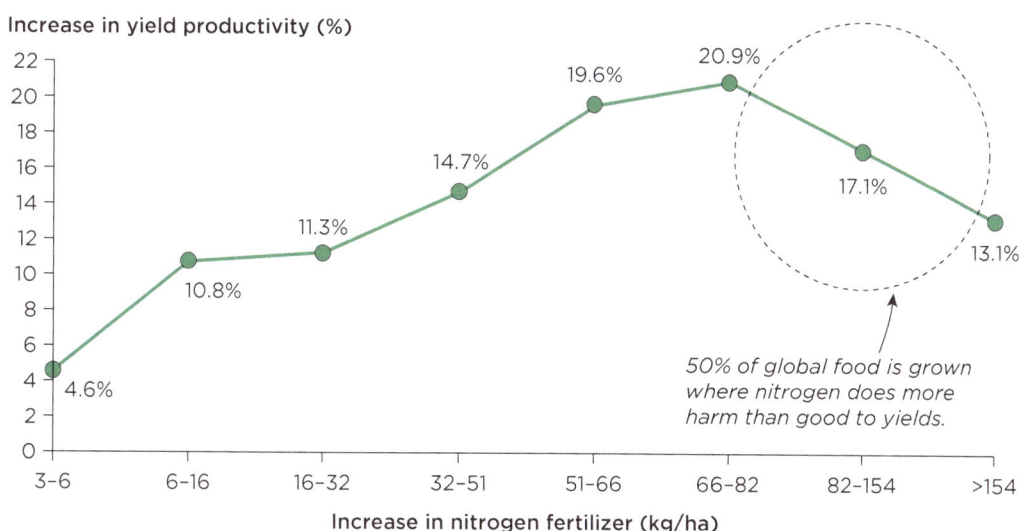

Source: Zaveri 2025.
Note: The figure shows point estimates of coefficients obtained for different quantiles of nitrogen fertilizer usage from the second to the ninth quantile relative to the omitted first quantile. kg/ha = kilograms per hectare.

The cost of nitrogen inefficiencies

Half of the nitrogen in fertilizers is lost to the surrounding environment in its multiple chemical forms, impacting human health and economic well-being.[2] The unique chemistry of the nitrogen cycle implies that a nitrogen atom once in reactive form can be readily converted to nitrites and nitrates that pollute the waterways, or volatized as ammonia or nitrogen oxides causing air pollution, or denitrified to nitrous oxide exacerbating ozone depletion. At every stage, nitrogen losses have adverse impacts on either health, the economy, and/or the environment.

Water

Nitrogen pollution in water contaminates drinking supplies, adversely affecting human health. Nitrate in drinking water is known to cause methemoglobinemia, or *blue baby syndrome*, which starves infants' bodies of oxygen—prompting the 10 parts per million safety standard. Evidence has suggested a host of other impacts at thresholds below the "safe standard." For example, causal analysis using health data has found stunted growth with cumulative exposure to nitrates in the early years, leading to impaired productivity in later life (Jones 2019; Zaveri et al. 2020). Other studies linked nitrate levels below the safety standard to thyroid disease and colorectal cancer. A systematic review involving more than 7 million subjects found that such levels are also associated with adverse reproductive and birth outcomes, including increased risk of preterm birth (Damania and Zaveri 2024; Lin et al. 2023; Ward et al. 2018). Overall, although there is clear evidence of adverse impacts, more research is required to establish causal relationships.

There are further impacts on water quality (refer to box 4.2, which discusses broader water quality challenges). Excess nitrogen leaches into surface water and groundwater, fueling eutrophication,[3] toxicity in aquatic systems, and drinking water contamination (Fowler et al. 2013; Oelsner and Stets 2019; Pennino et al. 2020). Eutrophication leads to harmful algal blooms, where algae—such as cyanobacteria— grow dense enough to pose health risks (Glibert et al. 2014). High algae densities reduce water clarity and promote toxic compound production in certain species (European Commission Joint Research Centre 2016). Cyanobacteria, which are visible to the human eye, release neurotoxins and hepatotoxins, such as microcystin and cyanopeptolin, which are harmful to humans, animals, and aquatic life (Damania et al. 2019). Large blooms can devastate ecosystems, often creating oxygen-depleted dead zones or *hypoxia*.

BOX 4.2
Tainted waters: The invisible crisis of water pollution

Water pollution is complex and pervasive, with impacts that evolve as countries grow. It spans a vast array of contaminants, from fecal bacteria and heavy metals to pharmaceuticals and plastics. Low-income countries grapple with pollutants of poverty, such as human waste and geogenic toxins like arsenic, while middle- and high-income countries increasingly face pollutants of prosperity, including nitrates, salts, and synthetic chemicals (Damania et al. 2019). Every year, more than 2 billion people still drink water contaminated with feces, placing billions at risk of disease and undermining global progress toward safe water for all (World Bank 2025).

Among these threats, heavy metals like lead and arsenic pose especially insidious risks. Naturally occurring arsenic affects millions in Latin America and South Asia, while industrial activities and aging infrastructure contribute to mercury and lead contamination. Lead is especially harmful to cardiovascular health and children's development (Crawfurd et al. 2024; Larsen and Sánchez-Triana 2023). These risks are entirely preventable by replacing lead pipes, implementing corrosion control, and enforcing lead-free procurement, which yield health and economic benefits at a benefit-cost ratio of 35:1 (Levin and Schwartz 2023).

Yet new contaminants are emerging faster than regulations can keep up. More than 90 percent of the 8.3 billion tons of plastic produced since the 1950s has not been recycled. As plastics degrade into micro- and nano-plastics, they infiltrate drinking water and food chains. Meanwhile, pharmaceuticals like antibiotics and hormones evade wastewater treatment, reentering the water cycle and contributing to antimicrobial resistance, which is a global health threat responsible for 700,000 deaths annually (Damania et al. 2019).

Tracking these threats is difficult because data gaps persist across much of the world. Unlike more routinely measured pollutants, like fine particulate matter in air, water quality data are sparse, inconsistent, and difficult to compare. As of 2022, about 73 countries were not reporting on water quality as part of Sustainable Development Goal 6, which aims to achieve clean water and sanitation for everyone (WHO/UNICEF Joint Monitoring Programme 2023). Emerging machine learning models can help fill gaps, but data must be transparent and credible to shape effective interventions.

The development toll of poor water quality is immense. In 2019, unsafe water, sanitation, and hygiene (WASH) contributed to 1.4 million deaths caused by diarrhea, acute respiratory infections, undernutrition, and soil-transmitted helminths (Wolf et al. 2023). Beyond health, contaminated water reduces labor productivity and impairs child development, with per capita economic costs of inadequate WASH services ranging from $3 to $49 depending on the country and context (Hutton and Chase 2017).

Even as countries develop, water pollution risks persist, just in new forms. The nitrates discussed in this chapter are one example. Salts are another emerging threat. Salinity in drinking water increases risks of miscarriage, preeclampsia, and infant mortality, yet there is no established health threshold in global drinking water standards (Damania et al. 2019). Salinity also depresses crop yields, with food losses equivalent to feeding 170 million people a day (Russ et al. 2020).

(continued)

These health and productivity impacts translate into macroeconomic losses. Pollution upstream acts as a drag on downstream economies. When biological oxygen demand, a common metric of pollution, exceeds 8 milligrams per liter, gross domestic product growth is reduced by a third (Damania et al. 2019).

Solving water quality requires prevention, data, and cost-effective interventions. Water quality is a "wicked problem"—multi-source and evolving. Yet targeted action works. Chlorinating water at the point of collection is among the most cost-effective health interventions, costing as little as $40 per disability-adjusted life year averted and reducing child mortality by 25 percent (Kremer et al. 2023). A mix of upstream prevention, nature-based solutions, and better monitoring could yield profound health and economic dividends.

New data-driven analysis has confirmed the link between nitrogen runoff and harmful algal blooms, offering global-scale evidence beyond previous country-specific or model-based studies. A global analysis combining satellite imagery with nitrogen fertilizer data revealed a strong positive relationship between upstream nutrient levels and the areal extent of blooms downstream (figure 4.4). Remote-sensing data have underscored the growing severity of the issue, with a 59.2 percent increase in bloom frequency and a 13.2 percent expansion in extent from 2003 to 2020 (Dai et al. 2023). The surge in harmful algal blooms in recent decades has spurred efforts to curb nitrogen runoff, including the establishment of global task forces (Basu et al. 2022; Gobler 2020).

Harmful algal blooms severely impact fish stocks and hence lead to losses in fisheries. The estimated empirical relationship shows that, in moderation, algal blooms increase catch, but they quickly become detrimental at higher intensities (refer to figure 4.5). Algal blooms at a harmful level affected a cumulative 24 million square kilometers (approximately 2.75 times the size of the United States) of coastal areas across the world. Using International Monetary Fund fish commodity prices, this implies approximately $12.21 billion to $18.59 billion in losses over 2003 to 2015. One important caveat is that these numbers serve as a lower bound as they exclusively capture damages to fisheries without enumerating other issues, such as health hazards to humans or diminished recreational use.

FIGURE 4.4 Effect of nitrogen use on algal bloom extent

Algal bloom (sq km)

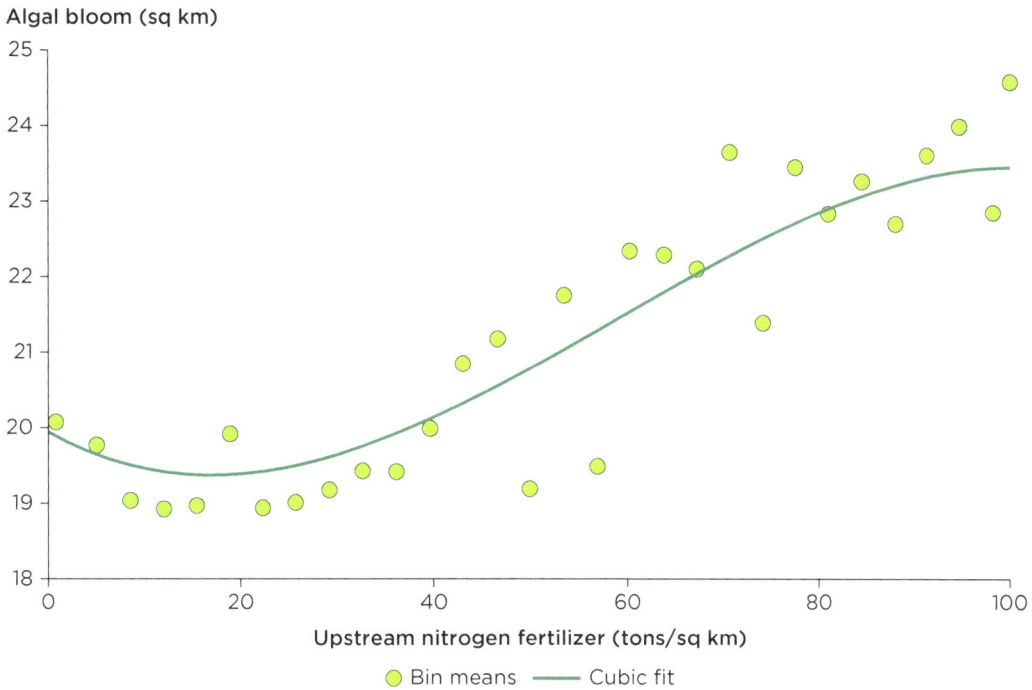

○ Bin means   ——— Cubic fit

Source: Original calculations based on data from Ho, Michalak, and Pahlevan 2019, and Lu and Tian 2017.
Note: The figure plots the output from a binned regression, using grid-level data on algal bloom extent and nitrogen fertilizer use. Each dot represents an equally-sized bin of observations (grouped over the x-axis). Within these bins, the average of the x- and y-variables is visualized in a scatterplot after controlling for gridcell and year fixed effects, and share of cropland. sq km = square kilometer.

FIGURE 4.5 Effect of algal bloom extent on fish catch

Change in tons of catch

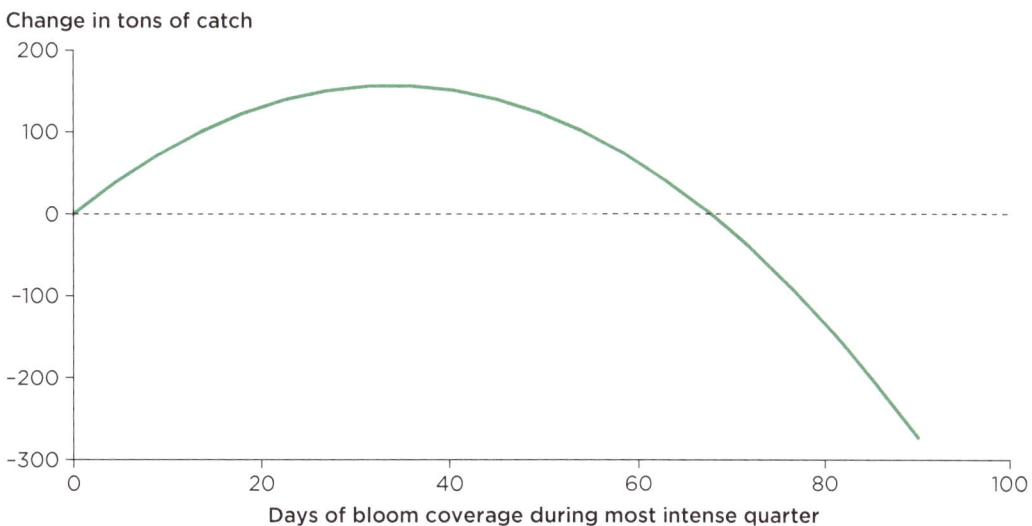

Days of bloom coverage during most intense quarter

Source: Original calculations based on data from Dai et al. 2023 and Watson 2017.
Note: The figure plots the relationship between quarterly bloom intensity (x-axis) and its impact on tons of fish caught (y-axis) after controlling for ocean temperature, fishing effort, and year fixed effects.

Legacy nitrogen, which can persist for decades after nitrogen inputs have ceased, is a key obstacle to improving water quality. *Legacy nitrogen* refers to nitrogen that has accumulated in soil and groundwater from past applications of fertilizer or livestock manure (Basu et al. 2022; Liu ct al. 2024). It can exist as soil organic nitrogen within the root zone of agricultural soil—known as *biogeochemical legacy*—or as dissolved nitrogen in groundwater—referred to as *hydrologic legacy* (Van Meter and Basu 2015). This stored nitrogen continues to leach into waterways, causing significant time lags between conservation efforts and observable improvements in water quality (McDowell et al. 2021; Van Meter, Van Cappellen, and Basu 2018).

Air

In the air, nitrogen compounds are major contributors to air pollution and ozone depletion. Nitrogen dioxide—from agriculture, industry, and vehicles—contributes to premature deaths from respiratory and cardiovascular diseases (EEA 2025; Sun, Lu, and Li 2024). Nitrous oxide, also known as laughing gas, is 300 times as potent as carbon dioxide at trapping heat, lingers for 114 years in the sky before disintegrating, and depletes the ozone layer (Kanter 2018; Tian et al. 2020). The climate impact of laughing gas is no joke, but it has largely been ignored in climate policies.[4] The first global nitrous oxide assessment (UNEP and FAO 2024) showed atmospheric concentrations rising faster than expected, with emissions up 40 percent since 1980—mostly from agricultural sources, such as synthetic fertilizers and manure. Nitrous oxide emissions also generate ammonia and nitrogen oxides that contribute to the formation of fine particulate matter, causing air pollution.

Managing nitrous oxide could yield major climate, air quality, and health benefits. The global assessment highlighted that ambitious abatement would avoid the equivalent of more than 6 years of carbon dioxide emissions from fossil fuels while delivering five times the ozone benefits of the 2007 accelerated hydrochlorofluorocarbon phase-out, the last significant ozone action under the Montreal Protocol (UNEP and FAO 2024). Adopting a sustainable nitrogen management approach that targets agricultural nitrous oxide emissions could also significantly reduce short-lived nitrogen compounds—such as nitrogen oxides and ammonia—rapidly improving air quality and averting 20 million premature deaths from air pollution (UNEP and FAO 2024).

Is nitrogen management worth the cost?

Overall, the costs of inefficient nitrogen use are staggering. Because nitrogen loss occurs through many pathways and causes diverse damages, estimates vary widely. Available estimates—based on willingness to pay for reducing the risks of nitrogen pollution or costs to ecosystems and health care services—are between $0.3 trillion and $3.4 trillion annually (Sutton et al. 2013). Applying estimates of the costs of abatement of nutrient runoff from the United States to the world (Rabotyagov et al. 2014) suggests that global abatement efforts could amount to approximately $10.1 billion per year. Although this figure is substantial, it does not represent an insurmountable challenge and is far lower than the willingness to pay to address the nitrogen damage. This may be a case that reflects the importance of prevention, emphasizing proactive measures to mitigate future costs.

There is considerable potential for country-specific policies to enhance nitrogen management and reduce harm to human health and the environment without jeopardizing food security. Satellite imagery and geospatial data have revealed that neighboring countries with similar environmental conditions and agricultural potential often have striking differences in nitrogen pollution, suggesting that policy choices—not environmental constraints—are the main drivers of inefficiency (Wuepper et al. 2020). It has been shown that if polluting countries matched the efficiency of their neighbors, nitrogen pollution could be reduced by 35 percent while increasing yield gaps by only 1 percent (Wuepper et al. 2020).

Cost–benefit calculations have further reinforced the case for action, showing that investing in nitrogen management yields benefits that far exceed the costs. It is often assumed that more pollution and the associated mortality and morbidity are an unavoidable cost of closing yield gaps, but this trade-off does not always exist. A recent global meta-analysis found that optimal farm-level nitrogen management practices[5] delivered societal benefits approximately 25 times greater than the implementation costs (Xia and Yan 2023). The benefits arose from enhanced crop yields, reduced premature mortality from respiratory diseases caused by air pollution, and the sustained services provided by unpolluted ecosystems (Gu et al. 2023) (figure 4.6). Overall, these practices could reduce total nitrogen losses by 30–70 percent while increasing yields by 10–30 percent and NUE by 10–80 percent, generating significant net economic gains.

The sources of these benefits vary across regions (figure 4.6): yield gains dominate in low-input settings like Sub-Saharan Africa, while reductions in pollution and health costs are the primary source of benefits in high-surplus regions like China and India. Yet, despite these clear advantages, reform remains uneven. In high–nitrogen surplus areas, farmers may lack the financial incentive to adopt better practices unless broader societal benefits are considered. Meanwhile, in low-nitrogen settings, limited access and affordability remain key barriers—even where yield gains offer high returns. These findings underscore the importance of tailoring policy support and investment to local nitrogen balances and capacity constraints. The next section outlines a range of policy solutions to do just that.

FIGURE 4.6 Costs and benefits of a selection of nitrogen management practices

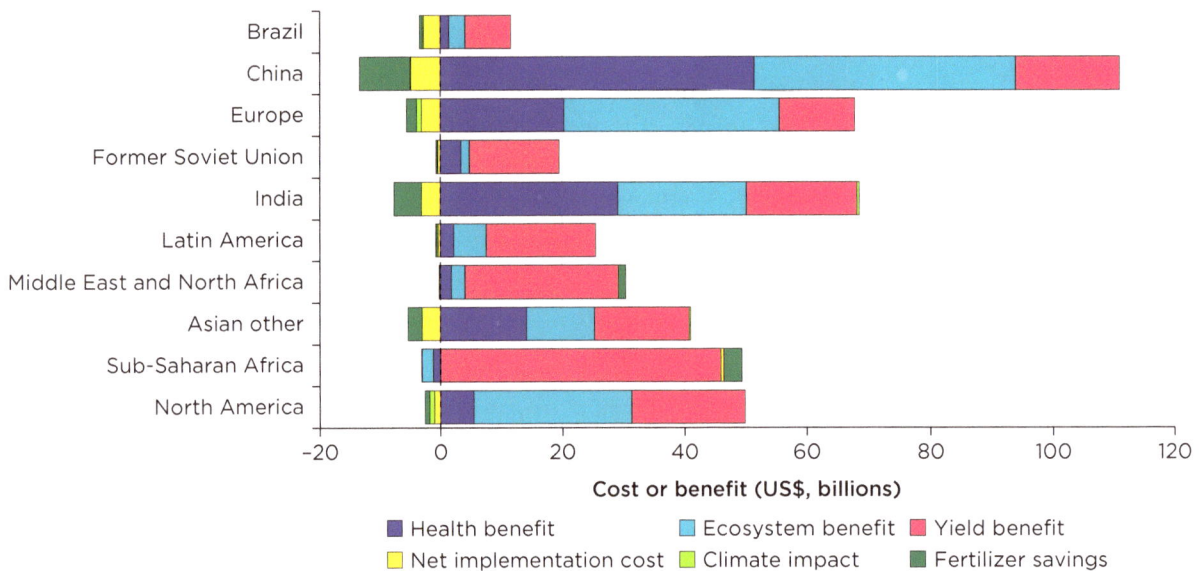

Cost or benefit (US$, billions)

Legend:
- Health benefit
- Ecosystem benefit
- Yield benefit
- Net implementation cost
- Climate impact
- Fertilizer savings

Sources: Xia and Yan 2023; adapted from Gu et al. 2023.
Note: Based on Gu et al. (2023), the figure shows the regional net costs and benefits of implementing a selection of nitrogen management practices globally. Using computational modeling, Gu et al. (2023) analyzed net implementation costs and regional benefits in 2015, accounting for yield gains, fertilizer savings, and avoided damage to climate, ecosystems, and human health. Negative values show net costs, and positive values represent net societal benefits.

Breaking the nitrogen fix: Cultivating on- and off-farm solutions

Nitrogen balances vary dramatically across the world, ranging from acute deficiencies in some regions to excesses in others. This highlights the dual challenge of ensuring that there is enough nitrogen to maintain food security while minimizing the harm caused by nitrogen waste. Even in advanced economies—such as the European Union, which has reduced its nitrogen waste over the past decades—progress has been uneven (Kanter et al. 2020).[6]

Given the complexity and variety of impacts, there is no simple solution to curbing nitrogen waste. Although it is possible to have an energy system without carbon, it is hard to imagine a food system without nitrogen. As the pressures of nitrogen loss mount, closing efficiency gaps and pursuing effective policies on- and off-farm that can decouple agricultural production from its environmental impacts become more urgent.

On the farm

The 4Rs of nutrient stewardship—applying the right nutrients, at the right rate, at the right time, and in the right place—can boost NUE, reduce waste, and improve yields. Yet farmers often struggle to optimize fertilizer use, especially where subsidies distort prices. Uniform application across fields is common, even when nutrient needs vary, leading to overuse, particularly when fertilizers are cheap (Damania et al. 2023; Duflo, Kremer, and Robinson 2011; Schultz 1964). Timing is just as important: nitrogen is easily lost through leaching, runoff, or volatilization. Better input management can significantly improve nitrogen efficiency through the 4Rs framework.[7]

Precision agriculture offers transformative tools to tailor nitrogen use to local conditions. A range of techniques can be used:

- *Manual techniques.* Low-cost approaches, such as seed priming, microdosing, leaf color charts, and chlorophyll meters can concentrate application in the vicinity of the plant, ensuring greater nitrogen uptake (Kanter et al. 2019).

- *Digital tools.* Advances in data analytics, mobile technology, and satellite data are revolutionizing nitrogen management at scale. Mobile phones now deliver real-time, site-specific fertilizer recommendations (Kanter et al. 2019; Singh, Ganguly, and Dakshinamurthy 2018),[8] while low-cost satellite sensors can predict crop yields at the individual field level, helping farmers decide where and when to apply fertilizer for maximum efficiency (Jain et al. 2019).[9]

Ethiopia's experience shows how precision tools can raise yields and profitability (box 4.3).

BOX 4.3
Optimized fertilizer disbursement in Ethiopia increases wheat yields and profitability

Targeted fertilizer management is key to unlocking Ethiopia's wheat production potential. Yields average just 2.6 tonnes per hectare—well below the 6–7 tonne potential—due in part to inefficient fertilizer use. Blanket recommendations ignore local soil and climate differences, leading to suboptimal application.

Machine learning is enabling site-specific fertilizer recommendations. In 2022, the World Bank–supported *Accelerating Impacts of CGIAR Climate Research for Africa* program partnered with Digital Green, the Ethiopian Institute of Agricultural Research, and the Ministry of Agriculture to pilot site-specific fertilizer recommendations (SSFRs). The algorithm, which was trained on 20,000 wheat observations and spatial data, tailored rates to local conditions.

The results were significant. Across 277 sites, national blanket recommendations increased yields by 16 percent, while more targeted district-level recommendations increased yields by 25 percent, boosting profits by $412 and $580 per hectare per season, respectively. As Ethiopia faces growing wheat demand, limited arable land, and climate challenges, SSFRs offer a path to higher productivity and food security.

Despite the public benefits, adoption of improved nitrogen practices is not guaranteed—education and extension are critical. Smallholders may distrust information or overapply nitrogen as insurance (Gars et al. 2025; Kanter et al. 2019). Barriers include limited knowledge, financial constraints, and reluctance to change. The complexity of nitrogen timing adds further uncertainty (Kanter et al. 2019). Strengthening local extension services is essential to bridge this gap (Dash 2019). Global examples highlight the impact of farmer training and education:

- *Bangladesh.* A simple rule-of-thumb training using leaf color charts reduced fertilizer use by 8 percent, without compromising yields, saving 180,000 tons of nitrogen fertilizer—worth $80 million, or 14 percent of the national input subsidy budget (Islam and Beg 2021).

- *China.* A decade-long training program reached 20.9 million farmers, boosting staple crop yields by 11 percent while cutting nitrogen use by 1.2 million tons—yielding $12.2 billion in returns (Cui et al. 2018). China also phased out nitrogen subsidies, redirecting funds to nutrient and manure management, and launched policies like the "Zero Growth Action Plan for Fertilizer Use," "Grain for Green," and "Double Reduction" to reduce fertilizer use without sacrificing grain output (Ji, Liu, and Shi 2020).

- *India.* Farmers adopted more balanced nutrient practices when provided with simplified tools and repeated extension services, reinforcing the importance of ongoing advisory support (Cole and Sharma 2017).

- *Viet Nam.* The World Bank–funded Vietnam Sustainable Agriculture Transformation Project, in partnership with the International Rice Research Institute, trained more than 800,000 rice farmers in optimizing fertilizer use. The initiative led to reductions in fertilizer use and improvements in yields and farmer incomes.[10]

Beyond optimizing nitrogen use, improving soil health lays the foundation for long-term fertility. As highlighted in spotlight 2, regenerative practices like conservation tillage, crop rotation, and cover cropping enhance soil structure and fertility. For example, legumes, such as beans, peas, and lentils, naturally fix nitrogen, complementing fertilizer use. Regenerative practices also improve water retention in the root zone—a critical factor in nutrient uptake (Kirkham 2005). By sustaining microbial activity and promoting stronger root systems, soil health practices that retain moisture, like cover cropping and mulching, build resilience and make fertilizers more effective.

Beyond the farm

On-farm efforts alone cannot solve nitrogen pollution. Improved farm management reduces nitrogen losses but does not address legacy nitrogen—stored nitrogen in soil and groundwater that leaches into waterways for decades (Basu 2025). Tackling this challenge requires off-farm solutions at the watershed level. Economic and policy forces beyond individual farms also shape on-farm nitrogen use and management, making it essential to engage input suppliers, consumers, and policy makers.

Wetlands: A natural solution for nitrogen removal

Wetlands and riparian buffers are cost-effective, yet often overlooked, nitrogen filters. Fluvial wetlands—connected to rivers and streams—remove both newly applied and legacy nitrogen before it enters downstream ecosystems (Cheng et al. 2020; Hansen et al. 2021; Bertassello et al. 2025) (box 4.4). Targeted wetland expansion is significantly more effective at removing nitrogen loads than traditional on-farm conservation measures (Cheng et al. 2020; Shen et al. 2025). Wetlands could also reduce reliance on costly wastewater treatment plants, which struggle to remove nitrogen at low concentrations. In China, restoring small wetlands could prevent the need for 800 new treatment plants, which would otherwise cost $8 billion annually—more than double China's $3.9 billion budget for water pollution control in 2022 (Shen et al. 2025). Wetlands provide added benefits: carbon sequestration, biodiversity support, flood control, and permanent nitrogen removal through plant uptake and denitrification, making them a fast-track solution for improving water quality.

Yet, the world has seen significant wetland loss in recent decades, mainly through agricultural drainage (Fluet-Chouinard et al. 2023). Farmers often resist wetland restoration, citing concerns about land fragmentation and inefficiencies, but subfield profitability analyses suggest that many low-yield depressions could be more profitable if restored to wetlands rather than cultivated (Brandes et al. 2016; Clare et al. 2021). Incentive programs that recognize these economic realities—by compensating farmers for nitrogen removal and habitat restoration—could tip the balance in favor of wetland conservation, providing both environmental and financial benefits.

BOX 4.4
Principles for wetland restoration to meet nitrogen removal goals

Size matters, but smaller can be smarter. Smaller wetlands are more effective nutrient filters per unit area due to higher sediment-to-volume ratios. Yet, they are disappearing at a faster rate. Prioritizing their restoration can boost nutrient retention with minimal impacts on land area.

Location matters, but pristine is not necessarily better. Restoration efforts should target nitrogen-rich zones, like agricultural hotspots, where benefits are greatest. For instance, a 10 percent increase in wetland area in these zones can cut nitrogen loads by up to 54 percent.

Connectivity matters, but not all connections are equal. Hydrologic connections bring in nitrogen-rich runoff, but too much flow during storms can flush nutrients out. Small, disconnected wetlands often retain nutrients better due to longer residence times. Balancing connectivity is crucial for nitrogen retention.

(continued)

Producers, consumers, and policy makers: Tackling the forces shaping nitrogen use

Input suppliers can play a vital role in reducing nitrogen losses. Performance standards could revitalize a fertilizer industry largely unchanged since the Haber-Bosch process, creating profitable opportunities for companies through patent-protected fertilizers, while also indirectly benefiting farmers by improving yields (Kanter 2018; Kanter et al. 2019). One example is efficiency-enhancing fertilizers like India's neem-coated urea, mandated since 2015, which gradually releases nitrogen, better aligning supply with crop demand (Government of India 2016). Other nitrification inhibitors warrant further research to determine their effectiveness under different conditions.

Consumers influence nitrogen demand through dietary choices. The EAT-*Lancet* Commission calls for shifting diets—particularly in high-consumption settings—away from foods considered unhealthy for humans and the environment (Willet et al. 2019). Plant proteins are far more nitrogen-efficient than animal-sourced foods, as livestock require large amounts of nitrogen-rich feed with only a small share converted into meat or dairy and the rest lost as waste and emissions (Poore and Nemecek 2018; Sutton et al. 2013). However, environmental impacts vary by livestock type and system, and consumption of animal-sourced foods remains low in many low-income, undernourished regions where increasing intake may be essential. A more universal strategy is reducing food loss and waste from farm to fork, which can ease pressure on land and fertilizer use without compromising nutrition.

Repurposing subsidies is key to reducing nitrogen overuse. In non–OECD (Organisation for Economic Co-operation and Development) countries, over 75 percent of agricultural policies promote nitrogen-intensive practices, often through *coupled* subsidies[11] tied to fertilizer use or output (Damania et al. 2023; Kanter et al. 2020). In contrast, *decoupled* subsidies, which are not linked to production, avoid altering incentives to change input or output levels, are less distortionary, and do not lead to harmful environmental spillovers (Zaveri 2025). As precision farming and new plant breeding gain importance, repurposing fertilizer subsidies toward these innovations and regenerative agriculture (refer to spotlight 2) could enhance productivity while reducing nitrogen pollution.

Fragmented nitrogen policies can lead to pollution swapping. Nitrogen's complex chemistry means reducing one pollutant can unintentionally worsen another—a phenomenon known as pollution swapping. For example, storing manure to limit nitrate runoff may increase ammonia emissions, and riparian buffers can reduce leaching but raise nitrous oxide (Hefting, Bobbink, and De Caluwe 2003). Nitrification inhibitors, meant to lower nitrous oxide and nitrogen leaching, can boost ammonia emissions by up to 25 percent if they are not paired with urease inhibitors (Kanter et al. 2020). In contrast, the EU Nitrates Directive, while designed to curb nitrogen runoff, also reduced ammonia and nitrous oxide emissions (Kanter 2018). Nitrogen policies remain unpredictable—some serendipitously deliver multiple benefits, and others shift problems elsewhere.

Nitrogen pollution must be tackled as a systemwide challenge. Recognizing this, the United Nations Environment Assembly[12] has urged cross-sectoral strategies that prevent trade-offs and address nitrogen pollution holistically. International organizations have set ambitious nitrogen reduction targets, but stronger national policies are needed to drive real progress. The European Commission's 2020 Farm to Fork Strategy[13] aims to halve nutrient losses by 2030, requiring a 20 percent cut in synthetic nitrogen fertilizer use. The Kunming-Montreal Global Biodiversity Framework adopted a similar goal. However, recent research has suggested that this will not be enough to meet the goal of halving nutrient losses, underscoring the need for more comprehensive reform to bridge the gap between ambition and action (Billen et al. 2024).

While humanity has become dependent on nitrogen, solutions exist to break the nitrogen fix. Achieving this requires coordinated policies that balance nitrogen's dual role as an agricultural necessity and a major pollutant, ensuring that they address the entire nitrogen cascade while minimizing unintended trade-offs.

Notes

1. *Planetary boundary* refers to the environmental limits within which humanity can safely operate.
2. There are several sources of nitrogen pollution, including fossil fuel combustion, industry, energy, transport, biomass burning, and wastewater, but the dominant source is the agriculture sector.
3. *Eutrophication* is when the nutrient imbalance feeds the growth of phytoplankton.
4. For example, it is not regulated by the Montreal Protocol on ozone-depleting substances, nor included in most countries' nationally determined contributions to reduce greenhouse gas emissions under the Paris Agreement.
5. These measures include enhanced-efficiency fertilizers, organic amendments (manure and straw), crop legume rotation, buffer zones, the 4R nutrient stewardship (right source, rate, time, and place), new cultivars, optimal irrigation, and tillage improvements.

6. https://www.eea.europa.eu/airs/2018/natural-capital/agricultural-land-nitrogen-balance.

7. https://www.tfi.org/insights/nutrient-stewardship/what-are-the-4rs/.

8. For example, UjuziKilimo, a Kenyan company, uses simple ground sensors to provide nitrogen recommendations via text messaging. A Nigerian company, Hello Tractor, connects farmers with tractor services via apps and text messaging (Kanter et al. 2019).

9. Data from satellite sensors such as the European Space Agency's SENTINEL-2 Multi Spectral Imager and the series of sensors from Planet's Dove satellites are being explored to bring monitoring to the individual-field level (Jain et al. 2019).

10. https://www.worldbank.org/en/news/feature/2024/05/14/greening-viet-nam-s-rice-bowl-a-mekong-delta-success-story?utm_source=chatgpt.com.

11. Subsidies can also discourage fertilizer innovation, crowd out commercial sales, and displace investment in agricultural research and development (Gulati and Banerjee 2015). Less often noted is the misdiagnosis of market failures, such as using subsidies to address high transport costs that would be better solved by infrastructure investment (Gautam 2015; Smale and Thériault 2018).

12. https://wedocs.unep.org/bitstream/handle/20.500.11822/39816/SUSTAINABLE%20NITROGEN%20MANAGEMENT.%20English.pdf?sequence=1&isAllowed=y.

13. https://food.ec.europa.eu/horizontal-topics/farm-fork-strategy_en.

References

Basu, N. 2025. "Legacy Nitrogen Meets Wetland Restoration: Solvable Problem or Fool's Paradise." Background paper for this report. World Bank, Washington, DC.

Basu, N. B., K. J. Van Meter, D. K. Byrnes, et al. 2022. "Managing Nitrogen Legacies to Accelerate Water Quality Improvement." *Nature Geoscience* 15 (2): 97–105.

Bertassello, L.E., N. B. Basu, J. Maes, B. Grizzetti, A. La Notte, and L. Feyen. 2025. "The Important Role of Wetland Conservation and Restoration in Nitrogen Removal Across European River Basins." *Nature Water*. https://www.nature.com/articles/s44221-025-00465-0.

Billen, G., E. Aguilera, R. Einarsson, et al. 2024. "Beyond the Farm to Fork Strategy: Methodology for Designing a European Agro-Ecological Future." *Science of the Total Environment* 908: 168160.

Brandes, E., G. S. McNunn, L. A. Schulte, et al. 2016. "Subfield Profitability Analysis Reveals an Economic Case for Cropland Diversification." *Environmental Research Letters* 11 (1): 014009.

Cassou, E., S. M. Jaffee, and J. Ru. 2018. *The Challenge of Agricultural Pollution: Evidence from China, Vietnam, and the Philippines*. Washington, DC: World Bank.

Cheng, F. Y., K. J. Van Meter, D. K. Byrnes, and N. B. Basu. 2020. "Maximizing US Nitrate Removal through Wetland Protection and Restoration." *Nature* 588: 625–30. doi:10.1038/s41586-020-03042-5.

Clare, S., B. Danielson, S. Koenig, and J. K. Pattison-Williams. 2021. "Does Drainage Pay? Quantifying Agricultural Profitability Associated with Wetland Drainage Practices and Canola Production in Alberta." *Wetlands Ecology and Management* 29 (3): 397–415.

Cole, S., and G. Sharma. 2017. "The Promise and Challenges of Implementing ICT in Indian Agriculture." India Policy Forum, National Council of Applied Economic Research 14 (1): 173–240.

Crawfurd, L., R. Todd, S. Hares, J. Sandefur, and R. Silverman Bonnifield. "The Effect of Lead Exposure on Children's Learning in the Developing World: A Meta-Analysis." *The World Bank Research Observer* 40 (2): 229–60. https://doi.org/10.1093/wbro/lkae010.

Cui, Z., H. Zhang, X. Chen, et al. 2018. "Pursuing Sustainable Productivity with Millions of Smallholder Farmers." *Nature* 555 (7696): 363–66.

Dai, Y., S. Yang, D. Zhao, et al. 2023. "Coastal Phytoplankton Blooms Expand and Intensify in the 21st Century." *Nature* 615 (7951): 280–84.

Damania, R., E. Balseca, C. de Fontaubert, et al. 2023. *Detox Development: Repurposing Environmentally Harmful Subsidies*. Washington, DC: World Bank. http://hdl.handle.net/10986/39423.

Damania, R., S. Desbureaux, A.-S. Rodella, J. Russ, and E. Zaveri. 2019. *Quality Unknown: The Invisible Water Crisis*. Washington, DC: World Bank.

Damania, R., and E. Zaveri. 2024. "Hidden Toxins: The Effects of Water Quality on Pregnancy and Infant Health." *BJOG: An International Journal of Obstetrics & Gynaecology* 131 (5): 535–37.

Dash, J. 2019. "Satellites and Crop Interventions." *Nature Sustainability* 2 (10): 903–04.

Duflo, E., M. Kremer, and J. Robinson. 2011. "Nudging Farmers to Use Fertilizer: Theory and Experimental Evidence from Kenya." *American Economic Review* 101 (6): 2350–90.

EEA (European Environment Agency). 2025. "Air Pollution in Europe: 2025 Reporting Status under the National Emission Reduction Commitments Directive." EEA, Copenhagen, Denmark. https://www.eea.europa.eu/en/analysis/publications/air-quality-status-report-2025.

Erisman, J. W., M. A. Sutton, J. Galloway, Z. Klimont, and W. Winiwarter. 2008. "How a Century of Ammonia Synthesis Changed the World." *Nature Geoscience* 1 (10): 636–39.

European Commission Joint Research Centre. 2016. *Algal Bloom and Its Economic Impact*. Brussels, Belgium: European Commission Publications Office. doi:10.2788/660478.

Fluet-Chouinard, E., B. D. Stocker, Z. Zhang, et al. 2023. "Extensive Global Wetland Loss over the Past Three Centuries." *Nature* 614 (7947): 281–86.

Fowler, D., M. Coyle, U. Skiba, et al. 2013. "The Global Nitrogen Cycle in the Twenty-First Century." *Philosophical Transactions of the Royal Society of London B: Biological Sciences* 368 (1621): 20130164.

Gars, J., R. Fishman, A. Kishore, Y. Rothler, and P. S. Ward. 2025. "Confidence and Information Usage: Evidence from Soil Testing in India." *American Journal of Agricultural Economics*. https://doi.org/10.1111/ajae.12513.

Gautam, M. 2015. "Agricultural Subsidies: Resurging Interest in a Perennial Debate." *Indian Journal of Agricultural Economics* 70: 83–105.

Gobler, C. J. 2020. "Climate Change and Harmful Algal Blooms: Insights and Perspective." *Harmful Algae* 91: 101731.

Government of India. 2016. *Economic Survey 2015–16, Volume I*. New Delhi: Department of Economic Affairs, Economic Division, Ministry of Finance.

Gu, B., X. Zhang, S. K. Lam, et al. 2023. "Cost-Effective Mitigation of Nitrogen Pollution from Global Croplands." *Nature* 613 (7942): 77–84.

Gulati, A., and P. Banerjee. 2015. "Rationalizing Fertilizer Subsidy in India: Key Issues and Policy Options." Working Paper 307, Indian Council for Research on International Economic Relations, New Delhi. https://www.academia.edu/14970315/Rationalizing_Fertilizer_Subsidy_in_India-Key_Issues_and_Policy_Options.

Hansen, A. T., T. Campbell, S. J. Cho, et al. 2021. "Integrated Assessment Modeling Reveals Near-Channel Management as Cost-Effective to Improve Water Quality in Agricultural Watersheds." *Proceedings of the National Academy of Sciences* 118 (28): e2024912118. doi:10.1073/pnas.2024912118.

Harou, A. P., M. Madajewicz, H. Michelson, et al. 2022. "The Joint Effects of Information and Financing Constraints on Technology Adoption: Evidence from a Field Experiment in Rural Tanzania." *Journal of Development Economics* 155: 102707.

Hefting, M.M., R. Bobbink, and H. De Caluwe. 2003. "Nitrous Oxide Emission and Denitrification in Chronically Nitrate-Loaded Riparian Buffer Zones." *Journal of Environmental Quality* 32 (4): 1194–1203.

Ho, J. C., A. M. Michalak, and N. Pahlevan. 2019. "Widespread Global Increase in Intense Lake Phytoplankton Blooms Since the 1980s." *Nature* 574 (7780): 667–70.

Huang, J., A. Gulati, and I. Gregory, eds. 2017. *Fertilizer Subsidies: Which Way Forward?* International Fertilizer Development Center and Fertilizer Association of India Report. https://hub.ifdc.org/handle/20.500.14297/2012.

Hutton, G., and C. Chase. 2017. "Water Supply, Sanitation, and Hygiene." In *Disease Control Priorities: Volume 7, Injury Prevention and Environmental Health*, 3rd ed., edited by C. N. Mock, R. Nugent, O. Kobusingye, and K. Smith, 171–90. Wahington, DC: World Bank.

Islam, M., and S. Beg. 2021. "Rule-of-Thumb Instructions to Improve Fertilizer Management: Experimental Evidence from Bangladesh." *Economic Development and Cultural Change* 70 (1): 237–81.

Jain, M., P. Rao, A. K. Srivastava, et al. 2019. "The Impact of Agricultural Interventions Can Be Doubled by Using Satellite Data." *Nature Sustainability* 2 (10): 931–34.

Ji, Y., H. Liu, and Y. Shi. 2020. "Will China's Fertilizer Use Continue to Decline? Evidence from LMDI Analysis Based on Crops, Regions and Fertilizer Types." *PLoS One* 15 (8): e0237234.

Jones, B. A. 2019. "Infant Health Impacts of Freshwater Algal Blooms: Evidence from an Invasive Species Natural Experiment." *Journal of Environmental Economics and Management* 96: 36–59.

Kanter, D. R. 2018. "Nitrogen Pollution: A Key Building Block for Addressing Climate Change." *Climatic Change* 147 (1): 11–21.

Kanter, D. R., F. Bartolini, S. Kugelberg, A. Leip, O. Oenema, and A. Uwizeye. 2019. "Nitrogen Pollution Policy beyond the Farm." *Nature Food* 1 (1): 27–32.

Kanter, D. R., O. Chodos, O. Nordland, M. Rutigliano, and W. Winiwarter. 2020. "Gaps and Opportunities in Nitrogen Pollution Policies around the World." *Nature Sustainability* 3 (11): 956–63.

Keeler, B. L., J. D. Gourevitch, S. Polasky, et al. 2016. "The Social Costs of Nitrogen." *Science Advances* 2 (10): e1600219.

Kirkham, M. B. 2005. *Principles of Soil and Plant Water Relations*. San Diego, CA: Academic Press.

Kremer, M., S. P. Luby, R. Maertens, B. Tan, and W. Więcek. 2023. "Water Treatment and Child Mortality: A Meta-Analysis and Cost-Effectiveness Analysis." Working Paper 30835, National Bureau of Economic Research, Cambridge, MA.

Kurdi, S., M. Mahmoud, K. A. Abay, and C. Breisinger. 2020. *Too Much of a Good Thing? Evidence That Fertilizer Subsidies Lead to Overapplication in Egypt (Vol. 27)*. Washington, DC: International Food Policy Research Institute.

Larsen, B., and E. Sánchez-Triana. 2023. "Global Health Burden and Cost of Lead Exposure in Children and Adults: A Health Impact and Economic Modelling Analysis." *The Lancet Planetary Health* 7 (10): e831–e840.

Levin, R., and J. Schwartz. 2023. "A Better Cost: Benefit Analysis Yields Better and Fairer Results: EPA's Lead and Copper Rule Revision." *Environmental Research* 229: 115738.

Lin, L., S. St Clair, G. D. Gamble, et al. 2023. "Nitrate Contamination in Drinking Water and Adverse Reproductive and Birth Outcomes: A Systematic Review and Meta-Analysis." *Scientific Reports* 13 (1): 563.

Liu, X., A. H. W. Beusen, H. J. M. van Grinsven, et al. 2024. "Impact of Groundwater Nitrogen Legacy on Water Quality." *Nature Sustainability* 7 (7): 891–900.

Lu, C., and H. Tian. 2017. "Global Nitrogen and Phosphorus Fertilizer Use for Agriculture Production in the Past Half Century: Shifted Hot Spots and Nutrient Imbalance." *Earth System Science Data* 9, 181–92. https://doi.org/10.5194/essd-9-181-2017.

McDowell, R. W., Z. P. Simpson, A. G. Ausseil, Z. Etheridge, and R. Law. 2021. "The Implications of Lag Times between Nitrate Leaching Losses and Riverine Loads for Water Quality Policy." *Scientific Reports* 11 (1): 16450.

Oelsner, G. P., and E. G. Stets. 2019. "Recent Trends in Nutrient and Sediment Loading to Coastal Areas of the Conterminous U.S.: Insights and Global Context." *Science of the Total Environment* 654: 1225–40.

Pennino, M. J., S. G. Leibowitz, J. E. Compton, R. A. Hill, and R. D. Sabo. 2020. "Patterns and Predictions of Drinking Water Nitrate Violations across the Conterminous United States." *Science of the Total Environment* 722: 137661.

Poore, J., and T. Nemecek. 2018. "Reducing Food's Environmental Impacts through Producers and Consumers." *Science* 360 (6392): 987–92.

Rabotyagov, S. S., T. D. Campbell, M. White, et al. 2014. "Cost-Effective Targeting of Conservation Investments to Reduce the Northern Gulf of Mexico Hypoxic Zone." *Proceedings of the National Academy of Sciences* 111 (52): 18530–5.

Richardson, K., W. Steffen, W. Lucht, et al. 2023. "Earth beyond Six of Nine Planetary Boundaries." *Science Advances* 9 (37): eadh2458.

Ritchie, H., M. Roser, and P. Rosado. 2013. "Fertilizers." *Our World in Data*. https://ourworldindata.org/fertilizers.

Russ, J. D., E. D. Zaveri, R. Damania, S. G. Desbureaux, J. J. Escurra, and A. S. Rodella. 2020. "Salt of the Earth: Quantifying the Impact of Water Salinity on Global Agricultural Productivity." Policy Research Working Paper 9144, World Bank, Washington, DC.

Schultz, T. W. 1964. *Transforming Traditional Agriculture*. New Haven, CT: Yale University Press.

Shen, W., L. Zhang, E. A. Ury, S. Li, B. Xia, and N. B. Basu. 2025. "Restoring Small Water Bodies to Improve Lake and River Water Quality in China." *Nature Communications* 16 (1): 294.

Singh, V., S. Ganguly, and V. Dakshinamurthy. 2018. "Evaluation of India's Soil Health Card from Users' Perspectives." CSISA Research Note, Vol. 12, International Food Policy Research Institute, Washington, DC.

Smale, M., and V. Thériault. 2018. "A Cross-Country Summary of Fertilizer Subsidy Programs in Sub-Saharan Africa." Feed the Future Innovation Lab for Food Security Policy Research Paper 169, Michigan State University, East Lansing, MI.

Smil, V. 2022. *How the World Really Works: A Scientist's Guide to Our Past, Present and Future*. London: Penguin UK. https://www.penguin.co.uk/books/319141/how-the-world-really-works-by-smil-vaclav/9780241989678.

Stewart, W. M., D. W. Dibb, A. E. Johnston, and T. J. Smyth. 2005. "The Contribution of Commercial Fertilizer Nutrients to Food Production." *Agronomy Journal* 97 (1): 1–6.

Sun, W., K. Lu, and R. Li. 2024. "Global Estimates of Ambient NO_2 Concentrations and Long-Term Health Effects during 2000–2019." *Environmental Pollution* 359: 124562.

Sutton, M. A., A. Bleeker, C. M. Howard, et al. 2013. *Our Nutrient World: The Challenge to Produce More Food and Energy with Less Pollution*. Edinburgh, UK: Centre for Ecology and Hydrology, on behalf of the Global Partnership on Nutrient Management and the International Nitrogen Initiative.

Tian, H., R. Xu, J. G. Canadell, et al. 2020. "A Comprehensive Quantification of Global Nitrous Oxide Sources and Sinks." *Nature* 586 (7828): 248–56.

UNEP and FAO (United Nations Environment Programme and Food and Agriculture Organization). 2024. *Global Nitrous Oxide Assessment*. Nairobi, Kenya: UNEP.

Van Meter, K. J., and N. B. Basu. 2015. "Catchment Legacies and Time Lags: A Parsimonious Watershed Model to Predict the Effects of Legacy Storage on Nitrogen Export." *PLoS One* 10 (5): e0125971.

Van Meter, K. J., P. Van Cappellen, and N. B. Basu. 2018. "Legacy Nitrogen May Prevent Achievement of Water Quality Goals in the Gulf of Mexico." *Science* 360 (6387): 427–30.

Ward, M. H., R. R. Jones, J. D. Brender, et al. 2018. "Drinking Water Nitrate and Human Health: An Updated Review." *International Journal of Environmental Research and Public Health* 15 (7): 1557.

Watson, R. 2017. "A Database of Global Marine Commercial, Small-Scale, Illegal and Unreported Fisheries Catch 1950–2014." *Scientific Data* 4 (1): 1–9.

WHO/UNICEF (World Health Organization/United Nations Children's Fund) Joint Monitoring Programme. 2023. *Progress on Household Drinking Water, Sanitation and Hygiene 2000–2022: Special Focus on Gender*. New York: UNICEF and WHO.

Willett, W., J. Rockström, B. Loken, et al. 2019. "Food in the Anthropocene: The EAT–Lancet Commission on Healthy Diets from Sustainable Food Systems." *The Lancet* 393 (10170): 447–92.

Wolf, J., R. B. Johnston, A. Ambelu, et al. 2023. "Burden of Disease Attributable to Unsafe Drinking Water, Sanitation, and Hygiene in Domestic Settings: A Global Analysis for Selected Adverse Health Outcomes." *The Lancet* 401 (10393): 2060–71.

World Bank. 2025. *Resilient Sanitation for People and Planet*. Washington, DC: World Bank.

Wuepper, D., S. Le Clech, D. Zilberman, N. Mueller, and R. Finger. 2020. "Countries Influence the Trade-off between Crop Yields and Nitrogen Pollution." *Nature Food* 1 (11): 713–19.

Xia, L., and X. Yan. 2023. "How to Feed the World While Reducing Nitrogen Pollution." *Nature* 613 (7942): 34–35.

Zaveri, E. 2025. "Fixing Nitrogen: Agricultural Productivity, Environmental Fragility, and the Role of Subsidies." Policy Research Working Paper 11050, World Bank, Washington, DC. http://hdl.handle.net/10986/42737.

Zaveri, E., J. Russ, S. Desbureaux, R. Damania, A.-S. Rodella, and G. Ribeiro. 2020. "The Nitrogen Legacy: The Long-Term Effects of Water Pollution on Human Capital." Policy Research Working Paper 9143, World Bank, Washington, DC.

Zhang, X., E. A. Davidson, D. L. Mauzerall, T. D. Searchinger, P. Dumas, and Y. Shen. 2015. "Managing Nitrogen for Sustainable Development." *Nature* 528 (7580): 51.

Zhang, X., T. Zou, L. Lassaletta, et al. 2021. "Quantification of Global and National Nitrogen Budgets for Crop Production." *Nature Food* 2 (7): 529–40.

CHAPTER 5

Parting the Smog: The State of Air Pollution around the World

"It was a foggy day in London, and the fog was heavy and dark. Animate London, with smarting eyes and irritated lungs, was blinking, wheezing, and choking; inanimate London was a sooty spectre, divided in purpose between being visible and invisible, and so being wholly neither."

—*Charles Dickens*, English Novelist (1812–1870)

Key messages

- As the second largest cause of death globally, air pollution leads to 7 million deaths each year (WHO 2023). The annual costs of indoor and outdoor air pollution have been estimated at $8 trillion.

- China's "war on pollution" has shown that, with concentrated effort, change is possible. Actions in China have included improving the monitoring and publishing of air pollution data, phasing out coal for cleaner-burning fuels, enforcing stricter emissions standards, relocating manufacturing, and expanding public transportation systems.

- Early trials of pollution markets in Gujarat, India, have shown promising results, with estimated benefits ($101.4 million per year) exceeding costs ($3.9 million per year) by a factor of over 25 (Greenstone et al. 2025).

- Cooperation in managing air pollution in shared airsheds could lower abatement costs from $2.6 billion to $278 million per microgram of pollutant per cubic meter of air (World Bank 2023b). These estimates from South Asia show that joint management of shared resources can drastically lower the costs of abating air pollution.

- Sources of air pollution are heterogeneous and context-specific, so solutions must be flexible. Air pollution markets can help combat pollution, and airshed agreements can be effective in areas where pollution is transboundary. Fossil fuel subsidy reform can also lead to multiple wins, freeing up budgetary space, reducing greenhouse gas emissions, cleaning the air, and saving hundreds of thousands of lives each year.

The deadly effects of air pollution

In the winter of 1952, a cold spell in London led to increased coal burning, which, combined with vehicle and industrial emissions, created dense smog trapped by temperature inversion. This five-day event, now known as the Great London Smog, caused concentrations of fine particulate matter ($PM_{2.5}$)[1] to reach a staggering level of 4,460 micrograms of pollutant per cubic meter of air ($\mu g/m^3$) (QUARG 1993), 300 times the current World Health Organization (WHO) guideline (15 $\mu g/m^3$) (WHO 2021). Initially blamed for 4,000 deaths, more-modern assessments suggest that more than 10,000 lives were lost (Stone 2002). The disaster prompted the United Kingdom to pass the Clean Air Act of 1956, which restricted the use of dirty fuels.

Although the Great London Smog catalyzed landmark reforms, air pollution today continues to cause millions of deaths annually, with far less attention. This chapter explores global air pollution trends, sources, and solutions, with a focus on $PM_{2.5}$, drawing from Sudarshan and Goswami (2024), a background paper for this report.

Global trends in air pollution

Air pollution, the second largest cause of death globally, is responsible for 7 million deaths each year and costs an estimated $8 trillion (Peszko et al. 2022; WHO 2023). This includes impacts from two categories of air pollution:

- *Ambient (outdoor) air pollution*, often generated by transport, energy generation, or economic activity, which caused more than 5.7 million deaths globally in 2020, with economic damages estimated at $4.5 trillion to $6.1 trillion annually, or 4.7 to 6.5 percent of global gross domestic product (GDP) (World Bank 2025).

- *Indoor air pollution,* which occurs within built environments, such as homes or workplaces.

Poor air quality causes respiratory and cardiovascular conditions, which increase infant mortality rates and reduce life expectancy. Globally, poor air quality is a greater threat to life expectancy than tobacco use, alcohol consumption, malnutrition, malaria, HIV/AIDS, or unsafe sanitation (Greenstone et al. 2024). There is also substantial evidence that air pollution diminishes cognitive abilities in a wide range of contexts—including language, mathematics, cognitive skills, and even sports refereeing—further depressing productivity (Archsmith, Heyes, and Saberian 2018; La Nauze and Severnini 2021; Zhang, Chen, and Zhang 2018).

Air pollution can substantially impact the economy through absenteeism and decreased cognitive performance. Evidence from Europe suggests that 1 $\mu g/m^3$ of $PM_{2.5}$ concentrations can lead to a 0.8 percent decrease in real GDP (Dechezleprêtre, Rivers, and Stadler 2019) due to diminished output caused by absenteeism, or lower productivity. Given this magnitude of impact, a 10 percent decrease in $PM_{2.5}$

concentrations would yield an additional €100 billion to €200 billion annually. Similar results on the impact of air pollution on growth have been found in India, where a 1 $\mu g/m^3$ increase in air pollution depresses GDP growth by 0.7 percentage points (Behrer, Choudhary, and Sharma 2023).

China's "war on pollution" has illustrated that, with concentrated effort, substantial improvements are possible. Globally, air pollution exposure per person is trending downward. This has largely been driven by China, where in the mid-2010s, the central government declared air pollution as a critical concern, and subsequently enacted a wide range of policies, such as $PM_{2.5}$ standards, carbon markets, and automated monitoring equipment (Greenstone et al. 2021; Greenstone et al. 2022). Box 5.1 details some of the measures taken by the Chinese government and their effects on air pollution. As a result of the drastic measures, global population-weighted $PM_{2.5}$ exposure decreased from a peak of 38.9 $\mu g/m^3$ (hazardous) in 2011 to 34.7 $\mu g/m^3$ (just below hazardous) in 2019 (Li et al. 2023). Figure 5.1 illustrates the effect these measures have had on global population-weighted exposure to air pollution.

BOX 5.1
China's "war on pollution"

In 2013, Beijing's fine particulate matter ($PM_{2.5}$) levels averaged 91 micrograms of pollutant per cubic meter of air—nearly triple the national standard—prompting public outcry. Earlier efforts had achieved only modest success, and pollution remained severe in major cities. In response, China launched the Air Pollution Prevention and Control Action Plan (2013–17), a mammoth $270 billion initiative, targeting three densely populated regions: the Beijing-Tianjin-Hebei area, the Pearl River Delta, and the Yangtze River Delta.

- The Beijing-Tianjin-Hebei Region cut $PM_{2.5}$ emissions by 25 percent by switching households from coal to electricity, phasing out dirty factories, and tightening industrial and power plant emissions standards (Greenstone and Schwarz 2018; Zhang et al. 2019).

- The Pearl and Yangtze River Deltas were similarly successful, reducing $PM_{2.5}$ by 15 and 20 percent, respectively (Greenstone and Schwarz 2018). This was accomplished through stricter vehicle and industrial emission controls and investments in renewables and electric vehicles.

This initiative was followed by the Blue Sky Protection Campaign (2018–20), which further reduced $PM_{2.5}$ and sulfur dioxide emissions by 18 and 15 percent, respectively, in China's most polluted regions. Like the air pollution action plan, the Blue Sky campaign focused on stringent industrial norms, promoting cleaner energy use, controlling vehicular emissions, and expanding public transportation systems to reduce reliance on private vehicles.

Together, these regional initiatives and air quality policies cut $PM_{2.5}$ in Beijing by over 40 percent between 2013 and 2018 (Myllyvirta 2018). As a result of substantial improvements in air quality, in 2017, there were 47,240 fewer deaths in China's 74 key cities than in 2013 (Huang et al. 2018).

Source: Sudarshan and Goswami (2024).

FIGURE 5.1 Population-weighted ambient PM$_{2.5}$, indexed to 2000, 2000–20

Population-weighted PM$_{2.5}$ concentrations (indexed to 2000)

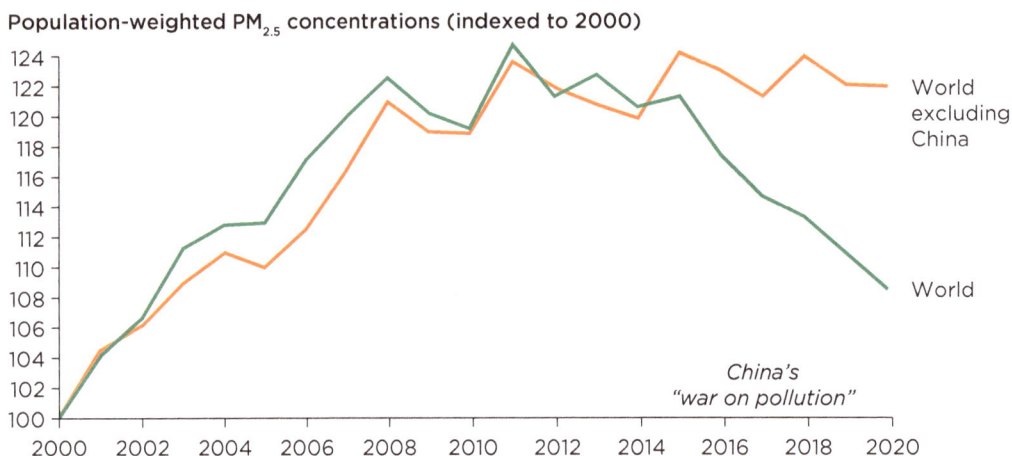

Sources: Original calculations based on PM$_{2.5}$ data from van Donkelaar et al. 2016 and population data from CIESIN 2018.

Note: The figure shows population-weighted PM$_{2.5}$ concentrations from 2000 to 2020, illustrating global trends in air pollution. From 2000 to the early 2010s, PM$_{2.5}$ concentrations rose steadily worldwide, peaking around 2013–14. However, following China's launch of its "war on pollution" in 2014–15, a significant divergence emerged. While global PM$_{2.5}$ levels excluding China remained high, China's aggressive air quality policies greatly pulled down the global average. PM$_{2.5}$ = fine particulate matter.

Unfortunately, China's experience has not been universal, and air quality has stagnated or worsened in most low-income countries (LICs) and lower-middle-income countries (LMICs) (figure 5.2). World Bank estimates suggest that 7.3 billion people globally are exposed to dangerous levels of PM$_{2.5}$ air pollution, with 80 percent living in LICs and LMICs (Damania et al. 2023; Rentschler and Leonova 2023). There are many causes of worsening air pollution levels in LICs and LMICs, including less stringent air quality regulations; the prevalence of older, more polluting machinery; fossil fuel subsidies; congested urban transport systems; and rapidly developing industrial sectors. High-income countries (HICs), in contrast, have seen consistent progress in reducing air pollution levels, and have largely decoupled growth from PM$_{2.5}$ emissions (refer to spotlight 1). They have done this through implementing strict environmental regulations, promoting cleaner energy sources, investing in public transportation systems, setting stricter vehicle emission standards, and prioritizing urban planning that encourages walking and cycling, all while transitioning industries toward cleaner technologies and practices.

FIGURE 5.2 **Population-weighted ambient PM$_{2.5}$, indexed to 2000, by country income group, 2000–20**

Population weighted PM$_{2.5}$ concentrations (indexed to 2000)

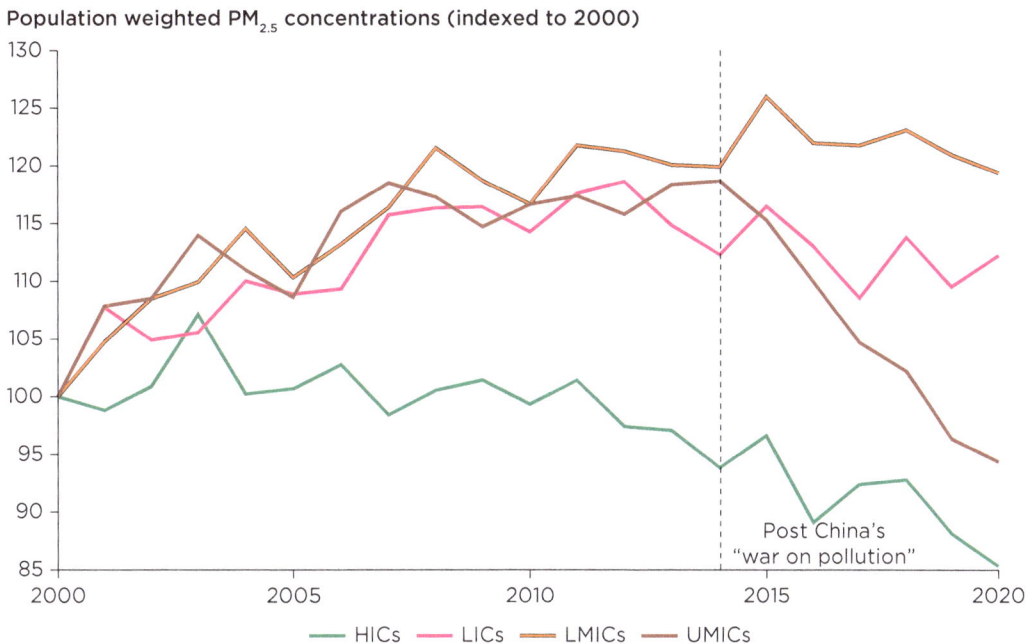

Sources: Original calculations, based on PM$_{2.5}$ data from van Donkelaar et al. 2016 and population data from CIESIN 2018.
Note: HICs = high-income countries; LICs = low-income countries; LMICs = lower-middle-income countries; PM$_{2.5}$ = fine particulate matter; UMICs = upper-middle-income countries.

While poorer countries tend to have higher levels of ambient air pollution, it is often the richer areas within them that are more polluted (Behrer and Heft-Neal 2024). Although this might seem paradoxical, pollution is often more severe in areas of higher economic activity, and those who are relatively poorer within a country are often not near the centers of economic activity. This also reflects patterns where jobs in polluted areas offer higher wages or other benefits, a concept rooted in Adam Smith's theory of compensating wage differentials (Smith 1776). Smith observed that workers will bargain for a higher wage in jobs that are unpleasant, dangerous, or otherwise unattractive. However, in some cases, pollution exposure reflects historical injustices such as redlining, rather than any compensatory advantage (Lane et al. 2022).

Sources of pollution

Sources of ambient air pollution are heterogeneous across countries, but some consistent trends exist (Sudarshan and Goswami, 2024). A substantial share of ambient particulate emissions in developing countries comes from the residential sector, largely from burning biomass or solid fuels for cooking or heating. Outside residential emissions, many sectors contribute, with the largest share usually coming from transportation (especially nitrogen dioxide), industry, and power generation (especially sulfur dioxide).

Poorer households face especially high levels of indoor air pollution, often relying on polluting biomass like firewood or charcoal due to limited access to cleaner fuels like liquefied petroleum gas (LPG) (Ferguson et al. 2020; Rao et al. 2021; World Bank 2011). In 2020, more than 3.2 million deaths were attributed to household air pollution, nearly half of all pollution-linked deaths worldwide, mainly in East Asia, South Asia, and Sub-Saharan Africa where solid fuel use is high (WHO 2023). Traditional cooking methods emit dangerous pollutants, such as $PM_{2.5}$, carbon monoxide, nitrogen oxide, and organic air pollutants, such as benzene and formaldehyde (Smith et al. 2013), with concentrations sometimes reaching over 100 times higher than WHO guidelines (WHO 2016). Improved cookstoves and heaters, and transitioning to better fuels like electricity, LPG, biogas, and ethanol can reduce emissions and health risks. However, adoption is constrained by affordability, infrastructure, and social factors.

As incomes rise, some polluting behaviors—such as burning biomass—decline, reflecting the economic concept of inferior goods. However, other pollution sources, such as fossil fuels for transport or industry, may grow with income. These shifting patterns are explored in box 5.2 through the lens of environmental Engel curves.

BOX 5.2
How consumption changes with income: Environmental Engel curves

Environmental Engel curves (EECs) show how the environmental footprint of household consumption changes with income. Adapted from the original Engel curve, a graph showing the relationship between income and demand, EECs can provide insights into the effectiveness and distributional effects of pollution control policies. EECs are upward-sloping for normal goods, indicating higher pollution with increased income, but may be downward-sloping if wealthier households shift to more-efficient, less-polluting goods.

Using harmonized household survey data from 64 countries, Ebadi and Rentschler (2025) analyzed how household consumption shares spent on 11 standard fuel types shifted across income levels. The study classified household energy types into three categories based on indoor air pollution intensity: high (charcoal, firewood, coal, and kerosene), medium (diesel, ethanol, gasoline, and oil), and low (electricity, liquefied petroleum gas, and natural gas).

In middle- and high-income countries, rising incomes tended to reduce reliance on high-polluting fuels and increase the use of cleaner energy (figure B5.2.1). This trend underscores the potential for wealthier households in these countries to adopt more environmentally sustainable energy consumption habits.

However, the transition did not occur in low-income countries, where budget shares of both high and low indoor air pollution intensities increased with income. This may have been because households in poor countries have a higher income elasticity for energy demand, and cleaner fuels may be less accessible. These patterns demonstrate the need to expand access to clean energy to enable sustainable consumption transitions.

(continued)

FIGURE B5.2.1 Household budget share, by energy type and country income group

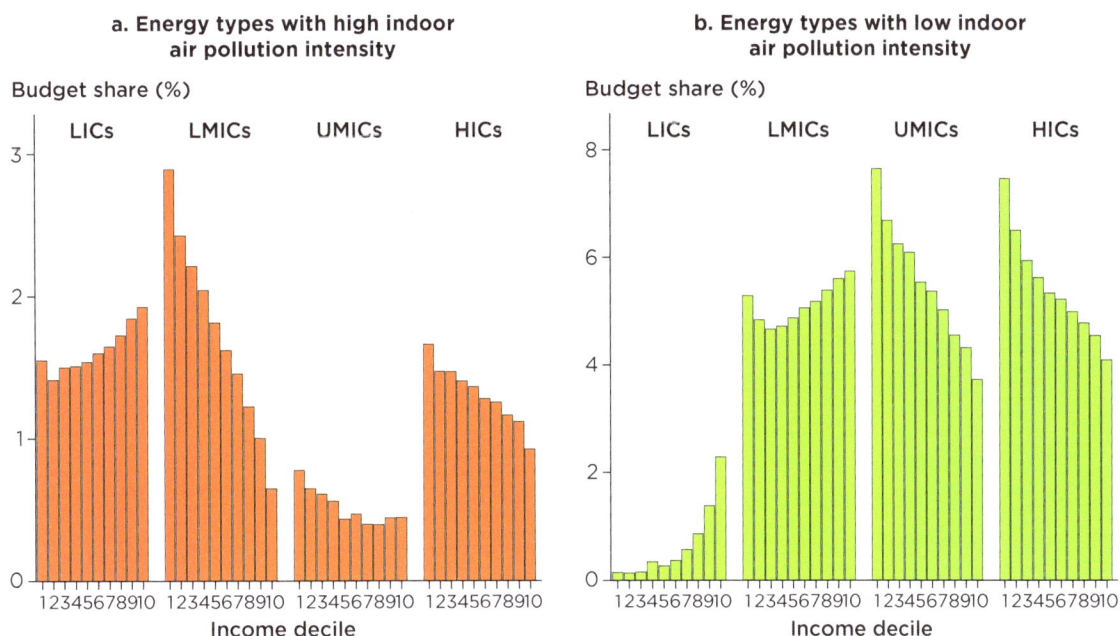

a. Energy types with high indoor air pollution intensity

b. Energy types with low indoor air pollution intensity

Source: Original calculations, based on data from the Climate Policy Assessment Tool, World Bank and International Monetary Fund; Ebadi and Rentschler (2025).
Note: High-polluting energy types (panel a) include charcoal, firewood, coal, and kerosene; low-polluting energy types (panel b) include electricity, liquefied petroleum gas, and natural gas. HICs = high-income countries; LICs = low-income countries; LMICs = lower-middle-income countries; UMICs = upper-middle-income countries.

Source: Ebadi and Rentschler (2025).

Clearing the air: Pitfalls and potential solutions

There is no single panacea to solve air pollution, as its sources are heterogeneous and policies must reflect local conditions and capacities. Nevertheless, addressing air pollution can yield large gains in life expectancy and unlock economic benefits. The World Bank's *Accelerating Access to Clean Air for a Livable Planet* report found that well-designed, cross-sectoral policy measures could halve the number of people exposed to $PM_{2.5}$ levels above 25 μg/m³, potentially preventing up to 2.1 million deaths annually and yielding economic benefits of $0.8 trillion to $6.2 trillion by 2040

(World Bank 2025). This section explores the key enablers and barriers to action, including the role of state capacity, the promise of market-based instruments, and the challenges posed by transboundary pollution and fossil fuel subsidies. These elements provide a roadmap for crafting policies that are effective, equitable, and context specific.

State capacity to enforce air pollution policies

Effective air pollution control depends not just on the presence of environmental regulations, but on a country's capacity to implement them. Sudarshan and Goswami (2024) showed that regions with severe air pollution, such as South Asia, often have a similar number of regulations as cleaner regions like Europe. Therefore, poor outcomes frequently reflect weak enforcement, misaligned incentives, and capacity shortfalls, rather than a lack of regulations and legal frameworks.

Enforcement gaps may stem from principal-agent problems, where regulators rely on information from polluters or intermediaries with conflicting incentives (Duflo et al. 2013; Stoerk 2016). Many environmental regulators also lack sufficient trained staff, a challenge exacerbated when pollution sources are numerous and decentralized (Ghosh et al. 2023). As box 5.3 shows, market-based instruments tend to be more effective and less susceptible to rent seeking than traditional command-and-control approaches, especially in low-capacity settings.

BOX 5.3
Controlling air pollution: Command-and-control or market-based instruments?

A meta-analysis conducted for this report evaluated 30 high-quality studies to explore the effectiveness of air pollution control policies, comparing 51 command-and-control (CAC) instruments and 10 market-based instruments (MBIs). All of the selected policies had statistically significant impacts on air pollutants, including carbon monoxide, nitrogen oxides, sulfur dioxide, and fine particulate matter ($PM_{2.5}$, and PM_{10}).

To compare the efficacy of CAC and MBIs, the analysis used Cohen's d, which measures the effect size and indicates the magnitude of difference between two groups. A higher Cohen's d value signifies a larger difference between the groups' means. Table B5.3.1 illustrates the differences in efficacy between CAC and MBIs, based on the estimates from the selected papers.

The CAC measures, while effective in the short term, tended to lose impact over time due to monitoring, enforcement, and other challenges. In some cases they even backfired, as in Mexico City's *Hoy No Circula* program, which restricted most drivers from using their vehicles one weekday a week, based on the last digit of their license plate, to reduce air pollution (Davis 2008). The policy led to increased vehicle purchases and higher emissions as people attempted to avoid the regulations.

(continued)

BOX 5.3

Controlling air pollution: Command-and-control or market-based instruments?
(continued)

The MBI measures, such as emissions trading or pollution taxes, became more effective over time. They offered flexibility, reduced emissions cost-effectively, and encouraged innovation, while also generating government revenue that could fund environmental programs or subsidize clean technology.

TABLE B5.3.1 Policy efficacy in reducing air pollution, by policy type and duration

Policy type	Less than a year	Duration above a year
CAC	−0.39	−0.12
	(0.24)	(0.31)
MBI	−0.21	−0.25
	(0.15)	(0.22)

Source: Original calculations based on estimates from 30 high-quality studies.
Note: The values in the table show the fraction of a standard deviation by which the policy reduced air pollution; the values in parentheses show the standard deviations of the policy impacts. CAC = command-and-control; MBI = market-based instrument.

Air pollution markets: An opportunity for developing countries

Air pollution markets, such as cap-and-trade systems, offer a cost-effective alternative to traditional command-and-control policies but are rarely adopted by developing countries. By setting an overall limit (cap) on total emissions and allowing firms to trade allowances, these market-based instruments (MBIs) incentivize cost-efficient pollution reduction and technological innovation. Unlike prescriptive rules, markets reward overperformance and create financial value for cleaner operations. While effective, implementing these systems requires careful regulatory design and skilled oversight. However, regulators can reduce overall staffing needs by relying on technologies such as real-time and continuous emissions monitoring systems, which also help reduce the falsification of data (Greenstone et al. 2022).

MBIs alter the enforcement and compliance landscape, reducing the burden on regulators and enabling effective implementation even in lower-capacity settings. In recent work, Greenstone et al. (2025) showed that emissions markets reduced pollution by 20 percent in India and remained effective over multiple years. The program was highly cost-effective, costing $3.9 million annually and delivering between $101.4 million and $847 million in annual benefits, depending on the conservativeness of the health benefits. Even at the lower bound, they appear to be a valuable tool in the fight against air pollution, with cost-benefit ratios of 1:26 to 1:215. Box 5.4 provides more context on the design and implementation of air pollution markets.

BOX 5.4

The success story of PM$_{2.5}$ markets in Gujarat, India

Governments in India are developing and trialing new policy approaches to curb fine particulate matter (PM$_{2.5}$) pollution. In 2019, the Government of Gujarat implemented the world's first emissions trading scheme for particulate pollution in Surat, a city that is home to emissions-intensive industries such as steel and cement production. The scheme operates on a cap-and-trade system, where the authorities set a total emissions allowance with purchasable permits entitling the holder to emit set quantities of PM$_{2.5}$. By requiring businesses to hold enough permits to cover their emissions, with fines for noncompliance, parts of the social cost of air pollution become internalized. The permit trade rewards less-polluting firms, as they require fewer permits or can resell their excess to less-clean firms. The project was conceptualized by the Ministry of Environment, Forests, and Climate Change in 2012 and executed with the help of the Gujarat Pollution Control Board and researchers from the University of Chicago and Yale University.

The initial results of the particulate trading scheme are promising: participating businesses have reduced their emissions by 24 percent. Automated monitoring systems that were in place before the scheme enabled consistent measurement of firms' emissions. The long-term effect of the policy on firms is an important consideration, as permits could significantly raise operating costs, which companies might pass on to consumers, or could lead to business closures. However, the trials were considered successful, and the authorities plan to extend the scheme within and outside Gujarat.

An opportunity for cooperation

Transboundary air pollution poses serious public health risks and is especially challenging in regions of the Indo-Gangetic Plain, where vast populations are affected. Managing it requires international cooperation, yet differences in environmental regulations, data standards, and political willingness often hinder progress. Economic priorities can also clash with environmental concerns, and disparities in technological capabilities and expertise can impede joint efforts.

Airshed approaches can help countries share regulatory resources and improve regional outcomes through coordinated monitoring, enforcement, and knowledge exchange. A successful model is the Geneva Convention on Long-Range Transboundary Air Pollution (CLRTAP), which has united 51 countries across Europe, North America, and beyond in reducing pollutants through eight protocols, including the 1998 Protocol on Heavy Metals, and the 1999 Gothenburg Protocol (Sudarshan and Goswami, 2024). These agreements have helped cut emissions of sulfur dioxide, nitrogen oxides, volatile organic compounds, and black carbon, making CLRTAP the first international agreement that explicitly addresses black carbon.

Distorted fossil fuel prices, distorted incentives

Air pollution is ubiquitous because polluting activities are underpriced. Despite well-documented harms from fossil fuels, governments around the world provide explicit subsidies—that is, direct financial transfers that lower the cost of using fossil fuels. Governments provide around $577 billion in direct fossil fuel subsidies (Damania et al. 2023), almost three times the global subsidies paid to the renewable energy sector (IRENA 2020).[2] Removing explicit subsidies could reduce $PM_{2.5}$ concentrations enough to prevent some 360,000 deaths between now and 2035. Yet, although this is a large number, it is only a small fraction of the overall deaths attributed to air pollution. Removing explicit fossil fuel subsidies is therefore a necessary first step, but it cannot solve the air pollution challenge alone. Furthermore, since political reality imposes a limit on how much energy prices can increase, governments will need complementary policies to ensure the availability and affordability of clean alternatives, address information and capacity constraints, and influence behavior (box 5.5).

BOX 5.5
Clean energy through policy support

Energy is a crucial driver of economic prosperity but has a significant environmental impact. Energy production affects air, water, and land, with fossil fuels contributing substantially to greenhouse gas (GHG) emissions. In 2019, the energy sector was responsible for 34 percent of global direct GHG emissions, with electricity and heat generation being the largest subsector (Dhakal et al. 2022). Carbon dioxide, primarily from fossil fuels, constitutes 74 percent of these emissions (Ge, Friedrich, and Vigna 2024). The energy sector also accounts for about 10 percent of global freshwater withdrawals (IEA, n.d.).

Recent energy policies have significantly altered the energy mix, efficiency, and emissions intensities, with decomposition analysis revealing their impacts on emissions. In lower-middle-income countries and upper-middle-income countries (UMICs), GHG emissions from electricity generation rose due to increased demand driven by economic growth and population rise (figure B5.5.1). However, in high-income countries (HICs) these increases were offset due to cleaner technologies and lower emissions intensities. As shown for other environmental pressures in spotlight 1, this analysis highlights that the scale effect, or increased electricity demand, was most pronounced in UMICs, where energy additions outpaced the shift to cleaner sources, leading to higher emissions. HICs successfully transitioned to less-polluting electricity generation, offsetting increased emissions from demand. While cleaner fuels are crucially important, energy efficiency has so far significantly offset emissions in several UMICs.

The energy transition will cut emissions and require a different configuration of resources. Technologies like biofuels, concentrated solar, and nuclear power have high water footprints. Renewables also require more land (Gross 2020), along with extensive infrastructure and critical minerals. Without safeguards, this can threaten forests and natural ecosystems. Effective policies can mitigate these risks, such as installing solar panels on rooftops or integrating them with agriculture, where partial shading can even boost crop yields.

(continued)

BOX 5.5
Clean energy through policy support *(continued)*

FIGURE B5.5.1 Decomposition analysis for electricity GHG and SO$_2$ emissions, by driver, 2004–17

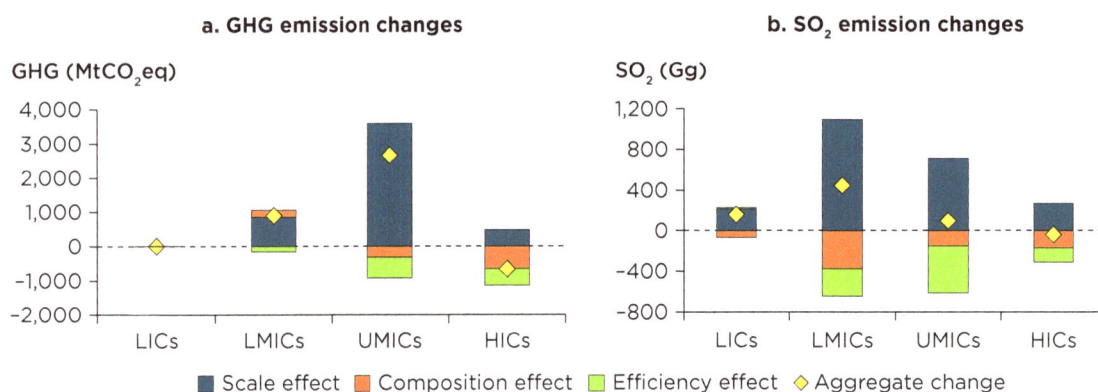

a. GHG emission changes

b. SO$_2$ emission changes

GHG (MtCO$_2$eq)

SO$_2$ (Gg)

■ Scale effect ■ Composition effect ■ Efficiency effect ◇ Aggregate change

Sources: Original calculations based on data from Global Trade Analysis Project databases.
Note: Efficiency is defined as the intensity of emissions per unit of value added, and composition is the electricity and heat generation distribution between different categories. Gg = gigagrams; GHG = greenhouse gas; HICs = high-income countries; LICs = low-income countries; LMICs = lower-middle-income countries; MtCO$_2$eq = million tons of carbon dioxide equivalent; UMICs = upper-middle-income countries; SO$_2$ = sulfur dioxide.

The World Bank's clean energy policy aims to facilitate a global shift to sustainable energy. It supports creating demand for energy efficiency through policy frameworks and overcoming financial and technical obstacles. The World Bank emphasizes mobilizing private investment to meet clean energy transition needs (World Bank 2023a). Developing countries face high electricity costs and fossil fuel dependency. To address this, the World Bank creates environments that attract private investment. Initiatives include India's sovereign green bond program and a private sector investment lab to expand risk mitigation and finance clean energy projects (World Bank 2023a, 2023b).

Conclusion

Air pollution is a more serious threat to health and economic growth today than at any time in human history, affecting more people than ever before. For the most part, it is not a diminishing problem. Despite this gloomy description, it can be tackled through comprehensive, context-specific air pollution policy that considers state capacity. The evidence for this lies in the success of China's approach and other countries across the world. The challenge now is to scale these successes globally, tailoring solutions to local contexts while unlocking the economic and health benefits of cleaner air.

Notes

1. PM$_{2.5}$ refers to particles that are less than 2.5 micrometers in diameter, which is approximately 5 percent of the size of a human hair. Due to their microscopic size, they can get lodged deep into the respiratory tract and absorbed into the circulatory system, which can lead to a wide range of health issues (for example, refer to WHO 2021).
2. Refer to https://www.oecd.org/en/topics/sub-issues/climate-finance-and-the-usd-100-billion -goal.html.

References

Archsmith, J., A. Heyes, and S. Saberian. 2018. "Air Quality and Error Quantity: Pollution and Performance in a High-Skilled, Quality-Focused Occupation." *Journal of the Association of Environmental and Resource Economists* 5 (4): 827–63. doi:10.1086/698728.

Behrer, A. P., R. Choudhary, and D. Sharma. 2023. *Air Pollution Reduces Economic Activity: Evidence from India*. Washington, DC: World Bank. http://documents.worldbank.org/curated/en /099710506302335471/IDU0536186520dc340403409fb001cfc637a9368.

Behrer, A. P., and S. Heft-Neal. 2024. "Higher Air Pollution in Wealthy Districts of Most Low- and Middle-Income Countries." *Nature Sustainability* 7: 203–12. doi:10.1038/s41893-023-01254-x.

CIESIN (Center for International Earth Science Information Network). 2018. *Gridded Population of the World, Version 4 (GPWv4): Population Count, Revision 11*. New York: CIESIN, Columbia University. doi:10.7927/H4JW8BX5.

Damania, R., E. Balseca, C. de Fontaubert, et al. 2023. *Detox Development: Repurposing Environmentally Harmful Subsidies*. Washington, DC: World Bank. http://hdl.handle.net/10986/39423.

Davis, L. 2008. "The Effect of Driving Restrictions on Air Quality in Mexico City." *Journal of Political Economy* 116 (1): 38–81. doi:10.1086/529398.

Dechezleprêtre, A., N. Rivers, and B. Stadler. 2019. *The Economic Cost of Air Pollution: Evidence from Europe*. Paris: OECD Publishing. doi:10.1787/56119490-en.

Dhakal, S., J. C. Minx, F. L. Toth, et al. 2022. "Emissions Trends and Drivers." In *Climate Change 2022: Mitigation of Climate Change. Contribution of Working Group III to the Sixth Assessment Report of the Intergovernmental Panel on Climate Change*, edited by P. R. Shukla, J. Skea, R. Slade, et al., 215–94. Cambridge, UK: Cambridge University Press. doi:10.1017/9781009157926.004.

Duflo, E., M. Greenstone, R. Pande, and N. Ryan. 2013. "Truth-Telling by Third-Party Auditors and the Response of Polluting Firms: Experimental Evidence from India." *Quarterly Journal of Economics* 128: 1499–1545.

Ebadi, E., and J. Rentschler. 2025. "Pollution Intensity of Consumption—Exploring the Environmental Engel Curve Based on Micro-Data for 89 Countries." World Bank, Washington, DC.

Ferguson, L., J. Taylor, M. Davies, C. Shrubsole, P. Symonds, and S. Dimitroulopoulou. 2020. "Exposure to Indoor Air Pollution across Socio-Economic Groups in High-Income Countries: A Scoping Review of the Literature and a Modelling Methodology." *Environment International* 143: 105748.

Ge, M., J. Friedrich, and L. Vigna. 2024. "Where Do Emissions Come From? 4 Charts Explain Greenhouse Gas Emissions by Sector." World Resources Institute, Washington, DC. https://www .wri.org/insights/4-charts-explain-greenhouse-gas-emissions-countries-and-sectors.

Ghosh, S., A. Mahajan, A. Karkun, S. Mathew, P. Dhawan, and B. Krishna. 2023. *The State of India's Pollution Control Boards*. Technical Report, Centre for Policy Research, New Delhi, India.

Greenstone, M., T. Ganguly, C. Hasenkopf, N. Sharma, and H. Gautam. 2024. "Air Quality Life Index Annual Update." Air Quality Life Index, Energy Policy Institute, University of Chicago, Chicago, IL.

Greenstone, M., G. He, R. Jia, and T. Liu. 2022. "Can Technology Solve the Principal-Agent Problem? Evidence from China's War on Air Pollution." *American Economic Review: Insights* 4: 54–70.

Greenstone, M., G. He, S. Li, and E. Y. Zou. 2021. "China's War on Pollution: Evidence from the First 5 Years." *Review of Environmental Economics and Policy* 15: 281–99.

Greenstone, M., R. Pande, N. Ryan, and A. Sudarshan. 2025. "Can Pollution Markets Work in Developing Countries? Experimental Evidence from India." *Quarterly Journal of Economics* 140 (2): 1003–60.

Greenstone, M., and P. Schwarz. 2018. "Is China Winning Its War on Pollution?" Energy Policy Institute, University of Chicago, Chicago, IL.

Gross, S. 2020. "Renewables, Land Use, and Local Opposition in the United States." Brookings Institution, Washington, DC. https://www.brookings.edu/wp-content/uploads/2020/01/FP _20200113_renewables_land_use_local_opposition_gross.pdf.

Huang, J., X. Pan, X. Guo, and G. Li. 2018. "Health Impact of China's Air Pollution Prevention and Control Action Plan: An Analysis of National Air Quality Monitoring and Mortality Data." *The Lancet Planetary Health* 2: e313–e323.

IEA (International Energy Agency). n.d. *Energy and Water.* Paris: IEA. https://www.iea.org/topics /energy-and-water.

IRENA (International Renewable Energy Agency). 2020. *Renewable Capacity Statistics 2020.* Abu Dhabi, United Arab Emirates: IRENA.

La Nauze, A., and E. Severnini. 2021. "Air Pollution and Adult Cognition: Evidence from Brain Training." Working Paper 28785, National Bureau of Economic Research, Cambridge, MA. doi:10.3386/w28785.

Lane, H. M., R. Morello-Frosch, J. D. Marshall, and J. S. Apte. 2022. "Historical Redlining Is Associated with Present-Day Air Pollution Disparities in U.S. Cities." *Environmental Science & Technology Letters* 9 (4): 345–50. doi:10.1021/acs.estlett.1c01012.

Li, C., A. van Donkelaar, M. S. Hammer, et al. 2023. "Reversal of Trends in Global Fine Particulate Matter Air Pollution." *Nature Communications* 14: 5349.

Myllyvirta, L. 2018. "Beijing's Air Pollution Just Got Worse Again." *Unearthed*, December 13.

Peszko, G., M. Amann, Y. Awe, G. Kleiman, and T. S. Rabie. 2022. *Air Pollution and Climate Change: From Co-Benefits to Coherent Policies.* International Development in Focus. Washington, DC: World Bank. doi:10.1596/978-1-4648-1835-6.

QUARG (Quality of Urban Air Review Group). 1993. *Urban Air Quality in the United Kingdom.* First Report of the Quality of Urban Air Review Group. London: QUARG.

Rao, N. D., G. Kiesewetter, J. Min, S. Pachauri, and F. Wagner. 2021. "Household Contributions to and Impacts from Air Pollution in India." *Nature Sustainability* 4: 859–67.

Rentschler, J., and N. Leonova. 2023. "Global Air Pollution Exposure and Poverty." *Nature Communications* 14: 4432. doi:10.1038/s41467-023-39797-4.

Smith, A. 1776. *The Wealth of Nations.* London: W. Strahan and T. Cadell.

Smith, K. R., H. Frumkin, K. Balakrishnan, et al. 2013. "Energy and Human Health." *Annual Review of Public Health* 34: 159–88. doi:10.1146/annurev-publhealth-031912-114404.

Stoerk, T. 2016. "Statistical Corruption in Beijing's Air Quality Data Has Likely Ended in 2012." *Atmospheric Environment* 127: 365–71.

Stone, R. 2002. "Counting the Cost of London's Killer Smog." *Science* 298 (5601): 2106–07. doi:10.1126/science.298.5601.2106b.

Sudarshan, A., and G. Goswami. 2024. "Solutions to Tackle Air Pollution." Background paper for this report. World Bank, Washington, DC.

van Donkelaar, A., R. V. Martin, M. Brauer, et al. 2016. "Global Estimates of Fine Particulate Matter Using a Combined Geophysical-Statistical Method with Information from Satellites, Models, and Monitors." *Environmental Science & Technology* 50: 3762–72. doi:10.1021/acs.est.5b05833.

WHO (World Health Organization). 2016. *Burning Opportunity: Clean Household Energy for Health, Sustainable Development, and Wellbeing of Women and Children*. Geneva: WHO. https://www.who.int/publications/i/item/9789241565233.

WHO (World Health Organization). 2021. *WHO Global Air Quality Guidelines: Particulate Matter (PM$_{2.5}$ and PM$_{10}$), Ozone, Nitrogen Dioxide, Sulfur Dioxide and Carbon Monoxide*. Geneva: WHO.

WHO (World Health Organization). 2023. *World Health Statistics 2023: Monitoring Health for the SDGs*. Geneva: WHO.

World Bank. 2011. *Household Cookstoves, Environment, Health, and Climate Change: A New Look at an Old Problem*. Washington, DC: World Bank. https://hdl.handle.net/10986/27589.

World Bank. 2023a. *Breaking Down Barriers to Clean Energy Transition*. Washington, DC: World Bank. https://www.worldbank.org/en/news/feature/2023/05/16/breaking-down-barriers-to-clean-energy-transition.

World Bank. 2023b. *Striving for Clean Air: Air Pollution and Public Health in South Asia*. South Asia Development Matters. Washington, DC: World Bank. doi:10.1596/978-1-4648-1831-8.

World Bank. 2025. "Accelerating Access to Clean Air for a Livable Planet." World Bank, Washington, DC. http://documents.worldbank.org/curated/en/099032625132535486.

Zhang, X., X. Chen, and X. Zhang. 2018. "The Impact of Exposure to Air Pollution on Cognitive Performance." *Proceedings of the National Academy of Sciences* 115 (37): 9193–97. doi:10.1073/pnas.1809474115.

Zhang, X., Y. Jin, H. Dai, Y. Xie, and S. Zhang. 2019. "Health and Economic Benefits of Cleaner Residential Heating in the Beijing–Tianjin–Hebei Region in China." *Energy Policy* 127: 165–78.

Part 2

Cities and Commerce

Part 2, comprised of two chapters, examines the environmental impacts of two prominent trends of the twentieth and twenty-first centuries: increasing urbanization and interconnectivity through trade.

CHAPTER 6

The Economics of Livable Cities

"Urbanization is not a problem to be solved; it is an opportunity to be embraced."

—*Edward Glaeser,* American economist (1967-)

Key messages

- By 2050, nearly two-thirds of the world's population will live in cities. So, without livable cities, there will not be a livable planet.

- Well-managed urbanization is an efficient way to organize economic activity, generating jobs and alleviating poverty while minimizing environmental impacts. Mismanaged urban growth, in contrast, limits economic opportunities, exacerbates inequality, and degrades natural resources such as forests, water, and air.

- Promoting compact city structures can be economically beneficial by creating agglomeration economies and reducing environmental impacts by reducing transportation needs. However, compactness must be coupled with strategic urban planning to avoid issues like congestion and pollution concentration.

- Investment in nature-based solutions and sustainable infrastructure is essential for maintaining livable cities. Such interventions offer multiple economic, environmental, and social benefits, yet require careful long-term planning, resources, and monitoring to be effective.

Introduction

Between 1960 and 2023, the urban share of the world's population climbed from 33.6 to 57.3 percent: by 2023, more than 4.6 billion people lived in cities.[1] This *great urbanization* has helped drive the expansion of global economic activity. Today, the world's cities account for over 80 percent of gross domestic product, nearly 90 percent of private sector job creation, and 83 percent of total household spending (Cooper and Fengler 2024; Duranton and Puga 2020; Zhang and Brown 2024). It has been projected that by 2050, nearly two-thirds of the world's population will live in cities, with the most rapid urbanization taking place in developing countries (Dodman et al. 2022). Therefore, effective *urbanization management* is more important than ever, to ensure and promote shared prosperity on a livable planet.[2]

This chapter provides an overview of the interdependence between cities and the natural capital upon which they depend. It defines a *livable city* as one that enhances prosperity for all residents while conserving Earth's key natural capital—land, water, and air. This may involve creating natural capital, such as urban forests and green roofs, or restoring degraded natural capital, such as wetlands. For fast-urbanizing developing countries, protecting existing natural capital will be crucial. Increasing both the provision of affordable housing and green spaces within well-planned compact areas and the coverage and quality of infrastructure and basic services is key to mitigating urban encroachment on natural lands and the adverse impacts of polluting activities.

Cities and natural capital: Land, air, and water

Cities provide an efficient way to organize economic activity in a way that accommodates the expanding global population. Studies have estimated that from 1992 to around 2015, urban built-up areas worldwide expanded at 3.2–3.5 percent annually (Huang et al. 2020; van Vliet 2019). That is around 350,000 to 380,000 square kilometers, comparable to the entire land area of Germany or Japan. Yet, in 2015, cities covered less than 1 percent of the world's land—around 653,000 to 713,000 square kilometers (Huang et al. 2020; Liu et al. 2020; van Vliet 2019)—and hosted 54 percent of the world's population, or around 4 billion people.[3] Seen through this lens, cities represent a much more efficient way to accommodate humanity than providing amenities across more dispersed rural settlements.

However, urban expansion places significant pressure on natural capital, affecting forests, air quality, and water systems. From 1992 to 2015, over 60 percent of urban expansion occurred on former croplands (Liu et al. 2020; van Vliet 2019). Although direct forest loss from urban growth has been relatively limited, the displacement of cropland to forested areas has led to substantial indirect deforestation, estimated at 171,000 to 318,000 square kilometers, nearly 10 times greater than the direct forest loss. Cities have also encroached into high-risk coastal and wetland areas. A global sample of around 10,000 cities analyzed for this report showed that during 2000–15, the populations of coastal cities[4] in East Asia and the Pacific, the Middle East and North Africa, and Sub-Saharan Africa grew at a rate that was 1.3 to 3 percent faster than inland cities in their regions. Meanwhile, the world has lost roughly 35 percent of its wetlands since 1970, mainly due to agriculture and urban expansion (Ramsar 2018). Without enough natural surface and adequate drainage systems, rapidly-growing cities face increasing flood risk.

Sinking cities reveal the growing dangers of groundwater overuse, as excessive extraction causes land subsidence with serious economic and environmental consequences (Josset et al 2024; Rodella, Zaveri, and Bertone 2023). This crisis is unfolding in places like Indonesia, the Islamic Republic of Iran, Mexico, and Viet Nam. Jakarta has sunk more than 3.5 meters since the 1980s, and parts of the Islamic Republic of Iran have subsided at 6 centimeters annually. Inadequate piped water supply has often forced residents to rely on unregulated groundwater, worsening subsidence and increasing flood and saltwater intrusion risks, especially in coastal regions like the Mekong Delta. Vulnerability depends on local geology, with clay-rich or sedimentary formations particularly prone to sinking when groundwater is drawn from beneath them. Remote sensing technologies, such as Interferometric Synthetic Aperture Radar, now allow cities to monitor elevation changes with precision, but these tools must be integrated with on-the-ground data to guide effective policy responses.

In developing countries, poorly planned rapid urbanization is associated with the proliferation of informal settlements and slums. Usually "hidden" in official statistics, these areas tend to sprawl beyond municipal boundaries, complicating infrastructure and service provision (Ellis and Roberts 2016).[5] Rapidly growing cities in Sub-Saharan Africa's resource-exporting countries, for example, have tended to show limited access to safe water and sanitation, high shares of slum residents, and high poverty rates (Ebeke and Etoundi 2017; Gollin, Jedwab, and Vollrath 2016). Inadequate services can easily lead to environmental degradation through open defecation, untreated wastewater discharge, and the open dumping and burning of solid waste (box 6.1), in addition to increased disease transmission (Adiga et al. 2018; Riley, Raphael, and Snyder 2020).

BOX 6.1
Solid waste management: A cross-cutting problem

Globally, more than 2 billion tonnes of municipal solid waste are generated each year (GPSC 2022), but urban solid waste composition and management practices vary. As cities develop, the share of organic waste—such as food and green waste from yards and gardens—tends to decrease, and the share of recyclables, such as plastic, glass, metal, paper, and wood, tends to increase (figure B6.1.1, panel a). Although cities in low-income countries openly dump most of their waste without proper treatment, the absolute volume of waste is much higher in cities in middle-income countries (figure B6.1.1, panel b). Cities in high-income and upper-middle-income countries landfill and incinerate significant volumes of waste, practices that, without gas control systems and combustion facilities, can be sources of toxic substances.

(continued)

BOX 6.1
Solid waste management: A cross-cutting problem *(continued)*

FIGURE B6.1.1 **Solid waste composition and management practices in cities, by country income group**

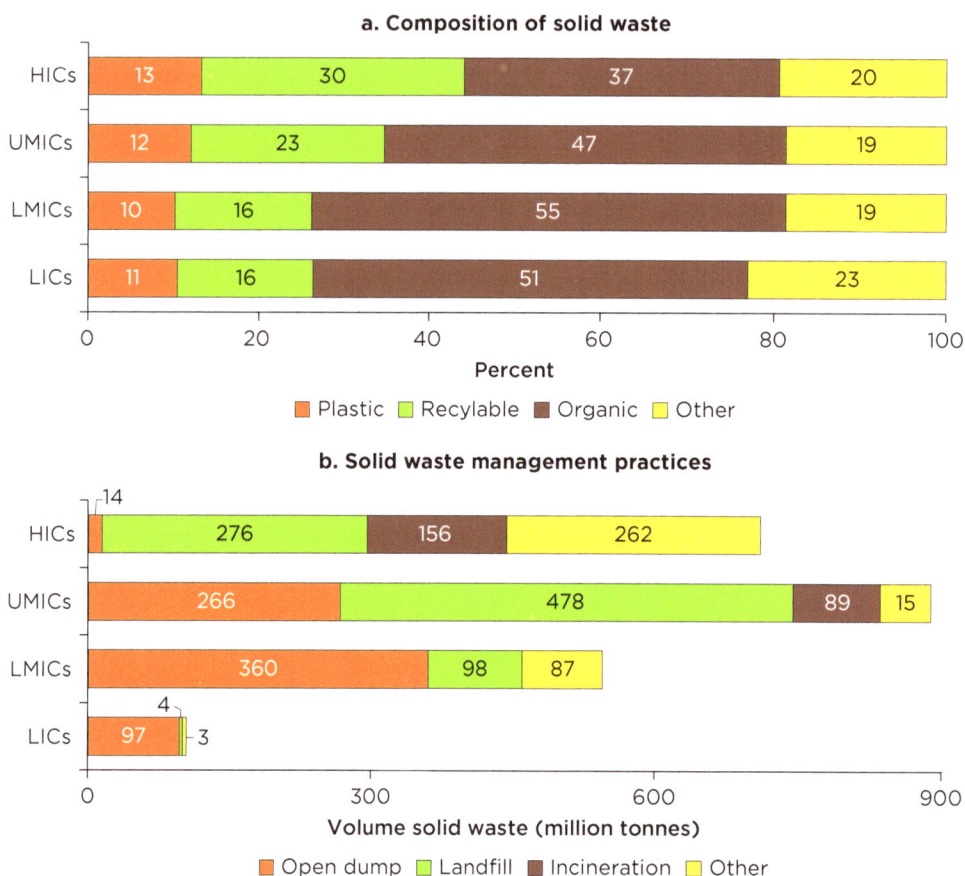

a. Composition of solid waste

Income group	Plastic	Recylable	Organic	Other
HICs	13	30	37	20
UMICs	12	23	47	19
LMICs	10	16	55	19
LICs	11	16	51	23

Percent

■ Plastic   ■ Recylable   ■ Organic   ■ Other

b. Solid waste management practices

Income group	Open dump	Landfill	Incineration	Other
HICs	14	276	156	262
UMICs	266	478	89	15
LMICs	360	98		87
LICs	97	4	3	

Volume solid waste (million tonnes)

■ Open dump   ■ Landfill   ■ Incineration   ■ Other

Source: Original calculations based on data from Kaza, Shrikanth, and Chaudhary 2021.
Note: Based on 61, 73, 97, and 58 cities in HICs, UMICs, LMICs, and LICs, respectively. In panel b, *other* include composting, recycling, and other advanced methods. HICs = high-income countries; LICs = low-income countries; LMICs = lower-middle-income countries; UMICs = upper-middle-income countries.

Poor waste management degrades the environment and public health. Open dumping and untreated landfills pollute soil and water, and incineration releases toxic pollutants over wide areas (UNEP 2024). Poor waste disposal also clogs drains, increasing flood risks and spreading waterborne diseases (Nepal et al. 2022). The sector contributes significantly to climate change, generating over 20 percent of global methane emissions and large amounts of black carbon (Bond et al. 2013; UNEP 2021). Health risks include elevated rates of birth defects, low birth weight, and cancers near landfills or incinerators. An estimated 0.4 million to 1 million deaths annually have been linked to mismanaged waste in developing countries (Williams, Gower, and Green 2019).

Promoting compact, vertical urban development is key to enhancing human livability while minimizing environmental encroachment. Expanding mid- and high-rise residential buildings in well-located areas can improve housing affordability and reduce urban land consumption (Jedwab, Barr, and Brueckner 2022). Higher-density urban areas yield several advantages:

1. *Cost efficiency.* A 1 percent increase in urban density correlates with a 0.17 percent decrease in local service costs (Ahlfeldt and Pietrostefani 2019).

2. *Public transportation access.* Denser areas enhance access to public transportation and waste management, improving air, water, and soil quality (Bibri, Krogstie, and Kärrholm 2020).

3. *Economic productivity.* A 1 percent increase in density can lead to a 4–6 percent ($74 to $111) rise in gross wages in developing countries (Grover, Lall, and Timmis 2023).

4. *Environmental benefits.* Dense cities shorten travel distances and reduce per capita energy consumption, reducing fine particulate matter ($PM_{2.5}$) emissions (Glaeser and Kahn 2010). However, compactness must be paired with urban planning to avoid congestion and pollution accumulation.

Urban green spaces are vital to city life, offering wide-ranging benefits, from cooler temperatures to cleaner air and better health. Across 500 cities, urban greenery has been shown to reduce surface temperatures by an average of 3 degrees Celsius during warm seasons (Li et al. 2024) and improve air quality by filtering pollutants and reducing the formation of secondary pollutants. Access to green spaces also has major health implications: in Europe, proximity to just 0.5 hectare of green space could prevent nearly 43,000 deaths annually (Barboza et al. 2021). Yet, access is highly unequal (box 6.2). Residents in high-income countries enjoy three times more access to green space than those in developing countries (Chen et al. 2022), and affluent neighborhoods tend to benefit the most. Urban design matters too—poorly planned greenery, such as roadside vegetation that restricts airflow, can worsen $PM_{2.5}$ exposure. Effective greening must therefore be paired with thoughtful design to ensure pollutant dispersion and equitable health gains (Alas et al. 2022; Venter et al. 2023).

BOX 6.2
Do welfare and thermal inequalities overlap?

Extreme heat has become more frequent and intense in cities, which already house 58 percent of the world's population and will host nearly 70 percent by 2050.[a] Urban areas are especially vulnerable due to the Urban Heat Island effect, driven by impervious surfaces, limited vegetation, and heat emissions. In East Asia, cities were up to 5.9 degrees Celsius (°C) hotter than surrounding rural areas (Roberts et al. 2023).

(continued)

BOX 6.2

Do welfare and thermal inequalities overlap? *(continued)*

Extreme heat episodes in cities have a wide range of negative consequences. These include reduced labor productivity and economic output, worse educational outcomes, and increased mortality and illness rates, and there have even been indications of heightened crime, violence, and car accidents (Roberts et al. 2023). Projections have suggested that by 2050, heat-related productivity losses could reduce real gross domestic product by 1.4–1.7 percent for the median city, and up to 11 percent for the most affected cities (Estrada, Botzen, and Tol 2017).

Understanding whether poorer neighborhoods are more exposed to increasingly severe, extreme heat in cities is crucial for designing equitable adaptation strategies. This report presents the results of new research by Avner et al. (2025) that examined data from roughly 10,000 cities to measure intra-city spatial welfare differences. This new evidence shows how average extreme heat levels vary across and within cities globally, and whether these are correlated with welfare inequalities. The research found the following:

1. Cities in high-income countries are, on average, 5.2–5.4 percent hotter than cities in low-income countries, even after controlling for a city's absolute latitude, baseline climate, population, and area. Hence, all else equal, cities in higher-income countries tend to suffer higher average extreme temperatures than cities in low-income countries.

2. Urban form matters: more impervious surfaces increase average extreme heat by about 8 percent, and more vegetation reduces it by roughly 5.6 percent.

3. Population density is positively correlated with heat exposure. Thus, cities are significantly hotter if either, for a given area, they are more populous or, for a given population, they have a smaller area.

4. Thermal inequality—differences across neighborhoods—is significantly lower in high-income countries, by about 11 percent on average, when controlling for a city's size, area, absolute latitude, and climate zone. This difference indicates that extreme levels of heat in different parts of cities in high-income countries are less marked than they are in cities in low-income countries.

5. Globally, there is no evidence of poorer neighborhoods being systematically hotter. Indeed, for the median city globally, the estimated gradient between temperature and the Relative Deprivation Index is slightly negative, indicating that poorer locations within the median city tend to be slightly cooler.

A city may be able to achieve a potentially sizable reduction in its average extreme heat level through greening policies that increase vegetation and reduce impervious surface areas. For example, if greening policies were able to increase vegetation in the Republic of Yemen to the median level for cities belonging to the same climate zone (arid), Al Mansurah's average extreme heat level would be expected to decrease from 53.1°C to 45.5°C, a substantial drop of 7.6°C. Similar impacts, but of slightly lower magnitude, can be found across other climate zones.

a. Data on the urban share of the global population are from the United Nations World Urbanization Prospects: 2018 Revision database (https://population.un.org/wup/).

Source: Avner et al. (2025).

Livable cities for prosperity and inclusiveness

Human capital and local productivity

Environmental amenities are key to urban livability, with direct effects on productivity, health, and human capital. Cleaner air, better sanitation, and reduced heat or water stress can boost both physical and cognitive performance. For instance, in Mexico City, a refinery closure that cut sulfur dioxide by 20 percent led to an increase of 1.3 work hours nearby (Hanna and Oliva 2015). Similarly, Beijing's anti-pollution policies yielded an estimated $380 million in annual productivity gains (Viard and Fu 2015). Livability improvements have also supported education and long-term development. Slum upgrades in Montevideo increased elementary school attendance (Zanoni, Acevedo, and Guerrero 2023), and reduced pollution in Brazil and Israel was linked to higher college entrance scores, with lasting career effects (Carneiro, Cole, and Strobl 2021; Lavy, Ebenstein, and Roth 2014).

Cities that improve environmental amenities can attract high-skilled workers, fueling higher wages and productivity. For example, estimates have shown that halving air pollution levels in Beijing could boost the city's average wage by 14.4 percent due to increased migration of highly skilled workers, which would benefit less-skilled workers through human capital complementarities and externalities (Khanna et al. 2021). In contrast, poor environmental conditions in urban areas contribute to brain drain as talent relocates to cleaner, more livable cities, often pulling firm productivity down in affected areas (Arntz, Brull, and Lipowski 2023; Tiebout 1956). Firms in polluted cities have resorted to offering wage premiums of up to 20 percent (Lai and Chitravanshi 2019).

Capital investment and job creation

With productive workers and a productivity-enhancing environment, livable cities are well positioned to attract investments and create jobs. A study found that in China, cities with better air and water quality attract more foreign direct investment inflows (Pisani et al. 2019), which can create more business opportunities and jobs for local residents (Saurav, Liu, and Sinha 2020). Livable cities also require technological innovation in sectors like sustainable transportation, waste (water) reduction, energy efficiency, and climate mitigation and adaptation, leading to more and better jobs across various industries. Consistent with evidence presented in chapter 9 of this report, green investments generally create more jobs, directly and indirectly, than less environmentally friendly alternatives. For example, mass transit creates 1.4 times as many jobs as road construction; electric vehicle charging infrastructure, twice as many jobs as manufacturing internal combustion vehicles; building energy efficiency, 2.8 times as many jobs as fossil fuels; and ecosystem restoration, 3.7 times as many jobs as oil and gas production (Jaeger et al. 2021).

Inclusiveness

Livable cities serve as a crucial pathway out of poverty, offering greater economic opportunities and amenities than rural areas. More-livable urban environments can facilitate faster upward mobility. For example, while the evidence may not be causal, Nakamura et al. (2023) found that urban residents living in low-flood-risk areas in Chile and Colombia have a higher probability of escaping poverty (figure 6.1), although exposure to harmful pollutants can hinder this. Enhancing livability through improved access to basic services can also significantly boost labor force participation, particularly among women, by reducing time spent on domestic chores (Duflo, Galiani, and Mobarak 2012). Slum titling initiatives in Buenos Aires and education and housing investments in Peru increased labor participation and earnings, while social housing projects in Cape Town resulted in an 18 percent rise in household income, primarily benefiting women (Franklin 2020). These findings suggest that enhancing urban livability fosters resilience among vulnerable groups and paves the way for shared prosperity within cities.

Developing livable cities for a livable planet

Several pathways can help urban development align with planetary livability and, ultimately, contribute to city prosperity and inclusiveness. This section explores actions that planners can take along four pathways: using urban planning to provide land and housing in safely located compact areas, improving living conditions in precarious settlements, integrating nature-based solutions (NbS) into mainstream urban planning, and incentivizing sustainable and inclusive solutions. The discussion does not examine crosscutting issues—such as financing schemes and governance—in any depth. Although vital for creating livable cities, these topics remain for future research.

FIGURE 6.1 Predicted probability of escaping poverty in Colombia and Chile, by flood risk

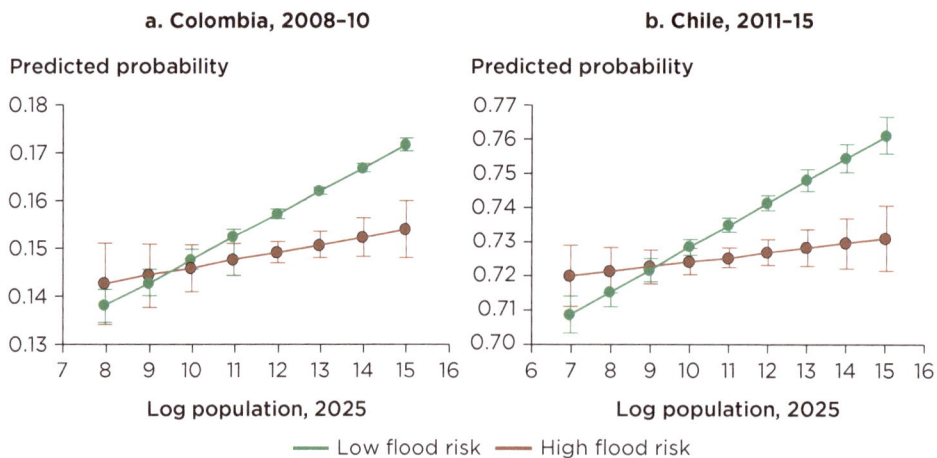

a. Colombia, 2008–10

b. Chile, 2011–15

Low flood risk — High flood risk

Source: Nakamura et al. 2023.

Using urban planning to provide land and housing in safely located compact areas

Reforming land tenure systems and easing land use restrictions can unlock developable land and expand the housing supply. In many developing countries, unclear or fragmented land ownership hampers property transfer and discourages formal investment, leading to inefficient land use and housing shortages (Bird and Venables 2020; Lall et al. 2021). Relaxing floor-area ratios (FARs), which are common in many cities, enables vertical development and reduces housing costs (World Bank 2013). Box 6.3 describes various approaches that the Indian government implemented to tackle the housing shortage for the urban poor.

Urban densification must also consider disaster risk. Integrating risk-sensitive urban planning, through updated land use maps, incentives, and emergency systems, can reduce exposure to hazards. Growth boundaries can help manage sprawl while steering new development away from hazardous areas.

BOX 6.3

Pradhan Mantri Awas Yojana-Urban: "Housing for All" in India

To address the chronic urban housing shortage problem, the Indian government launched Pradhan Mantri Awas Yojana-Urban (PMAY-U) in 2015. With a strong emphasis on empowering marginalized communities, the program has four approaches to implementation:

- Affordable housing-in-partnership is a resettlement approach where private developers receive a subsidy for building residential units for low-income households.
- In-situ slum redevelopment involves demolishing slums and building new housing in their place. Private developers receive a subsidy and a portion of the land for potential commercial development.
- Beneficiary-led construction is a self-upgrading model in which low-income households receive a subsidy to build or enhance their own housing, contingent upon land ownership.
- The credit-linked subsidy scheme gives households an interest subsidy on housing loans, which they can use to buy a new home (resettle) or enhance their existing home (redevelop).

PMAY-U has built more houses than any housing scheme in India. By 2023, it had approved 11.9 million new homes, completed 7.6 million, and provided slum dwellers with livable housing on rehabilitated land.

Source: World Bank 2023.

Improving living conditions in precarious settlements

Regularizing and upgrading existing settlements improves living conditions, empowers marginalized communities, and helps residents access better opportunities. Policy discussions of informal settlements often emphasize demolitions and resettlements to address their negative effects on public health, safety, and the environment. However, such approaches can be impractical, especially when residents choose informal settlements to gain better access to opportunities. For example, despite offering a path to homeownership, a government resettlement program in Manila, Philippines, led to a cycle of relocation and return for the urban poor, primarily because relocation offered limited access to jobs and services (Ajibade 2019).

Providing residents with land and property titles protects households from eviction and enables them to use their land as collateral for financial resources. This allows residents of informal settlements to invest in durable housing and self-protection against environmental hazards (Galiani and Schargrodsky 2010). Upgrading efforts are often most effective when they involve partnerships between government, nongovernmental organizations, and local communities, and use tailored approaches, such as in-situ upgrading and incremental housing that consider the unique characteristics and risks of each area.

Integrating NbS into mainstream urban planning

Creating, restoring, and preserving natural capital in cities requires long-term planning and consistent resources. Densification does not automatically create more green space. Instead, it often erodes green space through infill development (Haaland and van den Bosch 2015). Implementing NbS also faces barriers, including a lack of knowledge of local ecological conditions, challenges in acquiring land for large-scale projects, competing development priorities, and the need for sustained, cross-sectoral maintenance (Seddon et al. 2020).

To overcome these challenges, cities have used a mix of planning tools: zoning regulations to protect open space, design-based instruments like green infrastructure masterplans, quantitative targets for vegetation or stormwater management, and strategic land acquisition. These instruments help embed NbS into urban policy and make nature an intentional part of the built environment.

Still, a lack of evidence on the costs and benefits remains a key obstacle to scaling NbS (Chausson et al. 2020; Seddon et al. 2020; Tate et al. 2024). Many of the benefits—like improved air quality, flood reduction, and recreational value—are diffuse, long-term, or difficult to monetize. Meanwhile, operational costs are more visible. If evaluation focuses only on narrow outcomes, like cooling potential, green interventions may seem less cost-effective than gray alternatives. Broader assessments, which factor in energy savings, avoided emissions, stormwater fees, property values, and recreational values, can flip that conclusion (Crompton 2001; Sander, Polasky, and Haight 2010). A full accounting of benefits, trade-offs, and opportunity costs is essential for building a strong economic case (World Bank 2021).

Incentivizing sustainable and inclusive solutions

Cleaner mass transit can significantly reduce air pollution. In 58 cities, subway openings led to an average 4 percent drop in fine particulate matter (Gendron-Carrier et al. 2022). In Cairo, the phased opening of Metro Line 3 reduced particulate matter (PM_{10}) by 3 percent, and fuel subsidy removal added another 4 percent reduction (Heger et al. 2019). Where mass rapid transit systems are costly, alternatives like electric cable cars and bus rapid transit offer affordable options with substantial environmental benefits in developing countries (box 6.4).

Shifting people away from private vehicles is also key. Cities like Mexico City and São Paulo have been using demand-based parking fees and parking maximums, while Cairo's fuel reforms and taxi trade-in program—offering tax breaks, discounts, and loans—have cut transport emissions (C40 Cities 2015; ITDP 2014, 2017).

Governments can also harness private investment through incentives like density bonuses and tax breaks. In Kuala Lumpur, a 30 percent density bonus near transit zones has reduced sprawl, cleaned up brownfields, and met housing needs while increasing local tax revenues (CLC 2019).

BOX 6.4
Electric-powered bus rapid transit

Bus rapid transit offers major advantages over metro rail systems. It costs far less to build (at $1 million to $5 million per kilometer, compared to $15 million to $180 million per kilometer for metro rail systems) (ERIA 2013), and can be rolled out flexibly, adapting routes and frequency as demand shifts.

Electrifying bus rapid transit fleets eliminates tailpipe emissions and improves air quality (Severino et al. 2022). To support this transition, the Wuppertal Institute and SOLUTIONSplus[a] jointly developed the E-Bus Emission Assessment Tool,[b] which helps cities estimate emissions reductions by factoring in local energy mixes and bus fleets. For example, São Paulo was projected to see immediate benefits in terms of carbon dioxide emissions, thanks to clean energy generation in Brazil. In India, the volume and timing of carbon dioxide reduction will depend heavily on the growth of clean energy generation over the next 10–20 years. Both places are projected to experience immediate benefits in air quality, highlighting the need for transforming the transportation system and setting ambitious targets.

Expanding charging infrastructure is critical. Using existing assets—such as streetlights, lampposts, and utility poles—for chargers can cut installation costs by up to 70 percent (FLO 2021). Pairing this with light-emitting diode streetlight upgrades can boost energy efficiency and free up grid capacity for electric vehicles.

a. https://www.solutionsplus.eu/.
b. https://transformative-mobility.org/multimedia/e-bus-emissions-assessment-tool-e-beat/.

Conclusion

Enhancing planetary livability will hinge on how well the world manages urbanization in the next few decades. This chapter has outlined opportunities for building livable cities that enhance productivity, create jobs, and alleviate poverty while reducing human impacts on the environment. There is also a need to develop a robust system that allows cities to measure, monitor, and evaluate the status and progress of city livability. This will require cities to invest in capacity building to give government officials a clear understanding of the pros and cons associated with different interventions, so they can design effective projects, regulations, and incentives while monitoring progress. City leaders need to be cognizant of the effectiveness and affordability of emerging technologies related to construction materials, wastewater and solid waste treatment, NbS, clean energy, and so on, to facilitate wide uptake of sustainable solutions. They will also need to pursue multiple pathways in tandem, as their interlocking nature will require concerted efforts across sectors and jurisdictions over a long time horizon.

Notes

1. Based on data from the World Bank's World Development Indicators (https://wdi.worldbank .org/).
2. In this chapter, *urbanization management* refers to addressing underlying market and policy failures that undermine the efficient functioning of cities rather than directly attempting to target and, therefore, control the rate of urban population growth as a policy variable.
3. Based on data from the World Bank's World Development Indicators (https://wdi.worldbank .org/).
4. A *coastal city* is defined as having a geographic center within 100 kilometers of the nearest coastline.
5. Uncertainties in the number of slum dwellers complicate the design of targeted interventions and resource allocations. Recent studies have suggested that the United Nations estimate of around 1.1 billion people in 2020 may have undercounted the true number, as underlying national censuses lack spatial details and relevant nonmonetary measures (Breuer et al. 2024; Chen 2024; and Lucci, Bhatkal, and Khan 2018).

References

Adiga, A., S. Chu, S. Eubank, et al. 2018. "Disparities in Spread and Control of Influenza in Slums of Delhi: Findings from an Agent-Based Modelling Study." *BMJ Open* 8 (1): e017353. doi:10.1136 /bmjopen-2017-017353.

Ahlfeldt, G., and E. Pietrostefani. 2019. "The Economic Effects of Density: A Synthesis." *Journal of Urban Economics* 111: 93–107. doi:10.1016/j.jue.2019.04.006.

Ajibade, I. 2019. "Planned Retreat in Global South Megacities: Disentangling Policy, Practice, and Environmental Justice." *Climate Change* 157: 299–317. doi:10.1007/s10584-019-02535-1.

Alas, H., A. Stocker, N. Umlauf, et al. 2022. "Pedestrian Exposure to Black Carbon and PM$_{2.5}$ Emissions in Urban Hot Spots: New Findings Using Mobile Measurement Techniques and Flexible Bayesian Regression Models." *Journal of Exposure Science and Environmental Epidemiology* 32: 604–14. doi:10.1038/s41370-021-00379-5.

Arntz, M., E. Brull, and C. Lipowski. 2023. "Do Preferences for Urban Amenities Differ by Skill?" *Journal of Economic Geography* 23 (3): 541–76. doi:10.1093/jeg/lbac025.

Avner, P., E. Blanc, A. Castillo Castillo, M. Demuzere, and M. Roberts. 2025. "Twice Exposed? Do Welfare and Thermal Inequalities Overlap? Evidence From a Global Sample of Cities." Background paper for this report. World Bank, Washington, DC.

Barboza, E., M. Cirach, S. Khomenko, et al. 2021. "Green Space and Mortality in European Cities: A Health Impact Assessment Study." *The Lancet Planetary Health* 5 (10): E718–E730. doi:10.1016/S2542-5196(21)00229-1.

Bibri, S. E., J. Krogstie, and M. Kärrholm. 2020. "Compact City Planning and Development: Emerging Practices and Strategies for Achieving the Goals of Sustainability." *Developments in the Built Environment* 4: 100021. doi:10.1016/j.dibe.2020.100021.

Bird, J., and A. Venables. 2020. "Land Tenure and Land-Use in a Developing City: A Quantitative Spatial Model Applied to Kampala, Uganda." *Journal of Urban Economics* 119: 103268. doi:10.1016/j.jue.2020.103268.

Bond, T., S. Doherty, D. Fahey, et al. 2013. "Bounding the Role of Black Carbon in the Climate System: A Scientific Assessment." *Advancing Earth and Space Science* 118 (11): 5380–5552. doi:10.1002/jgrd.50171.

Breuer, J., J. Friesen, H. Taubenbock, M. Wurm, and P. F. Pelz. 2024. "The Unseen Population: Do We Underestimate Slum Dwellers in Cities of the Global South?" *Habitat International* 148: 103056. doi:10.1016/j.habitatint.2024.103056.

C40 Cities. 2015. "Cities100: Cairo—Taxi Trade-in Scheme Improves Air Quality." C40 Cities Case Studies. C40 Cities, New York. https://www.c40.org/case-studies/cities100-cairo-taxi-trade-in-scheme-improves-air-quality/.

Carneiro, J., M. Cole, and E. Strobl. 2021. "The Effect of Air Pollution on Students' Cognitive Performance: Evidence from Brazilian University Entrance Tests." *Journal of the Association of Environmental and Resource Economists* 8 (6): 1051–77. doi:10.1086/714671.

Chausson, A., B. Turner, D. Seddon, et al. 2020. "Mapping the Effectiveness of Nature-Based Solutions for Climate Change Adaptation." *Global Change Biology* 26 (11): 6134–55. doi:10.1111/gcb.15310.

Chen, B., S. Wu, Y. Song, C. Webster, B. Xu, and P. Gong. 2022. "Contrasting Inequality in Human Exposure to Greenspace between Cities in Global North and Global South." *Nature Communications* 13: 4636. https://www.nature.com/articles/s41467-022-32258-4.

Chen, W. 2024. "Underestimated Urban Slum Populations." *Nature Cities* 1: 727. doi:10.1038/s44284-024-00159-w.

CLC (Center for Livable Cities). 2019. *Affordable Housing: Profiles of Five Metropolitan Cities.* Singapore: CLC. https://www.metropolis.org/sites/default/files/resources/AffordableHousing-5Profiles-EN.pdf.

Cooper, M., and M. Fengler. 2024. *Cities Drive Global Prosperity—But the Way They Do That Is Changing.* Urban Transformation Stories, June 19. Geneva: World Economic Forum. https://www.weforum.org/stories/2024/06/how-cities-drive-global-prosperity/.

Crompton, J. 2001. "The Impact of Parks on Property Values: A Review of the Empirical Evidence." *Journal of Leisure Research* 33 (1): 1–31. doi:10.1080/00222216.2001.11949928.

Dodman, D., B. Hayward, M. Pelling, et al. 2022. "Cities, Settlements and Key Infrastructure." In *Climate Change 2022: Impacts, Adaptation, and Vulnerability. Contribution of Working Group II to the Sixth Assessment Report of the Intergovernmental Panel on Climate Change*. Cambridge: Cambridge University Press. doi:10.1017/9781009325844.008.

Duflo, E., S. Galiani, and M. Mobarak. 2012. *Improving Access to Urban Services for the Poor: Open Issues and a Framework for a Future Research Agenda*. Cambridge, MA: Abdul Latif Jameel Poverty Action Lab.

Duranton, G., and D. Puga. 2020. "The Economics of Urban Density." *Journal of Economic Perspectives* 34 (3): 3–26. doi:10.1257/jep.34.3.3.

Ebeke, C. H., and S. M. N. Etoundi. 2017. "The Effects of Natural Resources on Urbanization, Concentration, and Living Standards in Africa." *World Development* 96: 408–17. doi:10.1016/j .worlddev.2017.03.026.

Ellis, P., and M. Roberts. 2016. *Leveraging Urbanization in South Asia: Managing Spatial Transformation for Prosperity and Livability*. Washington, DC: World Bank. https://hdl.handle.net/10986/22549.

ERIA (Energy Research Institute for ASEAN and East Asia). 2013. "An Overview of Bus Rapid Transits in the World." In *Improving Energy Efficiency in the Transport Sector through Smart Development*, edited by I. Kutani, 5–25. ERIA Research Project Report 2013–27. Jakarta Pusat, Indonesia: ERIA. https://www.eria.org/RPR_FY2013_No.27_Chapter_2.pdf.

Estrada, F., W. W. Botzen, and R. S. Tol. 2017. "A Global Economic Assessment of City Policies to Reduce Climate Change Impacts." *Nature Climate Change* 7 (6): 403–06.

FLO. 2021. "How Leveraging Existing Utility and City Infrastructure Is Key to Accelerating Electric Vehicle Adoption." UtilityDive.org Sponsored Content, December 13. https://www.utilitydive.com /spons/how-leveraging-existing-utility-and-city-infrastructure-is-key-to-accelerat/611171/ https://www.utilitydive.com/spons/how-leveraging-existing-utility-and-city-infrastructure -is-key-to-accelerat/611171/.

Franklin, S. 2020. "Enabled to Work: The Impact of Government Housing on Slum Dwellers in South Africa." *Journal of Urban Economics* 118: 103265. doi:10.1016/j.jue.2020.103265.

Galiani, S., and E. Schargrodsky. 2010. "Property Rights for the Poor: Effects of Land Titling." *Journal of Public Economics* 94 (9–10): 700–29. doi:10.1016/j.jpubeco.2010.06.002.

Gendron-Carrier, N., M. Gonzalez-Navarro, S. Polloni, and M. A. Turner. 2022. "Subways and Urban Air Pollution." *American Economic Journal: Applied Economics* 14 (1): 164–96. doi:10.1257/app .20180168.

Glaeser, E., and M. Kahn. 2010. "The Greenness of Cities: Carbon Dioxide Emissions and Urban Development." *Journal of Urban Economics* 67 (3): 404–18. doi:10.1016/j.jue.2009.11.006.

Gollin, D., R. Jedwab, and D. Vollrath. 2016. "Urbanization with and without Industrialization." *Journal of Economic Growth* 21: 35–70. doi:10.1007/s10887-015-9121-4.

GPSC (Global Platform for Sustainable Cities). 2022. "Clean and Low-Carbon Cities: The Relationship between the Solid Waste Management Sector and Greenhouse Gases." GPSC Technical Brief, World Bank, Washington, DC. https://www.thegpsc.org/sites/gpsc/files/clean_and_low_carbon _cities_technical_brief.pdf.

Grover, A., S. Lall, and J. Timmis. 2023. "Agglomeration Economies in Developing Countries: A Meta-Analysis." *Regional Science and Urban Economics* 101: 103901. doi:10.1016/j.regsciurbeco .2023.103901.

Haaland, C., and C. van den Bosch. 2015. "Challenges and Strategies for Urban Green-Space Planning in Cities Undergoing Densification: A Review." *Urban Forestry & Urban Greening* 14 (4): 760–71. doi:10.1016/j.ufug.2015.07.009.

Hanna, R., and P. Oliva. 2015. "The Effect of Pollution on Labor Supply: Evidence from a Natural Experiment in Mexico City." *Journal of Public Economics* 122: 68–79. doi:10.1016/j.jpubeco.2014.10.004.

Heger, M., D. Wheeler, G. Zens, and C. Meisner. 2019. *Motor Vehicle Density and Air Pollution in Greater Cairo: Fuel Subsidy Removal and Metro Line Extension and Their Effect on Congestion and Pollution*. Washington, DC: World Bank. http://documents.worldbank.org/curated/en/987971570048516056.

Huang, Q., Z. Liu, C. He, et al. 2020. "The Occupation of Cropland by Global Urban Expansion from 1992 to 2016 and Its Implications." *Environmental Research Letters* 15 (8): 084037. doi:10.1088/1748-9326/ab858c.

ITDP (Institute for Transportation & Development Policy). 2014. "New São Paulo Master Plan Promotes Sustainable Growth, Eliminates Parking Minimums Citywide." Transport Matters Blog, July 7, 2014. https://itdp.org/2014/07/07/new-sao-paulo-master-plan-promotes-sustainable-growth-eliminates-parking-minimums-citywide-2/.

ITDP (Institute for Transportation & Development Policy). 2017. "Less Parking, More City: A Case Study in Mexico City." ITDP, New York. https://itdp.org/publication/less-parking-more-city-a-case-study-in-mexico-city/.

Jaeger, J., G. Walls, E. Clarke, et al. 2021. "The Green Jobs Advantage: How Climate-Friendly Investments Are Better Jobs Creators." Working Paper, World Resources Institute, Washington, DC. doi:10.46830/wriwp.20.00142.

Jedwab, R., J. Barr, and J. Brueckner. 2022. "Cities without Skylines: Worldwide Building-Height Gaps and Their Possible Determinants and Implications." *Journal of Urban Economics* 132: 103507. doi:10.1016/j.jue.2022.103507.

Josset, L., U. Lall, D. Prakash, and A. Dinar. 2024. "Public Health, Socioeconomic and Environmental Impacts of Urban Land Subsidence." Authorea Preprints. doi:10.22541/essoar.170808438.89045530/v2.

Kaza, S., S. Shrikanth, and S. Chaudhary. 2021. *More Growth, Less Garbage*. Washington, DC: World Bank. https://hdl.handle.net/10986/35998.

Khanna, G., W. Liang, A. Mobarak, and R. Song. 2021. "The Productivity Consequences of Pollution-Induced Migration in China." Working Paper 28401, National Bureau of Economic Research, Cambridge, MA. http://www.nber.org/papers/w28401.

Lai, C., and R. Chitravanshi. 2019. "Asia's Pollution Exodus: Firms Struggle to Woo Top Talent." Phys.*org*, March 31. https://phys.org/news/2019-03-asia-pollution-exodus-firms-struggle.html.

Lall, S., M. Lebrand, H. Park, D. Sturm, and A. Venables. 2021. *Pancakes to Pyramids: City Form to Promote Sustainable Growth*. Washington, DC: World Bank. https://hdl.handle.net/10986/35684.

Lavy, V., A. Ebenstein, and S. Roth. 2014. "The Impact of Short Term Exposure to Ambient Air Pollution on Cognitive Performance and Human Capital Formation." Working Paper 20648, National Bureau of Economic Research, Cambridge, MA. https://www.nber.org/papers/w20648.

Li, Y., J.-C. Svenning, W. Zhou, et al. 2024. "Green Spaces Provide Substantial but Unequal Urban Cooling Globally." *Nature Communications* 15: 7108. doi:10.1038/s41467-024-51355-0.

Liu, X., Y. Huang, X. Xu, et al. 2020. "High-Spatiotemporal-Resolution Mapping of Global Urban Change from 1985 to 2015." *Nature Sustainability* 3: 564–70. doi:10.1038/s41893-020-0521-x.

Lucci, P., T. Bhatkal, and A. Khan. 2018. "Are We Underestimating Urban Poverty?" *World Development* 103: 297–310. doi:10.1016/j.worlddev.2017.10.022.

Nakamura, S., K. Abanokova, H.-A. Dang, S. Takamatsu, C. Pei, and D. Prospere. 2023. "Is Climate Change Slowing the Urban Escalator out of Poverty?" Policy Research Working Paper 10383, World Bank, Washington, DC. https://openknowledge.worldbank.org/handle/10986/39626.

Nepal, M., B. Bharadwaj, A. K. Nepal, et al. 2022. "Making Urban Waste Management and Drainage Sustainable in Nepal." In *Climate Change and Community Resilience*, edited by A. K. E. Haque, P. Mukhopadhyay, M. Nepal, and M. R. Shammin, 325–38. Singapore: Springer. doi:10.1007/978-981-16-0680-9_21.

Pisani, N., A. Kolk, V. Ocelik, and G. Wu. 2019. "Does It Pay for Cities to Be Green? An Investigation of FDI Inflows and Environmental Sustainability." *Journal of International Business Policy* 2: 62–85. doi:10.1057/s42214-018-00017-2.

Ramsar. 2018. *Global Wetland Outlook 2018*. Ramsar Convention on Wetlands. Gland, Switzerland: Ramsar. https://www.global-wetland-outlook.ramsar.org/gwo-2018.

Riley, L., E. Raphael, and R. Snyder. 2020. "A Billion People Live in Slums. Can They Survive the Virus?" *The New York Times*, Opinion, April 8. https://www.nytimes.com/2020/04/08/opinion/coronavirus-slums.html.

Roberts, M., C. Deuskar, N. Jones, and J. Park. 2023. *Unlivable: What the Urban Heat Island Effect Means for East Asia's Cities*. Washington, DC: World Bank.

Rodella, A.-S., E. Zaveri, and F. Bertone, eds. 2023. *The Hidden Wealth of Nations: The Economics of Groundwater in Times of Climate Change*. Washington, DC: World Bank. http://hdl.handle.net/10986/39917.

Sander, H., S. Polasky, and R. Haight. 2010. "The Value of Urban Tree Cover: A Hedonic Property Price Model in Ramsey and Dakota Counties, Minnesota, USA." *Ecological Economics* 69 (8): 1646–56. doi:10.1016/j.ecolecon.2010.03.011.

Saurav, A., Y. Liu, and A. Sinha. 2020. "Foreign Direct Investment and Employment Outcomes in Developing Countries: A Literature Review of the Effects of FDI on Job Creation and Wages." In *Finance, Competitiveness and Innovation: In Focus*. Washington, DC: World Bank. http://documents.worldbank.org/curated/en/956231593150550672.

Seddon, N., A. Chausson, P. Berry, C. Girardin, A. Smith, and B. Turner. 2020. "Understanding the Value and Limits of Nature-Based Solutions to Climate Change and Other Global Challenges." *Philosophical Transactions of the Royal Society B* 375: 20190120. doi:10.1098/rstb.2019.0120.

Severino, A., G. Pappalardo, I. Olayode, A. Canale, and T. Campisi. 2022. "Evaluation of the Environmental Impacts of Bus Rapid Transit System on Turbo Roundabout." *Transportation Engineering* 9: 100130. doi:10.1016/j.treng.2022.100130.

Tate, C., N. Tran, A. Longo, et al. 2024. "Economic Evaluations of Urban Green and Blue Space Interventions: A Scoping Review." *Ecological Economics* 222: 108217. doi:10.1016/j.ecolecon.2024.108217.

Tiebout, C. 1956. "A Pure Theory of Local Expenditures." *Journal of Political Economy* 64 (5): 416–24. http://www.jstor.org/stable/1826343.

UNEP (United Nations Environment Programme). 2021. *Global Methane Assessment 2030: Baseline Report*. Nairobi: UNEP. https://www.unep.org/resources/report/global-methane-assessment-2030-baseline-report.

UNEP (United Nations Environment Programme). 2024. *Global Waste Management Outlook 2024*. Nairobi: UNEP. https://www.unep.org/resources/global-waste-management-outlook-2024.

van Vliet, J. 2019. "Direct and Indirect Loss of Natural Area from Urban Expansion." *Nature Sustainability* 2: 755–63. doi:10.1038/s41893-019-0340-0.

Venter, Z., A. Hassani, E. Stange, and N. Castell. 2023. "Reassessing the Role of Urban Green Space in Air Pollution Control." *Earth, Atmospheric, and Planetary Science* 121 (6): e2306200121. doi:10.1073/pnas.2306200121.

Viard, V., and S. Fu. 2015. "The Effect of Beijing's Driving Restrictions on Pollution and Economic Activity." *Journal of Public Economics* 125: 98–115. doi:10.1016/j.jpubeco.2015.02.003.

Williams, M., R. Gower, and J. Green. 2019. *No Time to Waste*. Teddington: Tearfund. https://learn.tearfund.org/en/resources/policy-reports/no-time-to-waste.

World Bank. 2013. *Planning, Connecting, and Financing Cities—Now: Priorities for City Leaders*. Washington, DC: World Bank. http://documents.worldbank.org/curated/en/512131468149090268.

World Bank. 2021. *Indonesia Vision 2045: Toward Water Security*. Washington, DC: World Bank. https://openknowledge.worldbank.org/entities/publication/beecd97a-0b5d-55f9-9d44-7f827b726e6c.

World Bank. 2023. *Enablers of Inclusive Cities: Enhancing Access to Services and Opportunities*. Washington, DC: World Bank. https://openknowledge.worldbank.org/handle/10986/40642.

Zanoni, W., P. Acevedo, and D. Guerrero. 2023. "Do Slum Upgrading Programs Impact School Attendance?" *Economics of Education Review* 96: 102458. doi:10.1016/j.econedurev.2023.102458.

Zhang, M., and L. Brown. 2024. *How to Build More Livable Cities for a Livable Planet*. Sustainable Cities Series. Washington, DC: World Bank. https://blogs.worldbank.org/en/sustainablecities/how-to-build-more-livable-cities-for-a-livable-planet.

CHAPTER 7

Trade: Friend or Foe of the Environment?

> "Out beyond ideas of wrongdoing and rightdoing,
> There is a field. I'll meet you there."
>
> —*Jalaluddin Rumi*, Persian poet (1207-1273)

Key messages

- Traded goods and commodities have a significant environmental footprint, accounting for 20–30 percent of global greenhouse gases and air pollution, 20 percent of global agricultural water use, and 26 percent of deforestation worldwide.

- However, evaluating the environmental impact of producing imported goods domestically reveals that trade has both pros and cons. It benefits the environment by reducing greenhouse gas emissions and global water usage. Conversely, it negatively impacts the environment by increasing global air pollution and deforestation.

- Efficiency and effective policies are key to reducing the environmental impacts of trade. Given trade's complex and multifaceted impact, maximizing efficiency significantly reduces its environmental footprint.

- Spotlight 3 in this report shows that investing in clean technologies and value chains can help lower-income countries diversify toward more rapid and sustainable development. As they have captured a growing share of green markets over the past two decades, developing countries demonstrate their potential to thrive in decarbonized value chains.

A balancing act

In the eighteenth and nineteenth centuries, the booming whaling industry allowed the inhabitants of New England in the United States to build their fortunes from whale oil, baleen, and blubber. Towns and economies thrived as ships embarked on long voyages to harvest whales from the Arctic to the Antarctic. The demand for whale products fueled unprecedented economic growth, powering lamps, machinery, and even fashion. However, the cost was staggering: overhunting led to the near-extinction of some species, such as right whales and sperm whales, and the collapse of marine ecosystems that relied on these giant creatures. As whale populations dwindled, the industry's long-term sustainability crumbled. Yet the drive for profit continued, pushing whales to the brink. By the time alternatives such as petroleum-based oil emerged, it was too late for many species.

The story of whaling is a stark reminder of how unchecked trade can sacrifice species and ecosystems, leaving a lasting imprint on the environment long after the profits fade. Global trade reached an all-time high of $33 trillion in 2024, marking an increase of more than five times since 2000 (UNCTAD 2024). Trade boosts economic growth by enhancing specialization, efficiency, and innovation, while enabling countries to focus on their comparative advantages. Without trade, countries would lack specialization and economies of scale and face inefficiencies and higher costs.

Although trade poses environmental challenges, it may also promote sustainability and innovation. On the downside, it can lead to pollution when damaging industries migrate to countries with laxer environmental regulations. On the upside, it can lead to green technology transfer, knowledge spillovers, and efficiency by enabling countries to specialize in their environmental comparative advantages. New evidence gathered for this report shows that the impacts are complex and vary with the environmental indicator under consideration and the nature of the goods that are traded. In cases where externalities remain uncorrected, by encouraging further production of polluting goods, trade could aggravate environmental damage. In other cases, where countries with a lower pollution footprint export to those with a higher pollution footprint, trade reduces total environmental damage. This chapter shows that, in general, trade reduces global water use and greenhouse gas (GHG) emissions but increases global fine particulate matter ($PM_{2.5}$) emissions and deforestation.

The environmental footprint of trade

Although international trade fosters economic growth, it may come at the expense of the environment. Trade-driven production is a driver of emissions and deforestation— especially for commodities such as palm oil, cocoa, and timber. Exports account for about one-third of global gross domestic product, implying that trade-related impacts on the environment are likely substantial. Around 15–30 percent of global emissions are embodied in traded goods, a share that has grown since 2004 (figure 7.1).

FIGURE 7.1 Share of direct emissions embodied in trade

Share (%)

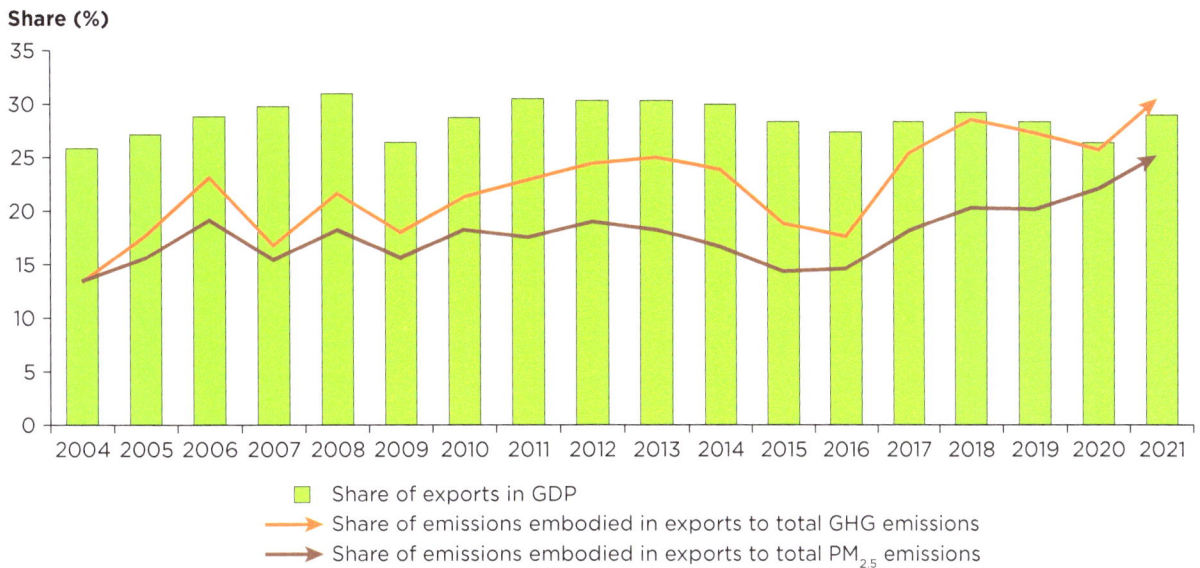

Legend:
- Share of exports in GDP
- Share of emissions embodied in exports to total GHG emissions
- Share of emissions embodied in exports to total PM$_{2.5}$ emissions

Sources: Original calculations, based on data from the World Development Indicators, World Bank; the Global Trade Analysis Project database; Ebadi and Aldaz-Carroll 2025.
Note: The shares represent the direct (scope 1) emissions embedded in exports, which originate from sources that are owned or controlled by firms. They do not include indirect (scopes 2 and 3) emissions, which arise from the production of purchased electricity, steam, heating, cooling, and other activities throughout the value chain. GDP = gross domestic product; GHG = greenhouse gas; PM$_{2.5}$ = fine particulate matter.

Agricultural trade has at times intensified pressures on natural resources. It has been projected that by 2050, demand for agricultural products will rise by 60–110 percent compared to their levels in 2005 (Zhao et al. 2023). Virtual water trade (VWT)—the water embedded in traded food—has nearly tripled since 1986, now representing 20 percent of global agricultural water use (Mekonnen et al. 2024). Groundwater depletion linked to trade is also rising, with 11 percent of global depletion between 2000 and 2010 attributed to VWT (Dalin et al. 2017). Trade also drives land use change. Around 26 percent of deforestation is tied to commodity exports (Abman and Lundberg 2020; Pendrill et al. 2019b).

Box 7.1 describes the emission and deforestation footprint of the transport sector, which is necessary for trade to occur. Shipping contributes significantly to air pollution, ocean plastic waste, and water pollution. In 2016, shipping accounted for 2 percent of global carbon dioxide (CO$_2$) emissions, and it has been projected that this share will potentially rise by 50–250 percent by 2050 (World Bank 2020). Road and rail transport also impact the environment, with rail being the lowest CO$_2$ emitter. Greening transportation can reduce the transport sector's environmental footprint.

BOX 7.1

Environmental footprint of the transportation sector

Transportation significantly impacts the environment through air pollution, greenhouse gas (GHG) emissions, water use, and land use. The transportation sector accounts for about 23 percent of global GHG emissions, with on-road diesel vehicles being significant polluters. During 2010–15, transportation was responsible for 11.6 percent of global population-weighted emissions of fine particulate matter ($PM_{2.5}$) (ICCT 2019). While regions with advanced emission standards, like the United States and the European Union, saw declines in $PM_{2.5}$ emissions, Southeast Asia and Sub-Saharan Africa experienced increases. Transportation modes vary in water consumption, with gasoline cars being the most water-intensive at 6.4 liters required to transport one passenger one kilometer (L/pkm), compared to urban electric trains at 3.4 L/pkm. The transportation sector also significantly affects land use, as infrastructure like roads and railways fragment ecosystems and reduce biodiversity, with roads alone covering more than 15 million square kilometers globally and contributing to deforestation and environmental challenges, especially in countries like the Democratic Republic of Congo, India, and Kenya.

Greening transportation involves multiple approaches, including supporting electric and hybrid vehicles, investing in efficient public transit, and developing infrastructure for walking and cycling. Electric vehicles (EVs) can significantly reduce air pollution compared to diesel and gasoline vehicles, with current life-cycle emissions over the lifetime of battery EVs already lower (by 19–69 percent) than those of comparable gasoline cars (ICCT 2021). *The Economics of Electric Vehicles for Passenger Transportation* underscored the economic benefits of EVs in developing countries (Briceno-Garmendia, Qiao, and Foster 2023). It also recommended the adoption of electric buses and smaller vehicles to improve urban mobility and reduce pollution, while providing policy guidance for accelerating EV adoption. Public transportation is crucial for reducing emissions, as buses and trains can lower air pollution by up to two-thirds per passenger, compared to private vehicles. Active mobility options like cycling and walking provide health benefits and a lower environmental footprint, complementing public transit systems.

Transitioning to green transportation technologies involves balancing environmental benefits with resource challenges. Transitioning to cleaner technologies can reduce air pollution but may increase water usage, as EVs have a higher water footprint due to cooling systems and mineral mining for batteries. Although alternative fuels like biofuels and hydrogen are promising, their production demands significant resources, raising environmental concerns about mineral extraction for EVs.

The World Bank has acknowledged these trade-offs and emphasized the need for cost-effective, sustainable infrastructure and efficient transport systems to promote economic growth and address climate change, especially in low- and middle-income countries. The World Bank's "Avoid-Shift-Improve" framework aims to reduce emissions by minimizing motorized transport, shifting to less emission-intensive modes, and enhancing emissions performance through efficient technologies and optimized routes (World Bank 2025). *Shrinking Economic Distance, Understanding How Markets and Places Can Lower Transport Costs in Developing Countries* examined the factors that keep transport costs high, delivery times long, and reliability low (Herrera Dappe, Lebrand, and Stokenberga 2024). Deregulating and increasing competition in the transport sector can enhance service quality and lower prices, while reducing empty running trucks and improving infrastructure can further decrease transport costs. Any reform agenda to develop efficient, high-quality freight transport and reduce economic distance should aim to foster efficient markets and places.

Although the statistics may be concerning, it would be misleading to conclude that trade is responsible for greater environmental degradation. Trade can be environmentally beneficial when exports flow from countries that utilize resources more efficiently and generate less pollution to the less resource-efficient importers. The remainder of this chapter explores this issue in more detail to determine whether trade has been beneficial or detrimental to air, land, and water resources.

The impact of trade on air quality, water, and forests

Simply looking at the environmental footprint of trade can be deceptive, as the emissions that are associated with trade need to be compared with a counterfactual scenario, such as one without trade (autarky). Without trade, countries would need to use their own resources and technologies to produce the goods they currently import, possibly resulting in a larger environmental footprint than that of trade with more resource-efficient exporters.

Varying impacts on emissions

Trade displaces emissions between centers of production and consumption. *Displaced emissions* are the difference between actual emissions embodied in trade and those in a hypothetical scenario of autarky. New analysis for this report estimated the volumes of displaced emissions at the global level, using the Global Trade Analysis Project database (box 7.2). The results show that restricting trade in countries where production is more polluting would lead to higher direct and indirect emissions,[1] which countries currently avoid through trade. Whether trade contributes to or reduces global emissions can be determined by aggregating the emission differential across all countries.

BOX 7.2
Displaced emissions at the global scale

The net effect of trade on country i's emissions (NE_i) equals the emissions embodied in exports (E_i^X) minus the emissions of the autarky scenario (E_i^R)—that is, the emissions from the goods consumed by a country if it were to produce them domestically:

$$NE_i = E_i^X - E_i^R = \left(E_i^X - E_i^M\right) - \left(E_i^R - E_i^M\right)$$

where E_i^X is calculated for each country by summing the pollution intensity of each sector multiplied by its exports. E_i^R is determined by summing the pollution intensity of each sector in

(continued)

BOX 7.2
Displaced emissions at the global scale *(continued)*

country *i* multiplied by its imports. E_i^M is determined by summing the pollution intensity of each sector in country *j*, multiplied by the imports of country *i*, using the pollution intensity of country *i*'s trade partners.

The analysis used the Global Trade Analysis Project (GTAP) database to estimate emissions embodied in global exports and imports (Aguiar et al. 2023). This was achieved by combining sector emissions from input-output tables in the GTAP database with GTAP trade data for 2004–21 covering 39 sectors across 133 countries, accounting for 96 percent of the global economy.

If the emissions embodied in exports (E_i^X) exceed the emissions of the autarky scenario (E_i^R), trade liberalization results in higher emissions for country *i*. Conversely, if emissions are greater in the autarky scenario, trade liberalization leads to lower emissions.

The first bracketed term ($E_i^X - E_i^M$) represents the emissions embodied in net exports, and the second ($E_i^R - E_i^M$) captures the emissions that country *i* avoids through imports. Aggregating across all countries, the net effect of trade on global emissions (NE_w) can be expressed as:

$$NE_w = \sum_i NE_i = \sum_i (E_i^X - E_i^M) - \sum_i \left(E_i^R - E_i^M \right) = \sum_i \left(E_i^M - E_i^R \right)$$

Since global exports equal global imports, the amount of emissions embodied in global net exports is zero. Therefore, the net effect of trade on global emissions is the total emissions embodied in imports minus the total emissions displaced by imports. This indicates that the net effect is influenced by the difference in pollution intensity between the importing and exporting countries.

Source: Ebadi and Aldaz-Carroll (2025).

The effect of trade on global emissions hinges on the differences in emissions intensity between importing countries and their exporting counterparts. If importing countries have lower pollution intensities than their trade partners, then trade increases global pollution. Instead, if importing countries have higher pollution intensities, then trade helps reduce global emissions by shifting production to nations with lower intensities.

Countries differ in the emissions embodied in the commodities they export. In 2021, high-income countries (HICs) were the leading exporters of GHG emissions embodied in traded commodities, accounting for 59 percent of global GHG emissions in exports. Around one-third of their exports went to developing countries. Because HICs have a lower emissions intensity in their production processes, the environmental impact was smaller than if these goods were produced domestically in developing countries. Conversely, upper-middle-income countries (UMICs) were responsible for 56 percent of global $PM_{2.5}$ emissions in exports, with over half directed to HICs. Due to UMICs' higher pollution intensity, these exports contributed to higher air pollution than if HICs produced the goods domestically.

Overall trade patterns reduce global GHG emissions but increase $PM_{2.5}$ emissions. From 2004 to 2021, trade consistently lowered global GHG emissions by 0.9–2.2 percent each year (figure 7.2)—largely because although HICs were the major exporters of GHG emissions, HICs exhibited lower GHG pollution intensities. In striking contrast, trade contributed to a 1 percent increase in global $PM_{2.5}$ emissions—driven by exports from UMICs to HICs with stricter environmental regulations.

FIGURE 7.2 Global direct emissions from trade

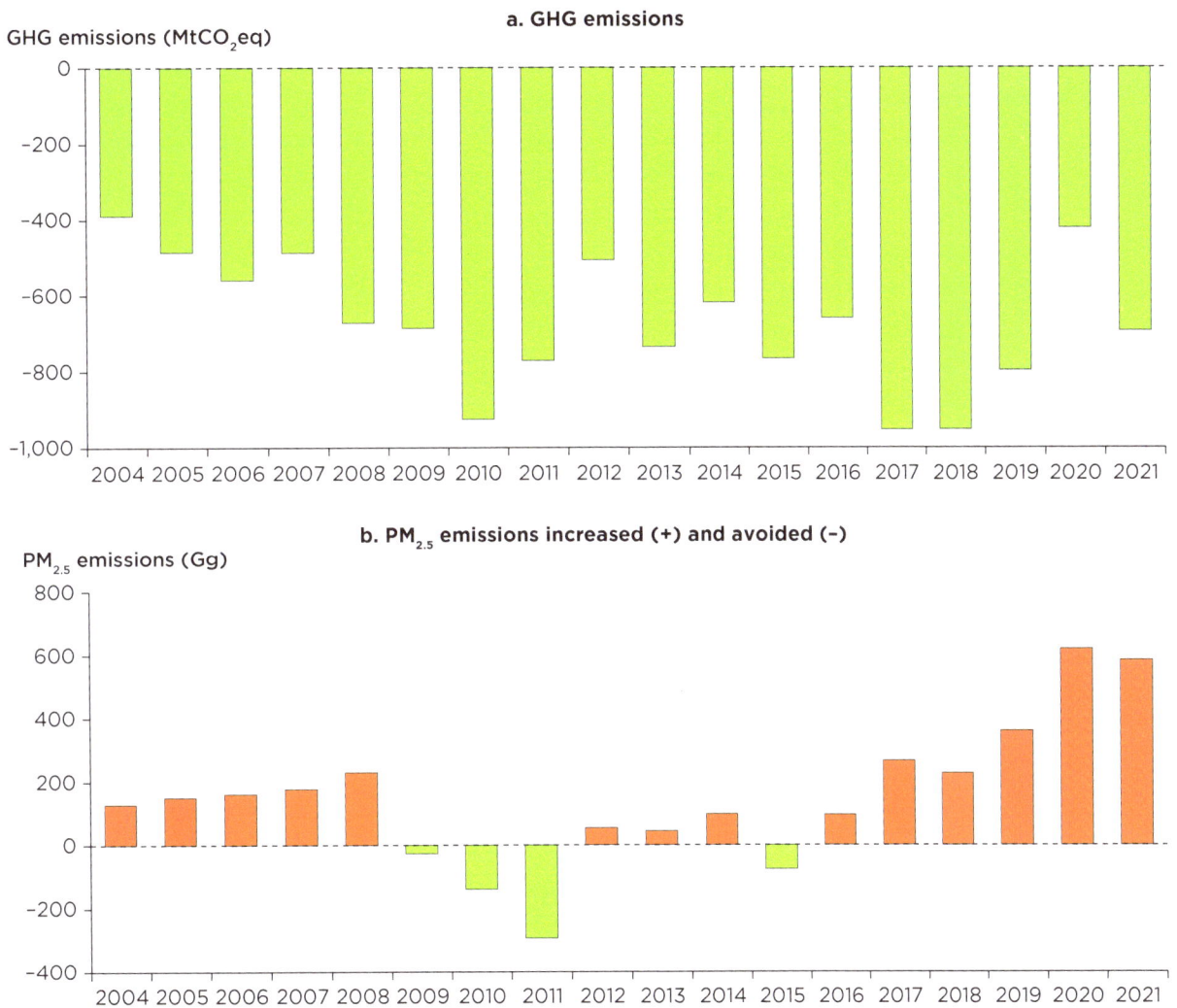

a. GHG emissions

GHG emissions ($MtCO_2eq$)

b. $PM_{2.5}$ emissions increased (+) and avoided (−)

$PM_{2.5}$ emissions (Gg)

Sources: Original calculations based on data from the Global Trade Analysis Project; Ebadi and Aldaz-Carroll 2025.
Note: Global emissions due to international trade equal total emissions displaced from imports minus total emissions embodied in imports. Gg = gigagrams; GHG = greenhouse gas; $MtCO_2eq$ = million tons of carbon dioxide equivalent; $PM_{2.5}$ = fine particulate matter.

The assessment suggests that trade has diverging impacts on aggregate air pollution emissions. Trade can reduce global emissions by allowing production to occur in countries that are more efficient due to specialization and economies of scale, using less energy and fewer resources. However, trade can also lead to *emissions leakage*, by shifting production—and therefore emissions—from countries with strict regulations to countries with fewer environmental controls. This is particularly evident with $PM_{2.5}$ emissions, where HICs have stronger environmental regulations. As a result, highly polluting firms may have incentives to shift production (Cherniwchan 2017; Levinson and Taylor 2008). The implication is that coordinated action and cooperation between countries is needed where emissions leakage dominates.

Does trade help prevent long-term water depletion?

Water-intensive agriculture tends to align with water abundance, and agricultural trade can therefore prevent long-term depletion in areas with less water. Early studies noted that trade could "save" local water resources in importing countries (Allan 1998). Global water savings represent the theoretical amount of water that would have been used if countries produced all goods domestically (as in autarky), compared to the actual water consumed under the current trade system. If trade liberalization encouraged water-abundant countries to export water-intensive goods while enabling water-scarce countries to import them, this would benefit both exporters (via trade revenues) and importers (through access to water-intensive products) while "saving" water (Allan 1996; Wichelns 2001).

Studies have consistently shown that, at the global level, the trade system facilitates more efficient water allocation, with water-intensive production concentrated in water-abundant regions. This efficiency translates into significant water savings— estimated at 352 cubic kilometers per year (Chapagain, Hoekstra, and Savenije 2006). These overall patterns align with the Heckscher-Ohlin model, which attributes comparative advantage to relative factor abundance (Carleton, Crews, and Nath 2023; Debaere 2014; Lai, Li, and Zhang 2023). A further example is that countries in the Middle East and North Africa rely heavily on virtual water imports, with Morocco importing 11 times more than it exports (World Bank 2023).

One of the main reasons for these patterns is that water input costs are significantly higher in areas with lower water resources and less rainfall. The physical scarcity of water strongly influences its effective input price, giving water-abundant regions a comparative advantage in water-intensive production (Carleton, Crews, and Nath 2023). Even in areas with no formal market to price water, the effective input price partially reflects its scarcity. In this context, the rising costs of physical scarcity partly compensate for the lack of functioning input markets by increasing the marginal cost of procurement where water is scarce.

Although water-intensive goods are generally produced in and traded from relatively water-abundant geographies, this does not ensure that water used in trade is *sustainable*. Unsustainable water use can occur when consumption exceeds local renewable availability, resulting in the loss of environmental flows that are essential for aquatic habitats and riparian biodiversity (Rosa et al. 2019). Unsustainable water use can also lead to the gradual depletion of groundwater reserves, which are critical for societal and ecosystem needs (Rodella, Zaveri, and Bertone 2023).

While most VWT uses sustainable sources, some trade relies on unsustainable water use, including nonrenewable groundwater extraction from deep aquifers (Dalin et al. 2017) or via the loss of environmental flows (Rosa et al. 2019). These local externalities can impact long-term water availability, ecosystem resilience, and societal well-being. Furthermore, this can pose risks to water-stressed trading partners that rely heavily on imports from regions dependent on unsustainable water sources, which may not remain viable in the long run. Some estimates have suggested that around 15–17 percent of unsustainable water is traded virtually (Mekonnen et al. 2024). Therefore, in some cases, complementary interventions—such as absolute limits to water use, effective irrigation pricing, or cap-and-trade systems that combine pricing with quantity controls—are required.

Ultimately, accomplishing sustainable water use requires a fundamental shift in how water is valued, recognizing it as a scarce and economically productive resource. This often, but not necessarily, requires using market forces and prices to guide water allocation decisions. This can be achieved through market-based mechanisms such as water pricing and trading schemes that allow users to buy, sell, or rent water rights, ensuring that water flows to higher-value uses. Evidence from Australia suggests that well-regulated water markets can generate high economic returns and incentivize conservation, although such success is contingent on robust governance. Establishing credible water trading systems remains a challenge as countries differ in legal traditions, institutional capacity, and socioeconomic conditions. However, if well-designed market-based solutions are achieved, then market trades can emerge in ways that deliver water to higher-value uses. When complemented with proper regulation and policies that secure essential allocations for the environment and the world's poorest, such a system could ensure sustainability and access for all.

Clearing the canopy: Links with deforestation

There are complex links between international trade and the exploitation of renewable resources, such as forests, fish, and wildlife, as they interact through multiple channels that differ from those associated with manufactured goods or nonrenewable resources. Among renewable resources, deforestation—which has implications for biodiversity, climate change, and livelihoods—is one of the most pressing global challenges.

Trade liberalization is often associated with increased agricultural production and the expansion of agricultural land at the expense of forests. When trade increases demand, it may lead to higher prices for commodities produced along the forest frontier, and this may result in an expansion of agricultural land and greater deforestation. Evidence has shown that global trade in soy, palm oil, beef, and other agricultural commodities is linked to deforestation in tropical regions. Indeed, 70 percent of deforestation in the Amazon rainforest is a consequence of cattle ranching and soy production, two heavily traded commodities (Pendrill et al. 2019a). Robust statistical approaches show that deforestation rates tend to peak after regional trade agreements, increasing net deforestation, particularly in developing tropical countries (Abman and Lundberg 2020).

Trade networks can further exacerbate deforestation risks, shifting the problem to countries with weak governance and regulatory frameworks. Global supply chains often obscure the origin of commodities, making it difficult to trace the environmental impact of imports. This lack of transparency makes policy implementation difficult. Adding to these challenges is the problem of displaced deforestation. Stricter environmental regulations in one country may shift deforestation to countries with laxer regulations. Although many HICs have increased their forest cover, this has often been at the expense of forests in developing countries (Pendrill et al. 2019a). This dynamic highlights the need for coordinated international efforts to tackle deforestation. Unilateral actions may inadvertently exacerbate the problem and will likely be ineffective if commodity production migrates to regions with laxer environmental protections.

Consumption patterns, especially in HICs, are ultimately responsible for a large proportion of trade-driven deforestation. Consumers in HICs benefit from products that are kept artificially cheap by unsustainable production patterns in the tropics (Brenton et al. 2024; WRI 2020). For example, there has been a significant increase in the consumption of foods from the forest frontier, such as beef and soy, with much of the soy destined for animal feed (figure 7.3). The land footprint of beef is approximately 20 times greater than that of legumes (Ritchie 2021). Over time, the trade in forest-risk commodities has increased.

Approaches to address the adverse effects of trade on forests have not been particularly effective. Traditional trade-related remedies have included tariffs, border tax adjustments on forest products, trade clubs, and bans on trade in certain commodities (such as rare timbers). However, these actions have not been economically efficient and some may even have had perverse effects, inducing further deforestation (Copeland, Shapiro, and Taylor 2022; Harstad 2024a). For example, when countries anticipate that there will be future trade-related penalties due to forest loss, there is an incentive to clear more land before negotiating a trade agreement. Once a country has cleared its forest, there may be no point in imposing trade sanctions, since "what has been done is done" (Harstad 2024b). The threat of sanctions lacks credibility and in the parlance of economics is not renegotiation-proof. This is an example of the *green policy paradox*, which arises when environmental policies—such as subsidies, conservation incentives, or offsets—fail to consider the unintended responses that they provoke.

FIGURE 7.3 Global deforestation and meat consumption, 1990–2020

Share of land covered by forests (%)   Per capita meat consumption (kg)

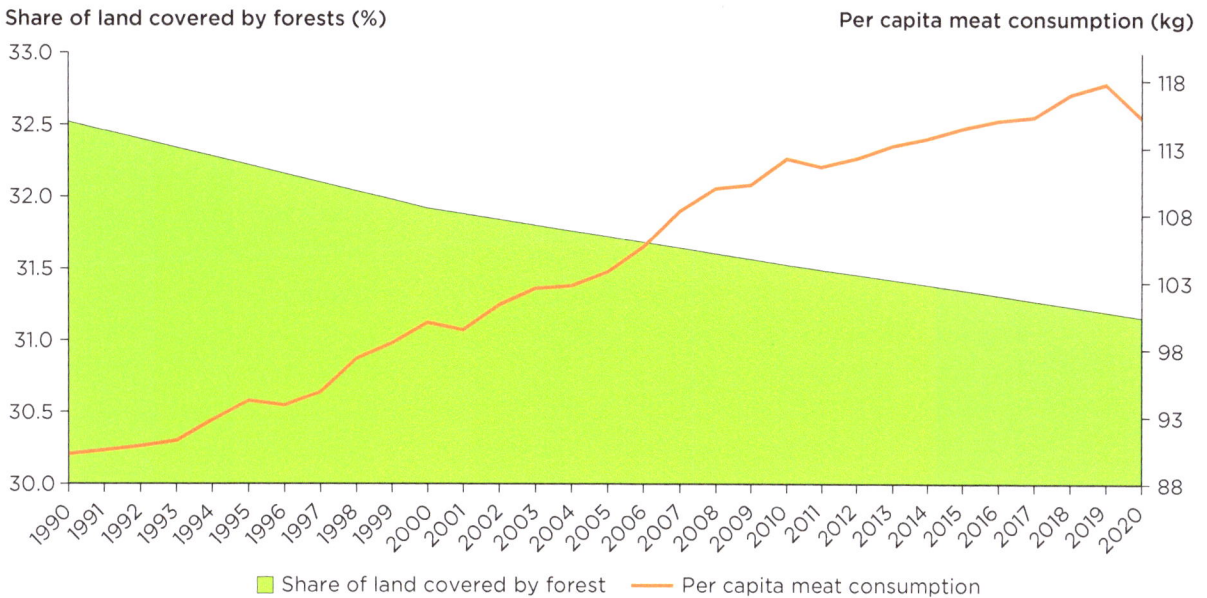

Share of land covered by forest — Per capita meat consumption

Source: Original calculations based on data from Our World in Data.
Note: kg = kilograms.

Well-crafted policies that leverage free trade can help mitigate deforestation and trade distortions, through contingent trade agreements (CTAs). This policy instrument links trade to environmental outcomes by establishing a framework in which tariffs are contingent on the conservation of critical resources, such as forest cover (Harstad 2024a). If the resource is preserved, goods can be exported tariff-free, but if monitoring reveals a decline in the resource, tariffs are imposed on imports. Empirical evidence supports the effectiveness of environmental provisions in regional trade agreements to limit deforestation (Abman, Lundberg, and Ruta 2024). The success of CTAs depends on the specifics of the agreement and the willingness of the parties to accept higher contingent tariffs. Preliminary simulations indicated that CTAs could be a potent tool for protecting tropical forests associated with globally traded commodities (Harstad 2024b). Unlike monetary compensation strategies, which face challenges like funding shortages and corruption, CTAs offer a promising alternative by avoiding the perverse incentives linked to direct payments.

Trade policies for environmental and economic gains

Well-designed trade policies can boost efficiency and stop environmental leakage, turning trade into a force for good. Although trade has a significant environmental impact, restricting trade could worsen environmental degradation—for example, by increasing water withdrawals. Instead of limiting trade, improving efficiency can reduce both the environmental footprint of production and the impact associated with trade. Trade policies can promote the transfer of clean technologies and enhance efficiency effects, and coordinating them can further prevent environmental leakage, increasing environmental gains while also maintaining the economic benefits of trade.

Efficiency: The key to safeguarding the environment

The impacts of international trade are both complex and multifaceted, so improving efficiency is key to shrinking the environmental footprint of trade. Trade offers significant environmental benefits, and countries can use trade policies to foster sustainability and innovation. By leveraging knowledge spillover and efficiency improvements, countries can significantly reduce pollution and resource consumption. Global value chains facilitate the transfer of clean technology and expertise, fostering innovation and making clean technologies and environmental goods more affordable (World Bank 2020). Enhancing efficiency can counteract trade-induced emissions while maintaining economic growth (box 7.3). Investing in clean energy could act as a major driver of economic growth, especially for low-income countries. Spotlight 3 shows that investing in green technologies and value chains can help lower-income countries diversify toward more rapid and sustainable development.

BOX 7.3

Decomposition analysis of the emissions embodied in trade

The environmental impacts of trade can emerge through three channels: the *scale* of commodities traded, the *efficiency* of production, and the *composition* of the traded goods. Trade boosts economic activity, leading to higher production and consumption, causing environmental degradation through increased resource extraction and pollution (scale effect). However, trade can also promote cleaner technologies and more efficient production, reducing environmental impacts (efficiency effect). Finally, comparative advantage enables countries to specialize in exporting goods they produce more efficiently and to import goods they produce less efficiently, shaping the mix of goods that countries trade (composition effect).

New research for this report found that emissions embedded in exports increased across all country income groups between 2004 and 2017. The analysis used data from input-output tables for 160 countries and regions from the Global Trade Analysis Project database (Aguiar et al. 2023) and techniques developed by Kaya and Yokobori (1997) and Ang (2004). The findings showed that the most substantial increase of greenhouse gas emissions (1,055 metric tonnes of carbon dioxide equivalent) was observed in upper-middle-income countries (UMICs), followed by high-income countries (HICs) and lower-middle-income countries (LMICs). The increase in low-income countries (LICs) was much smaller (figure B7.3.1). The fine particulate matter ($PM_{2.5}$) emissions embodied in exports increased in all country income groups, with a worldwide increase of around 1,465 gigagrams. UMICs contributed the most to this growth, followed by UMICs, HICs, and LICs.

(continued)

Decomposition analysis of the emissions embodied in trade *(continued)*

The efficiency effect—driven by reductions in the emissions intensity of exports—helped mitigate the overall pressure induced by the scale effect, constraining the growth in emissions, especially in UMICs and HICs.

FIGURE B7.3.1 Change in direct and indirect (scope 1, 2, and 3) emissions embodied in exports, by country income group, 2004–17

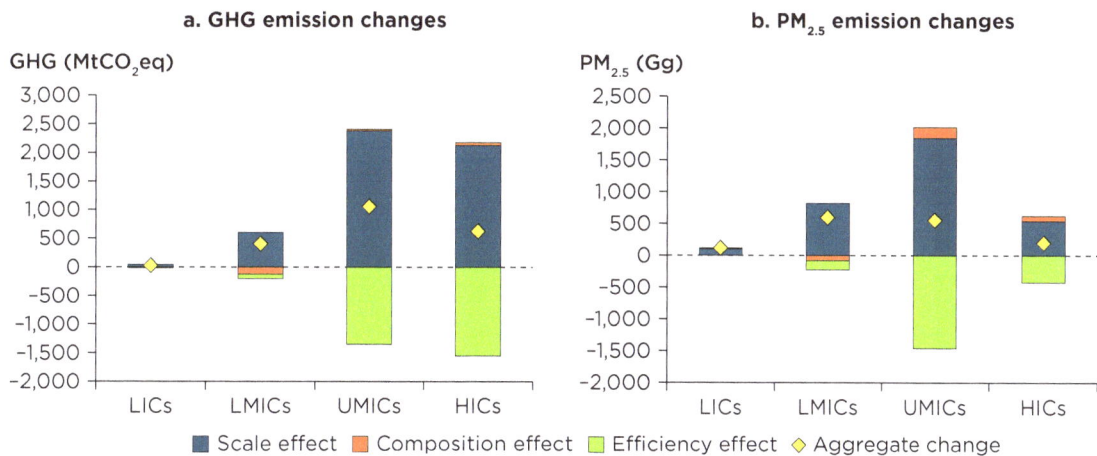

a. GHG emission changes

b. PM$_{2.5}$ emission changes

GHG (MtCO$_2$eq)

PM$_{2.5}$ (Gg)

■ Scale effect ■ Composition effect ■ Efficiency effect ◇ Aggregate change

Source: Original calculations based on data from GTAP databases.
Note: The composition effect highlights how shifts in the shares of exports across three main sectors—agriculture, manufacturing, and services—alter the emissions embodied in those exports. Gg = gigagrams; GHG = greenhouse gas; GTAP = Global Trade Analysis Project; HICs = high-income countries; LICs = low-income countries; LMICs = lower-middle-income countries; MtCO$_2$eq = million tons of carbon dioxide equivalent; PM$_{2.5}$ = fine particulate matter; UMICs = upper-middle-income countries.

Trade policies for environmental sustainability

Trade-related environmental issues are exacerbated when prices do not accurately represent the true economic value of natural resources, or the externalities generated by production or consumption. In such cases, market prices fail to reflect the true "shadow prices," or true economic values of resources. Although implementing a pollution tax[2] or quantity control measures would be the ideal or first-best solution, empirical work has suggested that although these are effective, they tend to be unpopular and harder to introduce (Carattini 2022). To correct such market failures, trade policies can offer a second-best solution.

The environmental effects of trade largely depend on whether production decisions consider the ensuing environmental impacts. For example, when air pollution causes health problems, political economy pressures can create strong incentives to regulate pollution. The US Clean Air Act and the EU Ambient Air Quality Directives are examples of how countries can regulate air pollutants to protect human health. Yet when pollution impacts are global—as is the case for GHG emissions or biodiversity loss—free-rider problems can emerge and collective action problems strengthen incentives to impose the cost of mitigating pollution on others. Environmental policies present both challenges and opportunities for developing countries' trade, and the World Bank Group actively supports these countries through various initiatives to leverage trade and investments for achieving environmental goals (box 7.4).

BOX 7.4
Navigating trade challenges and opportunities in developing countries

Environmental policies create both challenges and opportunities for developing countries' trade. *Climate Policies and Their Impact on Developing Countries' Trade* examined the challenges and opportunities that climate change and related policies, initiatives, and voluntary measures create for developing country trade (World Bank, forthcoming). It identified how developing countries can respond with international support, using trade and climate instruments, building on another study, *The Trade and Climate Change Nexus* (Brenton and Chemutai 2021). A working paper, "Global Ripple Effects: Knock-on Effects of EU, US, and China Climate Policies on Developing Countries' Trade," provided a comprehensive analysis of the challenges and opportunities for developing countries arising from the climate policies implemented by the three major economies—China, the European Union, and the United States (Aldaz-Carroll et al. 2024). This knowledge base from global analytics has been applied to country-specific work at the World Bank through Trade Competitiveness Diagnostics, contributions to Country Climate and Development Reports, and technical assistance to help countries such as Peru comply with new climate-linked trade regulations, like the EU Deforestation Regulation.

The World Bank Group, in collaboration with the World Economic Forum and the World Trade Organization, has launched the Action for Climate and Trade initiative to help countries leveraging trade and investments to achieve environmental goals (World Trade Organization 2023). The first pilots in Indonesia and Rwanda focused on concrete policy recommendations, including reducing tariffs for environmental goods, streamlining non-tariff measures, minimizing export barriers, addressing gaps in emissions accounting and developing quality infrastructure linked to due diligence, and establishing trade and climate strategies. These recommendations could be incorporated into World Bank Group financing instruments.

The World Bank Group helps countries in designing and implementing policies and initiatives through various actions. Development policy operations support trade policy reforms and address short-term balance of payments issues. Investment policy lending operations assist with subsidies, access to finance, technology transfers, and capacity building to enhance knowledge and human capital on climate change and trade. The World Bank Group also provides technical assistance and analytics for trade reforms and engages with other institutions for global advocacy and multilateral agreements on trade facilitation and international standards.

Environmental issues often extend beyond political borders, necessitating cooperation for effective policy making. Problems like air and water pollution, deforestation, and biodiversity loss may originate in one region but impact others, crossing borders or continents. Without addressing these cross-geographical effects, policies risk being ineffective or exacerbating environmental issues through environmental leakage, such as displaced $PM_{2.5}$ emissions or deforestation. Environmentally friendly trade policies can help by transcending political boundaries and removing incentives for environmental leakage.

Coordination is essential for achieving global environmental sustainability while preserving economic benefits. Cross-border environmental problems can lead to a free-rider issue, where countries depend on others to bear the costs. Coordination challenges arise across different jurisdictions, potentially resulting in loss of competitiveness and market access, as well as trade diversion. Despite these challenges, successful international policies could exist, such as "climate clubs" (Nordhaus 2015), where members agree on harmonized environmental policies and impose penalties on nonparticipating nations or offer favorable conditions to club members. Cooperation enhances environmental policies by aligning measurements for environmental indicators, ensuring transparent anti-leakage measures, and promoting green technology globally.

Notes

1. Higher production and economic activity often lead to an increase in scope 1 emissions (direct emissions from owned or controlled sources), scope 2 emissions (indirect emissions from the generation of purchased electricity, steam, heating, and cooling), and scope 3 emissions (all other indirect emissions that occur in a company's value chain), contributing to higher emissions.
2. A *Pigouvian tax* is a tax on negative externalities or activities that have adverse side effects on third parties, such as a tax on fossil fuels to reduce environmental pollution.

References

Abman, R., and C. C. Lundberg. 2020. "Does Free Trade Increase Deforestation? The Effects of Regional Trade Agreements." *Journal of the Association of Environmental and Resource Economists* 7 (1): 35–72.

Abman, R., C. Lundberg, and M. Ruta. 2024. "The Effectiveness of Environmental Provisions in Regional Trade Agreements." *Journal of the European Economic Association* 22 (6): 2507–48. doi:10.1093/jeea/jvae023.

Aguiar, A., M. Chepeliev, E. Corong, and D. van der Mensbrugghe. 2023. "The Global Trade Analysis Project (GTAP) Data Base: Version 11." *Journal of Global Economic Analysis* 7 (2): 1–37. doi:10.21642/JGEA.070201AF (Original work published December 19, 2022).

Aldaz-Carroll, E., E. Jung, M. Maliszewska, and I. Sikora. 2024. "Global Ripple Effects: Knock-on Effects of EU, US, and China Climate Policies on Developing Countries' Trade." Policy Research Working Paper 10988, World Bank, Washington, DC. https://documents1.worldbank.org/curated/en/099404411262427372/pdf/IDU1a2acff15188ab14ea51a7561bd72b8127c1d.pdf.

Allan, J. A. 1996. "Policy Responses to the Closure of Water Resources." In *Water Policy: Allocation and Management in Practice*, edited by P. Howson and R. C. Carter. London: Chapman and Hill.

Allan, J. A. 1998. "Virtual Water: A Strategic Resource." *Ground Water* 36 (4): 545–47.

Ang, B. 2004. "Decomposition Analysis for Policymaking in Energy: Which Is the Preferred Method?" *Energy Policy* 32: 1131–39.

Brenton, P., and V. Chemutai. 2021. *The Trade and Climate Change Nexus: The Urgency and Opportunities for Developing Countries*. Washington, DC: World Bank. http://hdl.handle.net/10986/36294.

Brenton, P., V. Chemutai, M. Maliszewska, and I. Sikora. 2024. *Trade and Climate Change: Policy Considerations for Developing Countries*. Washington, DC: World Bank.

Briceno-Garmendia, C., W. Qiao, and V. Foster. 2023. *The Economics of Electric Vehicles for Passenger Transportation*. Sustainable Infrastructure Series. Washington, DC: World Bank. http://hdl.handle.net/10986/39513.

Carattini, S. 2022. "Political Challenges of Introducing Environmental Tax Reforms in Developing Countries." World Bank, Washington, DC.

Carleton, T., L. Crews, and I. Nath. 2023. "Agriculture, Trade, and the Spatial Efficiency of Global Water Use." https://www.levicrews.com/files/p-wateruse_paper.pdf.

Chapagain, A. K., A. Hoekstra, and H. Savenije. 2006. "Water Saving through International Trade of Agricultural Products." *Hydrology and Earth System Sciences* 10: 455–68.

Cherniwchan, J. 2017. "Trade Liberalization and the Environment: Evidence from NAFTA and US Manufacturing." *Journal of International Economics* 105: 130–49.

Copeland, B. R., J. S. Shapiro, and M. S. Taylor. 2022. "Globalization and the Environment." In *Handbook of International Economics*, edited by G. Gopinath, E. Helpman, and K. NAFTA, 61–146. Amsterdam, Netherlands: North-Holland.

Dalin, C., Y. Wada, T. Kastner, and M. J. Puma. 2017. "Groundwater Depletion Embedded in International Food Trade." *Nature* 543 (7647): 700–704.

Debaere, P. 2014. "The Global Economics of Water: Is Water a Source of Comparative Advantage?" *American Economic Journal: Applied Economics* 6 (2): 32–48.

Ebadi, E., and E. Aldaz-Carroll. 2025. "Trade's Emissions Paradox: Cutting Greenhouse Gases, Raising Air Pollution." Policy Research Working Paper 11164, World Bank, Washington, DC.

Harstad, B. 2024a. "Contingent Trade Agreements." Working Paper 32392, National Bureau of Economic Research, Cambridge, MA.

Harstad, B. 2024b. "Trade and Trees." *American Economic Review: Insights* 6 (2): 155–75.

Herrera Dappe, M., M. Lebrand, and A. Stokenberga. 2024. *Shrinking Economic Distance: Understanding How Markets and Places Can Lower Transport Costs in Developing Countries*. Sustainable Infrastructure Series. Washington, DC: World Bank. http://hdl.handle.net/10986/42061.

ICCT (International Council on Clean Transportation). 2019. "A Global Snapshot of the Air Pollution-Related Health Impacts of Transportation Sector Emissions in 2010 and 2015." ICCT, Washington, DC. https://theicct.org/wp-content/uploads/2021/06/Global_health_impacts_transport_emissions_2010-2015_20190226.pdf.

ICCT (International Council on Clean Transportation). 2021. "Global Lifecycle Impacts of Passenger Cars: A Review of the State of the Art in LCA of Passenger Cars, with a Focus on the 2030s and Beyond." ICCT, Washington, DC. https://theicct.org/sites/default/files/publications/Global-LCA-passenger-cars-jul2021_0.pdf.

Kaya, Y., and K. Yokobori. 1997. *Environment, Energy, and Economy: Strategies for Sustainability*. Tokyo: United Nations University Press.

Lai, W., S. Li, and F. Zhang. 2023. "The Impact of Water Resources on Trade under a Changing Climate." doi:10.2139/ssrn.4627240.

Levinson, A., and M. S. Taylor. 2008. "Unmasking the Pollution Haven Effect." *International Economic Review* 49 (1): 223–54.

Mekonnen, M. M., M. M. Kebede, B. W. Demeke, et al. 2024. "Trends and Environmental Impacts of Virtual Water Trade." *Nature Reviews. Earth & Environment* 5 (12): 890–905.

Nordhaus, W. 2015. "Climate Clubs: Overcoming Free-Riding in International Climate Policy." *American Economic Review* 105 (4): 1339–70.

Pendrill, F., U. M. Persson, J. Godar, and T. Kastner. 2019a. "Deforestation Displaced: Trade in Forest-Risk Commodities and the Prospects for a Global Forest Transition." *Environmental Research Letters* 14 (5): 055003.

Pendrill, F., U. M. Persson, J. Godar, and T. Kastner. 2019b. "Deforestation Driven by International Trade: A Global Analysis of the Impact of Trade on Deforestation." *Global Environmental Change* 56: 1–11.

Ritchie, H. 2021. *Meat and Dairy Production*. Oxford, UK: Our World in Data.

Rodella, A.-S., E. Zaveri, and F. Bertone, eds. 2023. *The Hidden Wealth of Nations: The Economics of Groundwater in Times of Climate Change*. Washington, DC: World Bank. http://hdl.handle.net/10986/39917.

Rosa, L., D. D. Chiarelli, C. Tu, M. C. Rulli, and P. D'Odorico. 2019. "Global Unsustainable Virtual Water Flows in Agricultural Trade." *Environmental Research Letters* 14: 114001.

UNCTAD (United Nations Conference on Trade and Development). 2024. *The Impact of Trade on Sustainable Development*. Geneva: UNCTAD. https://unctad.org/system/files/official-document/ditcinf2024d3.pdf.

UNESCO (United Nations Educational, Scientific and Cultural Organization). 2015. *The United Nations World Water Development Report 2015: Water for a Sustainable World*. Perugia, Italy: United Nations World Water Assessment Programme.

Wichelns, D., 2001. "The Role of 'Virtual Water' in Efforts to Achieve Food Security and Other National Goals, with an Example from Egypt." *Agricultural Water Management* 49 (2): 131–51.

World Bank. 2020. *World Development Report 2020: Trading for Development in the Age of Global Value Chains (Vol. 1 of 2) (English)*. World Development Indicators | World Development Report. Washington, DC: World Bank. http://documents.worldbank.org/curated/en/310211570690546749.

World Bank. 2023. "The Road to Sustainable Transport: Transitioning to a Low-Carbon and Climate-Resilient Transport System." World Bank, Washington, DC. https://documents1.worldbank.org/curated/en/099753004192311879/pdf/IDU0114c90f5059af043410a34b0cf206b4ecebb.pdf.

World Bank. 2025. "The Global Facility to Decarbonize Transport (GFDT): Changing the Way We Move." World Bank, Washington, DC. https://www.worldbank.org/en/programs/global-facility-to-decarbonize-transport/approach.

World Bank. Forthcoming. *Climate Policies and Their Impact on Developing Countries' Trade*. Washington, DC: World Bank.

World Trade Organization. 2023. "WTO, World Bank, WEF Launch Joint Effort to Provide Tailored Trade and Climate Analysis." World Trade Organization, Geneva. https://www.wto.org/english/news_e/news23_e/envir_20apr23_e.htm.

WRI (World Resources Institute). 2020. "The Role of Trade in Deforestation." WRI, Washington, DC.

Zhao, L., Y. Lv, C. Wang, J. Xue, Y. Yang, and D. Li. 2023. "Embodied Greenhouse Gas Emissions in the International Agricultural Trade." *Sustainable Production and Consumption* 35: 250–59.

Using the Green Transition to Develop More Sophisticated Economies

The green transition is transforming global markets, creating new economic opportunities for developing countries. Many lower-income nations have historically relied on commodity exports but this dependence often leads to economic volatility, environmental degradation, and governance challenges. As demand grows for clean energy technologies, electric vehicles, and low-footprint manufacturing, developing countries have a chance to diversify their economies and integrate into higher-value green sectors. This spotlight examines how countries can strategically position themselves within green value chains, using comparative advantages to expand into clean energy, battery production, and other decarbonized industries.

While much of this report discusses the economic, health, and environmental challenges associated with environmental degradation, transitioning away from polluting and inefficient production methods also presents a key opportunity. Investing in clean energy contributes significantly to growth and accounts for about 10 percent of global gross domestic product growth (Cozzi et al. 2023). That is equivalent to $320 billion each year—or an economy the size of Czechia—from clean energy alone, and green sectors outside clean energy are likely even larger. Indeed, estimates suggest that by 2050, the transition toward net zero emissions could add $10.3 trillion to the global economy (Oxford Economics 2023). Capturing a large portion of that growth will be key to economic success in the twenty-first century (box S3.1).

BOX S3.1

Trade policies and green competitiveness: Asymmetry between countries

Creating global trade rules for green supply chains is key to advancing low-carbon industries and achieving worldwide decarbonization. But high-income countries (HICs) hold a clear advantage over developing countries when it comes to exporting green products. This disparity is highlighted by the Green Complexity Index (GCI), a measure introduced by Mealy and Teytelboym (2022) that assesses a country's ability to export complex green products based on their green productive

(continued)

Trade policies and green competitiveness: Asymmetry between countries *(continued)*

capabilities. The GCI shows whether a country has a relative advantage (positive score) or disadvantage (negative score) in exporting green products, compared to other nations.

A country's GCI is heavily influenced by its policies. Those that implement strong trade incentives and restrictions related to green products tend to perform better. In low-income countries, limited fiscal space and weak institutions often result in fewer trade incentives and less impact on the GCI, while trade policies in upper-middle-income countries show moderate positive effects, and HICs benefit the most, thanks to their ample fiscal space and stronger institutions. Figure SB3.1.1 illustrates how green trade policies affect a country's GCI.

FIGURE SB3.1.1 **Relationship between a country's GCI and the green trade policies it adopts**

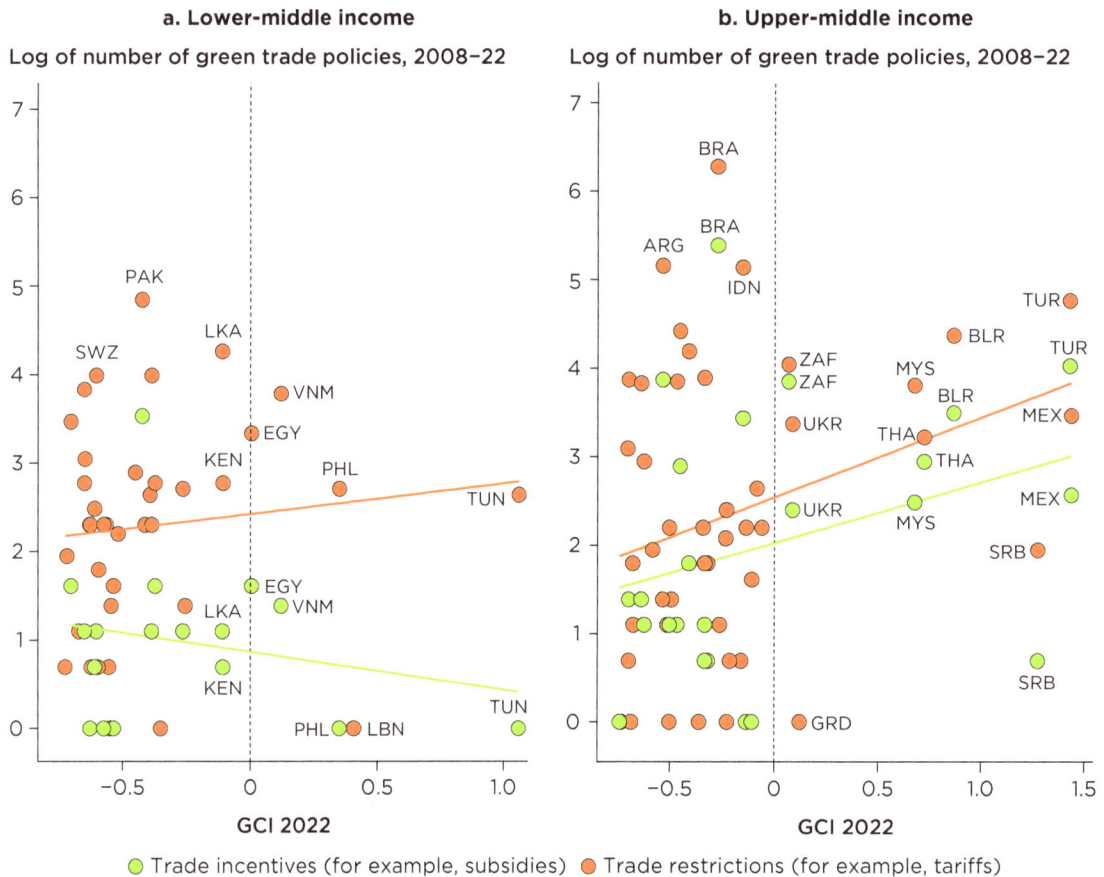

a. Lower-middle income

Log of number of green trade policies, 2008–22

b. Upper-middle income

Log of number of green trade policies, 2008–22

Trade incentives (for example, subsidies) ● Trade restrictions (for example, tariffs)

(continued)

Trade policies and green competitiveness: Asymmetry between countries *(continued)*

FIGURE SB3.1.1 **Relationship between a country's GCI and the green trade policies it adopts** *(continued)*

c. High-income

Log of number of green trade policies, 2008–22

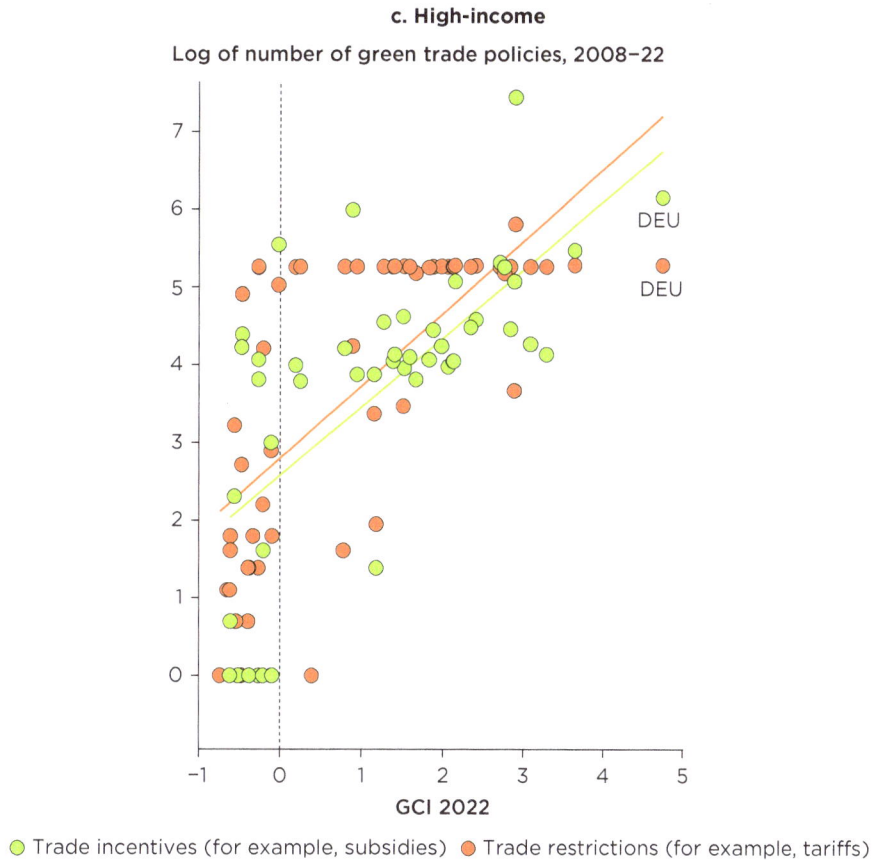

DEU

DEU

○ Trade incentives (for example, subsidies)   ● Trade restrictions (for example, tariffs)

Sources: Original calculations based on data from Global Trade Alert 2024 (https://data.globaltradealert .org/); Green Transition Navigator (https://green-transition-navigator.org/); Mealy and Teytelboym 2022.

Note: The y-axis uses data from Global Trade Alert 2024 on log trade policies that have impacted green products, as defined by Mealy and Teytelboym (2022); the x-axis is the GCI in 2022. For country abbreviations, see International Organization for Standardization, https://www.iso.org/obp/ui/#search. GCI = Green Complexity Index.

The data suggest that high-income countries (HICs) do not have an insurmountable advantage when it comes to green markets. In new analysis for this report, Martin (2024) explored trends in green transition–related business activities—that is, technologies and products that help the world economy transition to a lower emission equilibrium or adapt to the consequences of climate change. Martin (2024) found that the share of total green goods exported by HICs has declined sharply over the past 20 years, from 83.5 percent at the beginning of the century to 66.7 percent in 2018–22 (figure S3.1, panel a). Although the rise of China as a key exporter of green products explains a large proportion of this shift, the share of other developing countries in green exports increased modestly, from around 13 percent in 2000 to 17 percent in 2020. Low-income countries also dramatically increased their exports, although this was from a low base. At the same time, green exports have increased at a faster rate than total exports (figure S3.1, panel b), meaning that developing countries are capturing a bigger slice of a growing pie.

Gaining a piece of the growing green market share will be crucial for the economic prospects of developing countries, and there are reasons to think this will be possible. One way to measure green competitiveness is through a country's green revealed comparative advantage (RCA) in exports (Martin, 2024). A country's green RCA is the share of its exports in a green market segment relative to the global average share of exports in that segment.[1] A green RCA larger than 1 shows that a country's export share is larger than the world average, indicating that it probably has a comparative advantage in that green market segment and is more competitive in the green market, relative to other market segments in which it competes.

FIGURE S3.1 **Participation in decarbonized value chains**

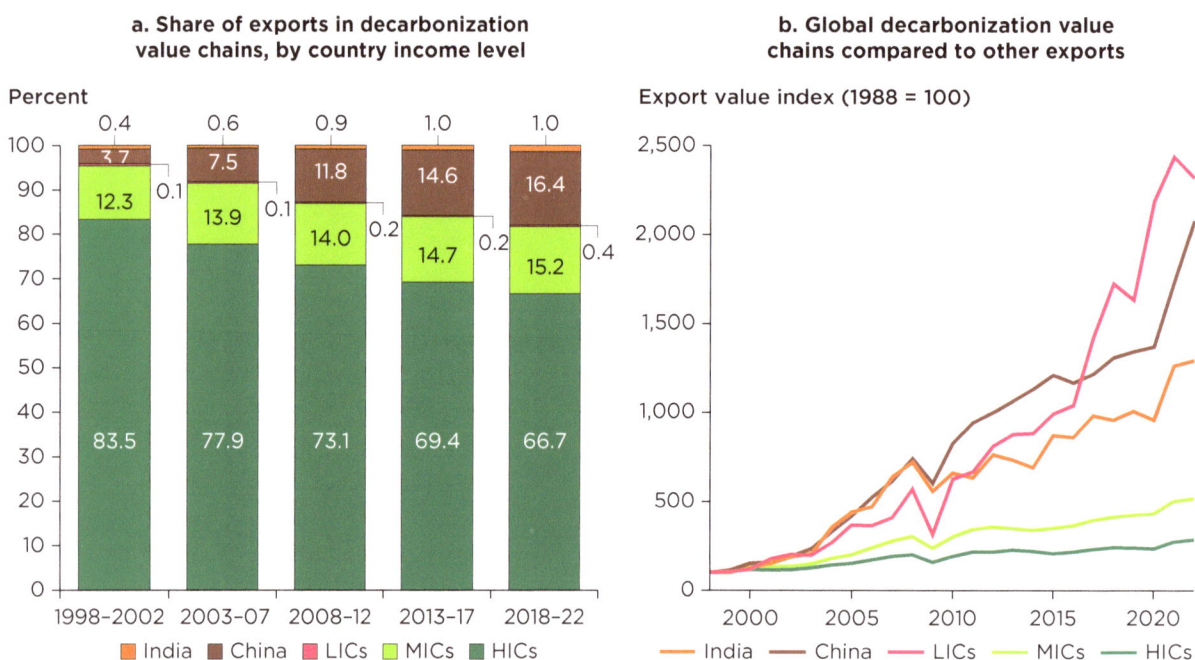

a. Share of exports in decarbonization value chains, by country income level

Percent

b. Global decarbonization value chains compared to other exports

Export value index (1988 = 100)

Source: Martin, 2024.

Note: The figure reports the evolution of decarbonization value chains, with panel a presenting 5-year averages of country group shares in these chains, excluding raw materials, and panel b comparing the value of the export market for decarbonization value chains to those for products with environmental benefits and total merchandise exports. In line with the definition in Rosenow and Mealy (2024), *decarbonization value chains* include electric vehicles, heat pumps, solar photovoltaics, wind turbines, and electrolyzers. Products with environmental benefits are defined in Mealy and Teytelboym (2022). HIC = high-income country; LIC = low-income country; MIC = middle-income country.

While richer countries are more likely to have a high green RCA, many developing countries perform higher than HICs (figure S3.2, panel a). The green RCA of poorer countries has increased faster than that of richer countries over the past 20 years (figure S3.2, panel b). Indeed, of the 20 countries with the fastest increases in RCA, only Romania is an HIC. This suggests that, under the right conditions, green industries can be a viable pathway to growth for developing countries.

FIGURE S3.2 Green export levels, by country

a. Relative comparative advantage

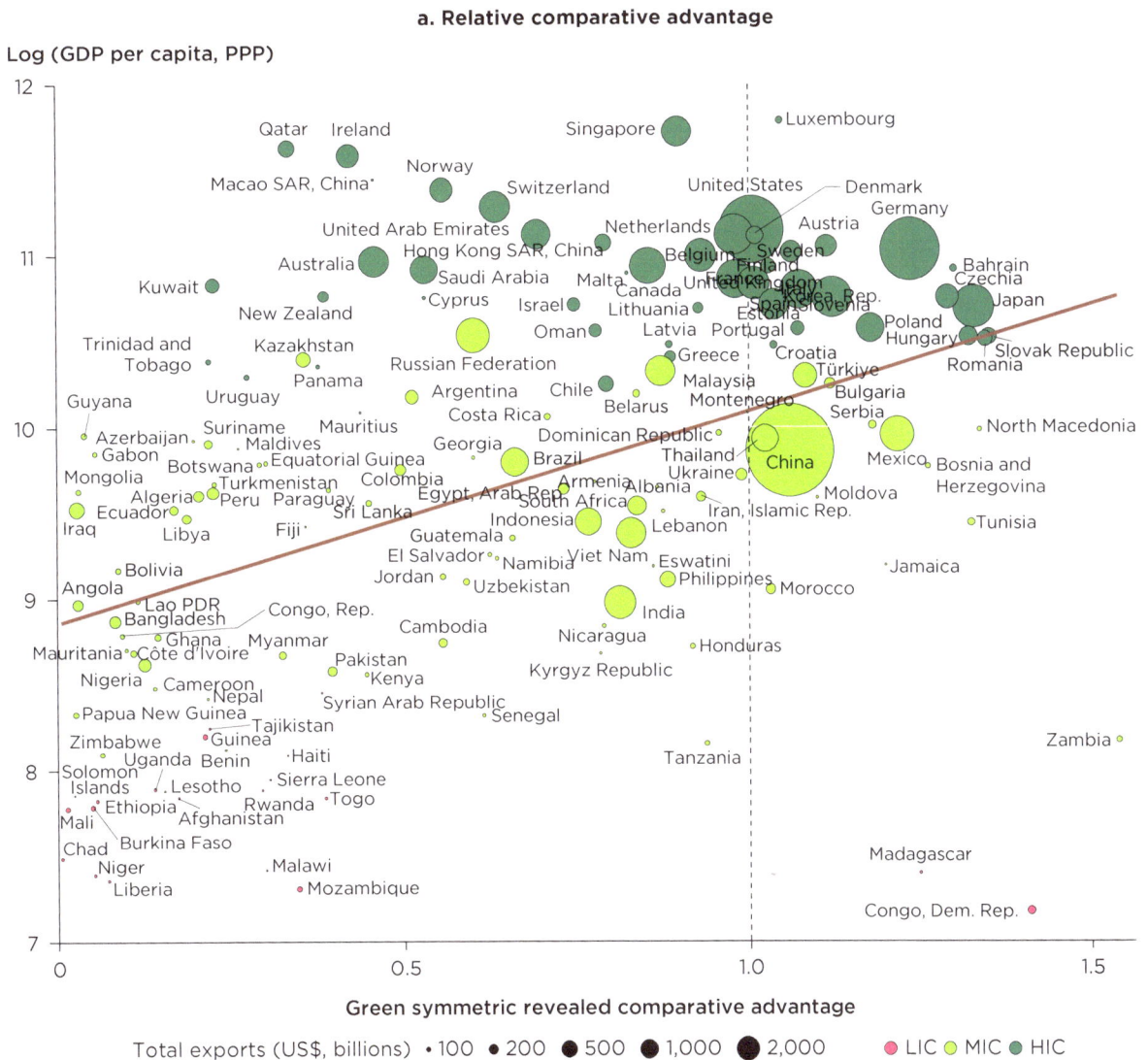

Log (GDP per capita, PPP)

Green symmetric revealed comparative advantage

Total exports (US$, billions) • 100 ● 200 ● 500 ● 1,000 ● 2,000   ● LIC ● MIC ● HIC

(continued)

Green export levels, by country *(continued)*

b. Change in relative comparative advantage

Log (GDP per capita)

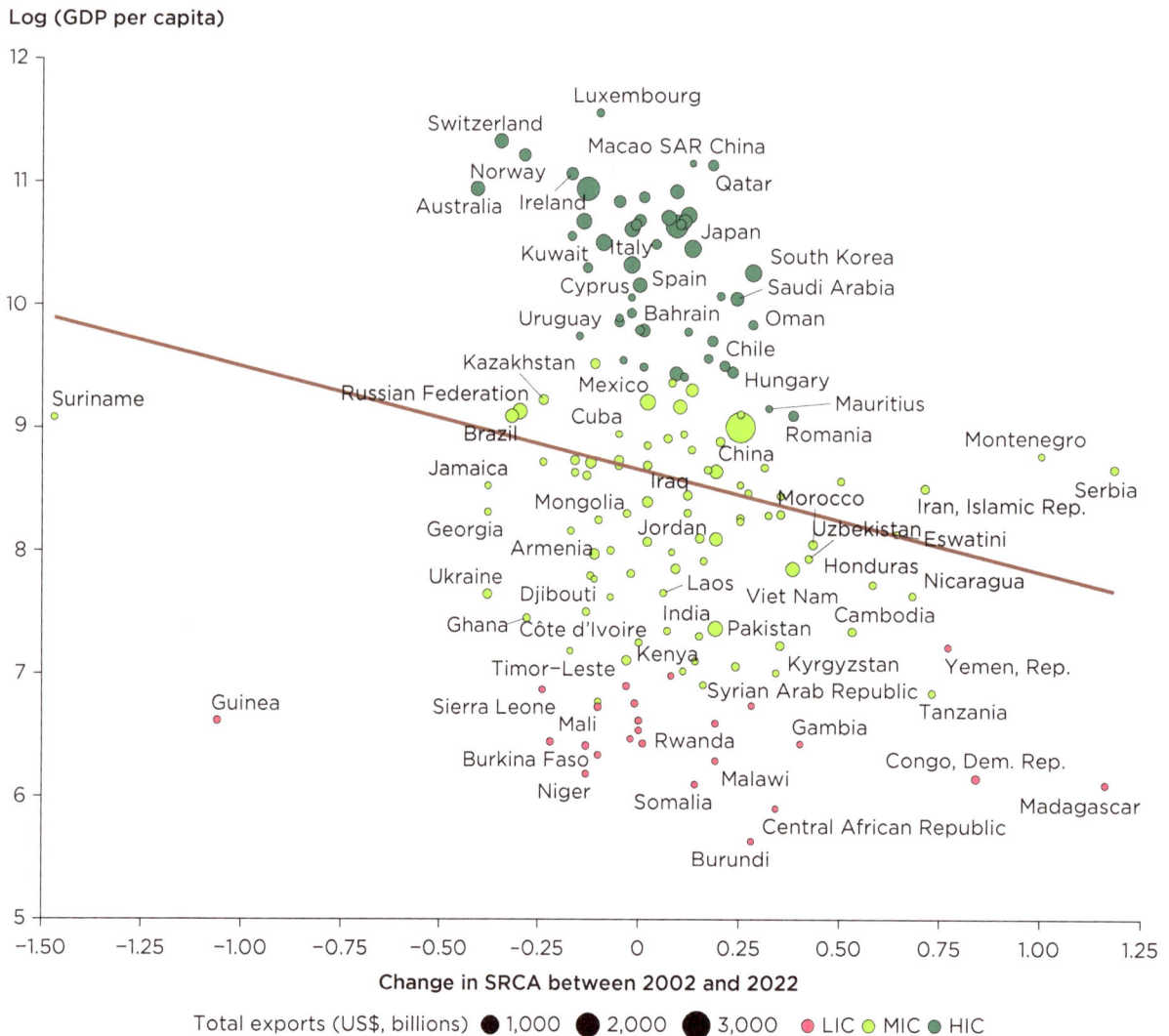

Source: Martin, 2024.

Note: Panel a shows a scatterplot of the symmetric RCA (SYRCA) for decarbonization value chain products (other than raw materials) between 2021 and 2022 versus a country's GDP per capita in 2022. $SYRCA = s_{ic}/(0.5 \times (s_{ic} + s_i))$, where s_i indicates the export share of good i by country c, and indicates the world's share of good i's exports in all goods exports. The SYRCA measure is bounded between 0 and 2 and symmetric around 1, and therefore helps avoid extreme values when countries export very few products. Panel b shows country-level changes in green comparative advantage, revealing that on average, compared to other countries, lower-income countries have improved their green RCA more over the past 20 years. GDP = gross domestic product; HIC = high-income country; LIC = low-income country; MIC = middle-income country; PPP = purchasing power parity; RCA = revealed comparative advantage; SRCA = symmetric revealed comparative advantage; SYRCA = scatterplot of the symmetric revealed comparative advantage.

Some developing countries may also find opportunities to invest in green technology and derive a deeper comparative advantage in green sectors. Since 2000, the share of academic research on "green" topics in developing countries has expanded considerably. This is important for many reasons. When research is conducted in developing countries, it is often better attuned to local environmental, social, and economic contexts. It is therefore more likely that technologies and policies are tailored to address the specific challenges they face. Perhaps more importantly, firms in lower-income countries tend to be smaller than those in richer countries, and are therefore more financially constrained, curtailing their ability to invest in research and development and making them more reliant on academic knowledge for their patented innovations. So, where academic capacities are strong, investing in their research can pay large dividends toward building a more competitive and greener private sector.

There are several important caveats to the green RCA approach. First, the data look exclusively at the export market and ignore goods produced for domestic consumption, which may have a different pattern. Second, the analysis ignores external influences—such as trade policies or subsidies—that may influence export decisions, so the "revealed" comparative advantage might not reflect an actual comparative advantage (trade is discussed more in chapter 7). Third, for countries with small export volumes, RCA calculations can be disproportionately affected by minor changes, leading to potential overestimation or underestimation of comparative advantage.

Note

1. $RCA_{ac} = \frac{s_{ac}}{s_a}$ where s_{ac} is the share of segment (for example, all green products) in country's exports and is the global share. A value larger than 1 indicates that country c has a relative comparative advantage in segment a. To avoid extreme outliers, a slightly modified symmetric version of RCA_a is used that is bounded between 0 and 2. Symmetric RCA is defined as $SYRCA_{ac} = \frac{2s_{ac}}{s_a + s_{ac}}$.

References

Cozzi, L., T. Gül, T. Spencer, and P. Levi. 2023. "Clean Energy Is Boosting Economic Growth." International Energy Agency, Paris. www.iea.org/commentaries/clean-energy-is-boosting-economic-growth.

Martin, R. 2024. "The Past, Present and Future of Green Comparative Advantage." Background paper for this report, World Bank, Washington, DC.

Mealy, P., and A. Teytelboym. 2022. "Economic Complexity and the Green Economy." *Research Policy, Special Issue on Economic Complexity* 51 (8): 103948. doi:10.1016/j.respol.2020.103948.

Oxford Economics. 2023. *The Global Green Economy Report 2023*. London: Oxford Economics. www.oxfordeconomics.com/wp-content/uploads/2023/11/The-Global-Green-Economy-Report-2023_FINAL_10MB-version.pdf.

Rosenow, S. K., and P. A. Mealy. 2024. "Turning Risks into Rewards: Diversifying the Global Value Chains of Decarbonization Technologies." Policy Research Working Paper 10696, World Bank, Washington, DC.

Part 3

Policies, Jobs, and Solutions for a Livable Planet

Part 3 examines solutions for a livable planet. It begins with a policy playbook and case studies of winning moves and costly mistakes, then examines digital solutions for sustainability. It looks at how green jobs can align economic growth with environmental goals, and concludes with the role of transition minerals required by new technologies.

CHAPTER 8

The Policy Playbook: Designing Effective Policies for a Livable Planet

> "The conservation of natural resources is the fundamental problem. Unless we solve that problem, it will avail us little to solve all others."
>
> —*Theodore Roosevelt*, American president and naturalist (1858-1919)

Key messages

- Effective environmental policy making requires having access to credible information and countering misinformation. History has shown that environmental reforms often gain traction only when public awareness and demand align with policy action. Strengthening independent data sources, enforcing transparency regulations, and leveraging digital tools for environmental monitoring can help drive accountability.

- Taking a systems approach to environmental policies will help avoid unintended consequences and maximize long-term benefits. Interventions designed to solve one problem can often create new challenges elsewhere, so effective policy making must consider interactions across sectors and disciplines, geographies, time, and communities to ensure that policies account for feedback loops and broader ecological and distributional impacts.

- Public investment plays a vital role in scaling green innovations. Many promising environmental technologies face a funding gap between early-stage development and commercial viability, known as the "Valley of Death." Public policies, including subsidies and research and development investments, are crucial for bridging this gap, ensuring that sustainable innovations can compete with entrenched high-emissions industries.

- Continuous evaluation strengthens environmental policy effectiveness. Building in regular monitoring and assessment helps policy makers adapt reforms over time, scale what works, and address emerging challenges before they escalate.

(continued)

- Tailoring reforms to institutional capacity increases the chances of success. Policies must be designed to match administrative capabilities, with simpler, incentive-compatible approaches in low-capacity settings and more complex, layered instruments where capacity is stronger.

Introduction

In the 1930s, the US government responded to the Dust Bowl, a severe drought that brought dust storms and soil erosion to millions of acres of farmland, with large-scale soil conservation efforts. These efforts included the promotion of kudzu (*Pueraria montana*), a fast-growing vine praised for its ability to stabilize degraded land (Forseth and Innis 2004). Millions of kudzu seedlings were planted across the southern United States under federal conservation programs, with farmers financially incentivized to participate. However, lacking natural predators in its new habitat, kudzu spread aggressively—smothering native vegetation, damaging infrastructure, and reducing biodiversity. By the 1950s, it was evident that the policy had backfired, and in 1970, the US Department of Agriculture classified kudzu as a noxious weed (Forseth and Innis 2004).

Kudzu remains a costly and persistent problem, particularly in the southeastern United States, highlighting the danger of environmental interventions that fail to anticipate long-term impacts. This case illustrates how well-meaning policies can backfire when ecological systems are not fully understood. It underscores the need for rigorous environmental assessments, especially when introducing non-native species, and for adaptive management strategies that can identify and respond to unintended consequences before they escalate into new crises.

Avoiding unintended consequences through a systems approach

To render development compatible with a livable planet, governments need to chart a path that does not address one challenge by creating others, as happened with kudzu. Policies are often designed with a single objective in mind, ignoring the many indirect consequences that follow. Through numerous examples, this report illustrates that natural resources are interconnected, both with each other and with different sectors in the economy. Mismanagement in one area—or even, in some cases, efficient management in only one area, space, or time period that ignores downstream or intertemporal impacts—can lead to harmful spillovers (box 8.1).

Navigating environmental policy in a second-best world

Environmental policies are often designed under the assumption of perfect information, fully functioning markets, and strong institutional capacity. Yet in practice, they are often implemented under constraints such as political resistance, market imperfections, and informational asymmetries. This creates a "second-best" context where ideal solutions may not be feasible. As Dercon (2024) observed, "policies that most economists think are unreasonable are often pursued." Recognizing these challenges, policy makers may have to adopt politically viable approaches that balance environmental goals with economic and governance realities.

Effective policy design in a second-best world requires aligning environmental objectives with existing political and economic incentives. In contexts where direct regulations or optimal taxation are politically infeasible, second-best strategies—such as indirect taxes, subsidies for cleaner technologies, or performance-based incentives—can achieve meaningful progress (Goulder and Parry 2008). Similarly, integrating environmental considerations into broader economic policies, such as trade or infrastructure investments, can generate co-benefits that make environmental action more palatable to decision-makers.

Flexibility and innovation are crucial for environmental policies to work under imperfect conditions. Given uncertainties in enforcement and market responses, adaptive policies—such as gradually increasing pollution pricing, leveraging technology to improve monitoring, and fostering voluntary agreements—can help build momentum for stronger long-term action (Rodrik 2008). Although second-best policies may not fully resolve environmental challenges, they can serve as stepping stones toward more comprehensive solutions, ensuring progress even in less-than-ideal circumstances.

Effective environmental policies must take a systems approach to avoid unintended consequences and maximize long-term benefits. Interventions designed to solve one problem can often create new challenges elsewhere if they fail to consider complex interconnections across ecological, economic, and social systems. Access to reliable *information* and *technological innovation* is foundational to sound policy making, to enable governments to track environmental changes in real time, improve enforcement, and drive public demand for action. Policy makers must also create *enabling policies* that prioritize linkages—recognizing how actions in sectors, geographic communities, or timeframes influence others—to ensure that interventions create synergies rather than conflicts. Finally, *policy evaluation* is critical for improving adaptivity and response, to ensure that policies are working as intended and adapting to changing realities on the ground.

Information and innovation

Information has historically been a catalyst for environmental change by making invisible threats visible and mobilizing public demand. In 1962, Rachel Carson's

Silent Spring exposed the hidden dangers of industrial chemicals, transforming obscure environmental risks into a national concern. Written in accessible language but rooted in scientific rigor, Carson's work galvanized public opinion, spurred congressional debates, inspired grassroots activism, and contributed to the creation of the US Environmental Protection Agency and the ban on DDT. More than just a book, *Silent Spring* became a clarion call for the power of information, highlighting how raising awareness can shift national priorities and drive systemic change. *Silent Spring* underscored the potential of information dissemination as a catalyst for environmental protection.

More recently, monitoring and information can be credited for helping spur what is perhaps the quickest and most dramatic environmental turnaround in modern history. As discussed in chapter 5 of this report, the past 10 years have seen a dramatic decline in air pollution in China, attributable to the country's "war on pollution." Introducing automatic air pollution monitoring was a major—and some argue, the critical enabling—component of this war (Greenstone 2022). Before the "war on pollution," general monitoring and enforcement of air pollution regulations were weak, particularly at the local level. The high cost of verifying local information, combined with the short-term economic gains associated with high pollution levels, created perverse incentives. To solve this conflict of interest and information asymmetry (known in economics as a *principal-agent problem*), China's central government implemented an automatic pollution monitoring system, which allowed for real-time data sharing and increased the accuracy of monitoring data.

China's automatic air pollution monitoring system had an almost immediate impact (Greenstone 2022). The new information created an upswell of public awareness of the harms of air pollution. Online searches for pollution avoidance technologies, such as face masks and air filters, immediately and permanently increased. In addition, because data were publicly available, local officials had strong incentives to reduce air pollution levels in their cities. These two effects led to both the creation of demand and the enabling conditions to manage the trade-offs associated with reducing pollution.

Silent Spring and China's "war on pollution" both demonstrated that information can be a catalyst for driving demand for environmental change. People cannot be expected to demand change if they are not aware of environmental hazards and their impacts, which can be distant, abstract, or slow to take effect. History has shown that environmental conditions rarely improve unless and until the public demands it. In an era where people are constantly bombarded with misinformation and disinformation (box 8.2), providing *compelling*, *credible*, and *consistent* information on environmental conditions will be key to achieving and maintaining a livable planet. Digital technologies can play a crucial role in providing transparent, accurate, local, and actionable information to drive change. Spotlight 5 discusses new technologies and applications for addressing environmental challenges.

Information in a world full of misinformation

In the digital age, misinformation is a significant barrier to environmental action. False or misleading information can obscure real hazards, misguide priorities, and reduce urgency. Social media platforms, driven by engagement-based business models, create echo chambers that amplify misinformation (Acemoglu, Ozdaglar, and Siderius 2024). Outrageous or misleading content often performs better, making it harder for accurate information to gain traction and incentivizing platforms to promote sensationalism over science.

Companies also often employ disinformation and greenwashing tactics to downplay environmental harm (Ford et al. 2025). These tactics include softening language (for example, calling sewage overflows "diluted rainwater"), exaggerating the costs of environmental improvements to delay action, and misdirecting blame—such as blaming consumers rather than infrastructure failures for sewage overflows. Companies also distract from major pollution issues by spotlighting minor sustainability efforts or casting doubt on scientific evidence, echoing tactics once used by the tobacco industry.

Countering environmental misinformation requires actions across three fronts: regulatory oversight, corporate accountability, and improved public access to credible data. Tools like content labeling, algorithmic transparency, and limits on monetizing misleading content can curb online misinformation (Acemoglu et al. 2024). Corporate disclosure rules, such as third-party verification of environmental claims and penalties for deceptive marketing, can reduce greenwashing (Ford et al. 2025). Governments can also require science-based proof for sustainability claims, as done in consumer protection laws. Public education, media literacy, and open-access environmental monitoring platforms can help people recognize disinformation and ensure that reliable data remain accessible. Tackling misinformation through coordinated regulatory, corporate, and informational strategies is key to restoring public trust and enabling action.

New innovations and technologies, such as automatic monitoring systems, offer opportunities for information sharing to create demand for environmental change and reduce environmental challenges. For these innovations to contribute meaningfully to sustainability, they must be supported by deliberate policy interventions that incentivize green technology development and ensure broad access to reliable environmental data. Acemoglu et al. (2023) noted that even when green innovations are profitable, they need additional policy support to counter path dependence and lock-ins, which make existing (dirtier) technologies more profitable than the new and cleaner innovations. In this context, merely correcting the environmental externality through taxes or other policy instruments will not induce a shift to cleaner innovations. Early and front-loaded support is vital to escape the "Valley of Death"—a stage that often exists between early development and commercial uptake of a new technology (Kauffman 2024).

Policies such as loan guarantees, early-stage subsidies, and procurement commitments for green technologies have played a key role in enabling major environmental innovations that would not have succeeded in purely market-driven environments. For example, the Danish and US governments played an essential role in building bridges over "Valleys of Death" to catalyze the commercialization of the following transformative environmental technologies:

- *Electric vehicles (EVs)*. The US Department of Energy provided a $465 million loan to a prominent US producer of EVs in 2010, enabling the company to scale production of EVs and establish manufacturing facilities. This public support was instrumental in the country's ability to drive widespread adoption of EVs (Kao 2013).

- *Solar photovoltaic cells (Bell Labs, USA)*. The first silicon solar cells were developed at the US-based Bell Labs in 1954, but their widespread adoption was driven by the US National Aeronautics and Space Administration's (NASA's) procurement of solar panels for satellites and government research funding, which led to improved efficiency and cost reductions (Solar Energy World 2011).

- *Wind research programs (Denmark, 1970s–1990s)*. The Danish government played a pivotal role in early wind energy innovation by funding research programs and testing facilities and developing prototypes in the 1970s–1990s. State-backed initiatives led to breakthroughs in horizontal-axis wind turbine designs, improved blade aerodynamics, and increased energy efficiency. These innovations laid the foundation for Denmark's leadership in the wind industry and influenced modern turbine technology worldwide (Karnøe 1990).

These examples highlight how government policies have accelerated the adoption of green innovations by reducing market uncertainties and ensuring that sustainable technologies receive the necessary support to compete with entrenched, carbon-intensive industries.

Policy enablers to prioritize linkages

Effective policy making requires a holistic approach that considers the nexus of interactions across sectors, geographies, time, and communities. Environmental, social, and economic systems are deeply interconnected—so, actions in one area often have ripple effects elsewhere. Policies must therefore integrate multiple disciplines, address spatial challenges that can often extend beyond national borders, anticipate long-term consequences by accounting for feedback loops and future uncertainties, and address harmful impacts on vulnerable communities. A complementary package of policies can strengthen policy outcomes and address unintended adverse impacts.

Linkages across sectors and disciplines

Policies must break out of silos and consider how different sectors interact to create both systemic challenges and opportunities. Climate change, pollution, and resource depletion are not isolated issues. They are shaped by interlinked systems like energy, agriculture, transportation, trade, and finance. Effective solutions require cross-sectoral coordination to avoid unintended harm and instead build synergies that drive sustainability and economic resilience.

Green water illustrates this interconnectedness. As described in chapter 2, nearly half of global rainfall stems from terrestrial moisture recycling—evapotranspiration from vegetation. Deforestation disrupts this cycle, affecting water availability for downwind and downstream sectors like agriculture. Likewise, trophic cascades (chapter 3) shows how the loss of a single species, like vultures or sparrows, can trigger public health and food security crises, costing tens of billions of dollars in damages. Chapter 4's discussion of the nitrogen cascade reveals how fertilizer subsidies, which are meant to boost yields, can degrade soil, pollute air and water, and damage fisheries through algal blooms.

Such cross-sector trade-offs are everywhere. Transitioning to clean energy improves air quality but may displace fossil fuel workers. The rapid rise of artificial intelligence, while promising efficiency gains, demands vast amounts of energy and water (refer to box 8.3). Managing these trade-offs requires integrated policies that minimize harm and amplify co-benefits. Reinvesting fossil fuel subsidies into renewable energy jobs programs, redesigning infrastructure for resource efficiency, or aligning land-use policies with conservation goals can help balance economic, social, and environmental objectives, ensuring that long-term sustainability is not sacrificed for short-term goals.

The transition to cleaner energy will require a host of so-called "transition minerals," which also has the potential to lead to trade-offs. Some of these minerals have been found in places with elevated risks in regions struggling with poverty, conflict, and weak regulatory oversight. Estimates have suggested that deforestation rates are higher where these minerals are extracted, compared to conventional mining sites (refer to Katovich and Rexer (2025) and spotlight 6 in this report). Nevertheless, at this stage, the overall material burden of transition minerals remains relatively low, and good governance can help mitigate localized harm from transition minerals.

BOX 8.3
The thirsty cloud: The AI boom and its growing appetite for water

The rapid rise of artificial intelligence (AI) poses a new and emerging challenge for the sustainable use of water. Generative AI systems, like OpenAI's ChatGPT, rely on energy-intensive data centers that also consume vast amounts of freshwater for cooling and generating electricity offsite. The irony is that while AI holds immense potential to shrink humanity's environmental footprint—by improving climate models, tracking endangered species, or even finding ways to cut water use—it is also leaving a significant environmental footprint of its own. The International Energy Agency projected that data

(continued)

BOX 8.3

The thirsty cloud: The AI boom and its growing appetite for water *(continued)*

center electricity demand would double between 2022 and 2026, reaching the equivalent of Japan's total annual consumption (IEA 2024). Water demand for cooling is following a similar trajectory: in 2022 alone, Google's water use rose by 20 percent, and Microsoft's rose by 34 percent (Crawford 2024).

Training and running large language models like Generative Pre-trained Transformer 3 (GPT-3) can use millions of liters of water. One estimate suggests that Just 10–50 queries can consume up to 500 milliliters, and GPT-4 and future generations will likely consume even more (Li et al. 2023). Total AI-related water demand may reach 6.6 trillion liters by 2027—six times Denmark's annual water withdrawals (Li et al. 2023).

The concern is not just the amount of water, but where the water is used. Many data centers are located in regions already facing water scarcity. A new geospatial analysis for this report used globally available subnational data sets on data centers and water scarcity indicators. More than 6,000 data centers across Asia, Europe, North America, and South America were extracted and geo-coded using OpenStreetMap. These data were aggregated at the administrative level and combined with a geographical map of water scarcity (Mekonnen and Hoekstra 2016) to determine the share of data centers subject to water scarcity.

Strikingly, the results in figure B8.3.1 reveal that more than a third of the sampled data centers are in water-scarce regions where the blue-water-scarcity index is greater than 1 (meaning net water withdrawals exceed available blue water). This finding underscores the need to balance the scaling of AI with sustainable water management, particularly as tensions over competing demands intensify.

FIGURE B8.3.1 **Number of data centers, by water scarcity quartile**

Number of data centers

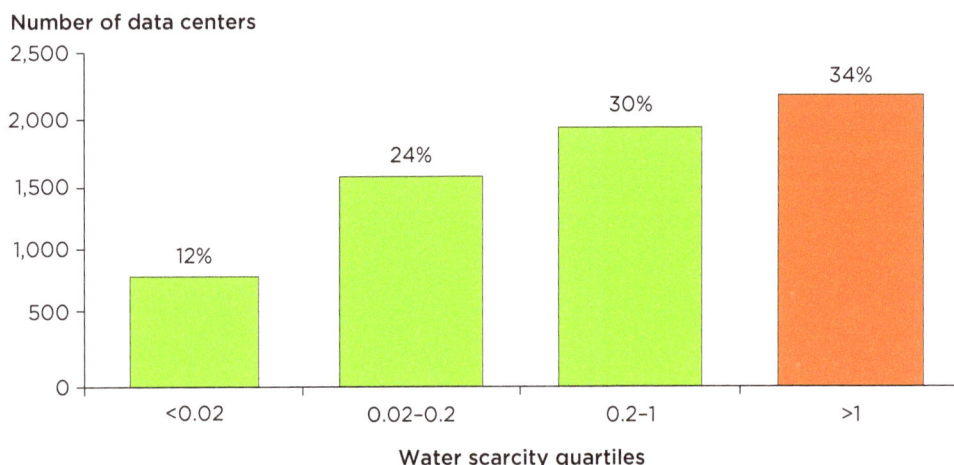

Water scarcity quartiles

Sources: Original calculations using lists of data centers web-scraped and extracted from CloudScene, a market intelligence firm, and geolocated to level 2 administrative boundaries using OpenStreetMap. *Note:* Regions with data centers are grouped into water-scarcity quartiles using the blue-water-scarcity index as defined by Mekonnen and Hoekstra (2016).

(continued)

Synergies across sectors can amplify outcomes, making integrated policies far more effective than isolated interventions. Agricultural subsidy reform, for example, is more successful at reducing land conversion when paired with extension services that promote sustainable practices (WWF 2021). Emissions pricing mechanisms—such as pollution taxes or trading schemes—also deliver greater reductions when combined with renewable energy subsidies and energy efficiency standards (Stechemesser et al. 2024). Policy mixes that align taxes with fossil fuel subsidy reform or building code changes consistently outperform stand-alone measures (Stechemesser et al. 2024). These examples highlight the need for collaboration across ministries to design holistic policies that *create reinforcing benefits and manage trade-offs*. Without coordination, policies risk being fragmented or counterproductive—such as promoting renewables while subsidizing fossil fuels, or building hydropower in locations that degrade natural habitats and destroy wildlife. A whole-of-government approach would ensure that reforms are more impactful, equitable, and durable over time.

Linkages across geographies

Many environmental problems transcend political boundaries, making cooperation essential for effective policy making. Air and water pollution, deforestation, and biodiversity loss often originate in one jurisdiction but cause harm in another, whether across city lines, national borders, or entire continents. Similarly, rivers and watersheds span multiple jurisdictions, meaning that upstream activities—such as agriculture, mining, and wastewater discharge—can degrade water quality for downstream users, or lead to reductions in terrestrial moisture recycling downwind (chapter 2). Without coordinated action, policies implemented at one level may be undermined by inaction elsewhere, leading to policy leakage, resource misallocation, and environmental injustices.

Linkages across time

Environmental policy design requires long-term vision that accounts for the evolution of systems over time. The consequences of today's policy decisions often unfold over decades, meaning that short-sighted approaches can lead to unintended consequences or missed opportunities. Feedback loops can either reinforce or undermine policy goals. For example, a pollution tax may curb emissions and drives innovation in cleaner technologies, while efficiency improvements can sometimes backfire by increasing overall consumption. Policy makers must therefore anticipate both effects to maximize benefits and mitigate unintended consequences.

Avoiding lock-ins is equally important, as investments in infrastructure and regulations can entrench unsustainable pathways that are costly to reverse. To prevent this, governments should prioritize no-regret policies, ensuring flexibility under various future scenarios. At the same time, policies must be dynamically adaptable, evolving with scientific advances and technological shifts. Regular reassessments—such as updating carbon pricing, refining regulatory standards, or leveraging real-time environmental monitoring—help ensure effectiveness over time.

Linkages across communities

Policies must acknowledge that environmental damages may disproportionately impact some communities and may exacerbate income inequalities. Poorer households may take nearly a decade longer to recover from extreme weather, while pollution, deforestation, and land degradation further undermine their livelihoods (Damania et al., 2025). Poorer households are also more likely to live in areas that experience deforestation and land degradation, which reduce agricultural productivity and heighten food insecurity. Addressing these disparities requires inclusive environmental policies that integrate social protection, critical infrastructure, and targeted investments to prevent further marginalization (box 8.4). Challenges such as elite capture and deeply rooted social norms can impede the effectiveness of representation.

BOX 8.4

Community-led models for service delivery

Various models of service delivery influence how vulnerable communities access essential environmental services. Centralized models, where national governments provide infrastructure and services, can lead to inefficiencies, especially in remote or underserved areas. Decentralized models grant more authority to local governments, potentially improving service responsiveness but sometimes exacerbating inequalities if local governance is weak. Community-driven development approaches have emerged as an alternative, empowering local groups to manage resources and infrastructure.

Several countries, including India and Thailand, have implemented a community-based approach for mangrove management. In India, joint forest management programs have been established in which village institutions collaborate with government forest departments to manage mangrove ecosystems, resulting in effective conservation and enhanced community welfare (Singh et al. 2010). The success of both initiatives has depended on strong institutional support, capacity building, and safeguards against elite capture to ensure that marginalized communities truly benefit from decision-making power. In Thailand, programs initiated in the 1980s have successfully maintained more than 2,000 square kilometers of mangrove cover (Mohammed 2004), with local communities involved in managing and rehabilitating mangrove forests, decentralizing authority from the government to the community level.

Policy evaluation

Evaluation is essential for adaptive policy in a changing world. In an era of environmental uncertainty, geopolitical shifts, and rapid technological change, fixed policy designs are a liability. What works today may falter tomorrow. Independent evaluation provides the evidence and feedback needed to recalibrate programs as contexts evolve. Through real-time learning, portfolio reviews, and structured feedback loops (formal processes that collect, analyze, and apply information to adjust policies over time), evaluation helps identify what is working, for whom, and under what conditions. This is not about assigning blame—it is about improving results. In low-capacity settings, for instance, evaluations have helped redirect overly complex policy instruments toward more feasible, community-based solutions. In high-capacity contexts, evaluations have illuminated where regulatory layering delivers diminishing returns. Without structured learning and course correction, even the best-intentioned reforms risk stagnation or drift. Evaluation is not a luxury. It is a form of institutional memory that makes policy smarter, more resilient, and more accountable over time.

Evaluation also ensures that success is recognized, scaled, and sustained. Too often, promising results remain trapped in pilot projects or isolated reforms. Evaluation helps surface what works and why, so that governments can scale up effective interventions with confidence. This is particularly vital for environmental policy, where outcomes often take years to materialize and causal chains are complex. Rigorous impact evaluations, process reviews, and comparative studies provide the analytical backbone for informed scaling decisions. When evaluation is embedded in the policy cycle—not as an afterthought but as an enabler—success is not just celebrated, it is institutionalized. That is critical for moving from experiments to systems change.

Policy instrument choice under uncertainty

Selecting the most effective policy instrument is crucial to environmental policy success. Pollution taxes and cap-and-trade systems have strengths and trade-offs. Taxes offer a predictable price signal, but they do not guarantee emissions reductions, as polluters may choose to pay rather than cut emissions. In contrast, cap-and-trade systems enforce hard limits and create financial incentives for firms to reduce emissions at lower costs than their rivals, making them well-suited for pollutants with clear ecological thresholds. However, cap-and-trade systems can suffer from price volatility and speculative distortions, which may deter long-term investment (Goulder and Schein 2013).

In practice, the choice depends on the environmental goal and policy context. For gradual reductions, such as cutting single-use plastics, taxes may be more effective, while hard caps are preferable for critical thresholds like carbon or freshwater. Regardless of the tool, the precautionary principle remains essential: when activities pose a high risk to human or environmental health, action should be taken even if full scientific consensus is not yet available (Kriebel et al. 2001). The Montreal Protocol illustrated this—by phasing out ozone-depleting substances before the exact rate of ozone loss was fully understood, it prevented far worse damage (Green 2009). Applying this principle today to emerging risks like microplastics, synthetic biology, or deep-sea mining could prevent future crises.

Tailoring policy reforms to capacity

Countries need to tailor reforms to fit their institutional contexts. In low-capacity settings, several priorities emerge: (1) keep reforms simple and incentive-compatible, (2) frame policies around development co-benefits, (3) empower local institutions, (4) leverage technologies that leapfrog enforcement and infrastructure gaps, and (5) sequence reforms to build trust and institutional capacity. In high-capacity countries, other complementary levers are feasible: (1) layering complex policy instruments over time, (2) supporting breakthrough innovations to overcome the "Valley of Death," (3) enabling global technology diffusion, and (4) unlocking new markets.

The following subsections explore how these reforms can be practically applied to create momentum toward a more livable planet.

Reform strategies for low-capacity settings

In low-capacity settings, policies must be simple, incentive-compatible, and adaptive. Where administrative systems are weak and formal enforcement is limited, success hinges on designing interventions that are feasible to implement and politically sustainable. For instance, in Bangladesh, a leaf color chart program optimized nitrogen fertilizer use and increased yields through a simple visual guide requiring no laboratories or formal regulation (Islam and Beg 2021). This led to a reduction in nitrogen application and increases in yields, without the need for administratively heavy regulations.

Framing reforms around co-benefits—such as jobs, health, and livelihoods—broadens support. Policies that align with local development priorities enjoy greater public buy-in and political traction. In Thailand, the community-led mangrove restoration initiative in Pred Nai village succeeded not because of top-down regulation, but because it supported fisheries, protected coastlines, and generated income through sustainable resource use and micro-credit schemes (Nature-based Solutions Initiative, n.d.). Linking environmental goals to tangible, near-term benefits makes reform more resilient in politically constrained settings.

Local institutions and communities can fill enforcement gaps and build legitimacy. Where state capacity is weak, devolving authority to trusted actors can enhance compliance and reduce transaction costs. In rural Nepal, community forest user groups manage large swaths of public forest, improving access for households while supporting environmental recovery. Research has shown that proximity to community-managed forests increases rural land values, reflecting both their economic and ecological value. Such forests are often better maintained, with greater biomass and regeneration, than state-managed alternatives (Nepal, Nepal, and Berrens 2017). When communities have a direct stake in natural resources, conservation outcomes can often be achieved through social trust and participation, even in the absence of strong bureaucratic enforcement.

Technology can help leapfrog infrastructure gaps in low-capacity settings. In the Democratic Republic of Congo, communities use ForestLink to report illegal logging via smartphones, improving oversight where formal enforcement is weak (Rainforest Foundation UK 2021). In Pakistan, biometric ID systems linked to social registries have enabled rapid emergency cash transfers during devastating floods—demonstrating how digital tools can strengthen resilience when traditional infrastructure fails (refer to spotlight 5). Furthermore, innovations can automate enforcement, improve transparency, and reduce corruption—through, for example, smart contracts or blockchain (Damania et al. 2019).

Reforms should be designed to layer and evolve over time. In low-capacity settings, this means initiating reform with politically feasible, low-cost interventions that build trust and institutional capability, laying the groundwork for more ambitious policies. For instance, the Arab Republic of Egypt's energy subsidy reform began with gradual price adjustments and public communication strategies (World Bank 2021), which helped ease fiscal pressures and encouraged greater private investment in clean energy, with solar and wind generation growing almost threefold between 2014 and 2019 (Hallegatte et al. 2024). Sequencing allowed for initial "quick wins" that bolstered public confidence and institutional learning, enabling more complex reforms over time.

Unlocking ambition in high-capacity settings

In high-capacity settings, incremental, long-term investments enable deeper and more durable reform. When governments have strong institutions, regulatory coherence, and technical expertise, a wider set of instruments becomes feasible. Effective strategies often combine taxes, standards, labeling, and pricing mechanisms over time. The US sulfur dioxide cap-and-trade program began with a narrow focus on large emitters and expanded in scope and ambition as monitoring and trading infrastructure developed (Schmalensee and Stavins 2013). Similarly, the European Union has systematically built a comprehensive climate policy framework that blends carbon pricing through the Emissions Trading System, regulatory mandates under the "Fit for 55" package, and finance mechanisms such as EU-issued green bonds (European Commission 2021; IEA 2023).

Public investment and policy support are essential to overcoming the "Valley of Death" in clean innovation. Many breakthrough environmental technologies fail not because they lack promise, but because they cannot bridge the gap between early-stage development and commercial viability. Entrenched incumbents and infrastructure lock-in often prevent cleaner alternatives from gaining traction. Strategic public support—such as research and development (R&D) funding, demonstration programs, and procurement commitments—can de-risk early-stage technologies and attract private investment. For instance, the first silicon solar cells were developed at Bell Labs in 1954, but it was NASA's procurement for space missions and sustained government R&D funding that enabled their eventual cost reductions and commercial viability (Solar Energy World 2011). These cases illustrate that targeted public intervention can transform promising technologies into scalable solutions. The challenge lies in avoiding the known pitfalls of rent-seeking for protectionist policies that result in support for unviable initiatives.

High-capacity countries can unlock new markets, sharpen their competitive edge, and deliver better, more affordable products with lower environmental footprints. History has shown how this process can unfold: the leap from the horse and buggy to internal combustion engines revolutionized transport, and today the shift from gasoline-powered vehicles to electric ones is similarly reshaping mobility.

Early public investments played a decisive role in scaling EV technologies, driving down costs while raising performance standards. For instance, in the United States, a $465 million government loan enabled a pioneering EV manufacturer to scale production, catalyzing industrywide transformation in battery range, design, and autonomous driving capabilities (Kao 2013). These cleaner, more efficient vehicles are no longer niche alternatives—they are setting new benchmarks in performance, safety, and energy use.

Technology diffusion from high- to low-capacity countries is essential for global progress. High-capacity countries can support this process through concessional finance, technical assistance, and open-source innovations. For example, solar mini-grids in Sub-Saharan Africa have scaled with donor-backed pilots and technologies adapted from advanced economies (ESMAP 2022). To be effective, diffusion must be paired with local capacity building and institutional support to ensure sustainable outcomes.

Matching policy integration with financial innovation

Realizing the potential of environmental assets for long-term growth and resilience will require better-aligned financing instruments. A variety of financial mechanisms—ranging from green bonds and sustainability-linked loans to payments for ecosystem services—are emerging to mobilize capital toward nature (box 8.5). However, public finance and donor flows still dominate the landscape, underscoring the need to crowd in private sector investment. Encouragingly, the number of financial instruments tied to nature has grown rapidly over the past decade, and innovative vehicles like debt-for-nature swaps and resilience bonds have increasingly been deployed to meet both environmental and development goals. Nevertheless, uptake has remained limited, particularly in low- and middle-income countries, where capacity constraints, lack of creditworthiness, and regulatory gaps impede access to these tools.

BOX 8.5
Instruments for financing environmental assets

A livable planet requires large-scale investment in environmental sustainability—investment that is often difficult to mobilize, especially where national budgets are constrained or the global benefits of environmental assets outweigh local incentives. In response, a growing ecosystem of financing instruments has emerged to bridge this gap, aligning the interests of countries, investors, and the global community. These instruments fall into several categories:

Green bonds. These fixed-income instruments are used to raise funds for environmentally beneficial projects, such as renewable energy, energy efficiency, and sustainable infrastructure.

(continued)

BOX 8.5
Instruments for financing environmental assets *(continued)*

Sustainability bonds. Similar to green bonds but broader in scope, these support both environmental and social objectives—for example, combining clean energy investments with affordable housing initiatives.

Sustainability-linked bonds. These link borrowing costs to performance on environmental or social metrics. For instance, firms may receive lower interest rates if they meet specific emissions reduction targets.

Environmental impact bonds. These are a form of outcome-based financing where returns are tied to project success. One example is the Chesapeake Bay environmental impact bond, which funds water quality improvements.

Debt-for-nature swaps. These agreements forgive a portion of a country's debt in exchange for commitments to conservation investments.

Payment for environmental services. These incentive schemes pay landowners or local stewards to maintain ecosystem services like carbon storage or water purification.

Wildlife bonds. Instruments such as the Rhino Bond channel capital into species protection, with investor returns contingent on achieving conservation outcomes.

Tokenization of natural assets. Digital tokens representing fractional ownership of ecosystems (for example, forests) are being piloted to attract private investment in conservation.

Alongside these global efforts, the World Bank is innovating to provide better support to countries that take on the burden of delivering global public goods:

The *Framework for Financial Incentives* (FFI) grew from the World Bank's Evolution Roadmap, which set out the World Bank's vision for ending poverty on a livable planet. It aims at helping to reshape the world's aid architecture, catalyzing innovative investment in and reforms for global public goods by middle-income countries. The FFI rewards countries that scale up national projects where the benefits extend beyond their borders. For example, a water project in one country could be expanded to support shared rivers, watersheds, and aquifers, strengthening irrigation and agricultural outputs along with social and environmental gains in neighboring countries. The FFI combines the powerful leverage of the World Bank's balance sheet with targeted concessionality, offering maximum impact for the lowest-cost investment. All countries that borrow from the International Bank for Reconstruction and Development (IBRD) are eligible for FFI incentives, provided their projects address global challenges with cross-border effects. For approved projects, IBRD incentives will increase the volume (amount) of lending, extend the tenor (the time required for repayment) of a loan, and/or discount the price of lending.

The World Bank is also expanding concessional lending to low-income countries through the International Development Association (IDA). IDA's well-established Regional Window is becoming the Global and Regional Opportunities Window (GROW). Over the next three years, GROW will make up to $13.5 billion available for country allocations on IDA's concessional terms, funding operations that address global challenges and strengthen regional integration. Somalia is an initial

(continued)

recipient of GROW funding through its integration into the Horn of Africa digital network. The design of the FFI draws on more than two decades of experience gained from the Regional Window, which has road tested innovations that solve national development problems while producing positive cross-border effects. Demand for such initiatives has been consistent, and the window has grown throughout each IDA replenishment since it was introduced in 2003.

Together, these instruments reflect a growing commitment to mobilize capital for the global good—ensuring that countries acting in service of the planet are supported, and that private and public capital flows toward a more sustainable future.

To address these barriers, countries must take a systems approach to investment planning. This includes building track records of nature-positive investments to lower perceived risk, integrate environmental assets into national accounts to clarify their economic value, and strengthen institutions that can manage blended finance platforms. Importantly, financial reform should not merely aim to mobilize more money, but also to steer capital toward outcomes that matter most for long-term environmental health and inclusive growth. These financing solutions form an essential complement— ensuring that ambition is matched with the necessary resources and instruments to drive results.

Conclusion: Building systems for effective environmental policy

Bold, evidence-based policies that recognize environmental interconnections, leverage reliable information, and integrate technological advancements are essential for securing a livable planet. Given the complexity of environmental challenges, isolated interventions—however well-intentioned—risk unintended consequences unless they account for cross-sectoral, geographic, and temporal linkages. Policies must also recognize the critical role of information in shaping both public awareness and demand for action, and in enabling governments to make informed decisions. Real-time environmental monitoring, transparent and accessible data, and countering misinformation are powerful tools for accountability and effective decision-making. China's air quality improvements, for example, were not achieved through information disclosure alone. Success also depended on strict enforcement protocols, financial incentives for clean energy adoption, and coordinated public campaigns to shift households' behavior. These efforts were embedded in a strong governance framework with clear mandates and capacity across levels of government.

The example of China's air quality improvements highlights the importance of building institutional foundations that allow environmental interventions to succeed. While each country's context differs, common enablers include (1) reliable, actionable data systems; (2) capable bureaucracies with enforcement power; and (3) channels for public participation and accountability. Embedding these from the start, rather than as afterthoughts, can dramatically increase the odds of success.

Spotlight 4 explores policy successes and failures, contrasting cases that embraced systems thinking with those that did not. Spotlight 5 examines how technologies—from satellite monitoring to AI-powered analytics—can improve decisions and mobilize action. Finally, annex 8A presents examples of key interventions. Together, these examples show how policies can remain effective in evolving environmental and technological landscapes.

Annex 8A. Examples of key interventions

Cross-cutting solutions

- **Information:** Build environmental monitoring systems, invest in real-time monitoring and data access, and mandate public reporting.
- **Enabling policies:** Align policies to prevent pollution swapping and environmental leakage—"problem shifting" across sectors or locations. Repurpose harmful subsidies.
- **Implement evaluation:** Regular program evaluations to strengthen transparency and accountability.

Green water management

Investment	Policy reform
Carry out landscape-scale forest restoration and natural regeneration to enhance soil moisture, groundwater recharge, and evapotranspiration—while delivering low-cost co-benefits for biodiversity, water security, and resilience.	*Prioritize native and mixed-species forests* over monoculture plantations in afforestation policies to enhance drought buffering, rainfall recycling, and soil moisture retention.
Provide finance for the adoption of regenerative agricultural practices—such as no-till farming, mulching, pitting, contouring, terracing, cover cropping, and crop rotation—to retain soil moisture, reduce erosion, improve infiltration, and enhance soil health.	*Ensure that land use planning and management policies* recognize green water functions and services.
Strengthen soil monitoring systems—combine in-situ sensors, remote sensing, and digital platforms to assess key indicators like soil moisture, organic matter, and infiltration rates, and link to tools such as soil health cards and mobile advisory apps to deliver location-specific guidance.	*Encourage payments for ecosystem services schemes* that support green water by promoting land uses and vegetation cover that retain water in soils and promote infiltration.

Integrated blue and green water management

Investment	Policy reform
Modernize irrigation to reduce waste and increase efficiency.	*Incorporate green water sources* into blue water management and planning.
	Design policies to reduce water pollution, including command-and-control or economic instruments.

Nitrogen management

Investment	Policy reform
Promote precision agriculture using the 4Rs of nutrient stewardship—applying the right nutrients at the right rate, right time, and right place using both digital tools (for example, satellite data and mobile advisories) and low-cost methods (for example, leaf color charts and chlorophyll meters) to tailor nitrogen use, improve yields, and reduce waste.	*Repurpose subsidies* from generic fertilizer application toward specific measures for improving soil health and efficiency-enhancing inputs.
Use farmer training and extension to support adoption of efficient nitrogen practices, particularly in smallholder systems.	*Apply whole-farm nitrogen caps (for example, as in Denmark) and nutrient trading schemes (for example, like for the Chesapeake Bay)* where there is excess usage.
Carry out wetland restoration to remove both current and legacy nitrogen—residual nitrogen in soil and groundwater that continues to leach into waterways—while providing co-benefits like carbon storage and biodiversity. Effectiveness depends on strategic design, targeting, and scaling.	*Prioritize investments in fertilizer access,* soil testing, and balanced nutrient application to close yield gaps where there is low usage.

Biodiversity

Investment	Policy reform
Expand collaborative management partnerships between state and non-state actors, particularly in regions with high human pressure.	*Facilitate the use of financial instruments* that encourage biodiversity conservation such as conservation tenders/auctions, credits, and sustainability-linked bonds.

Air quality—reducing particulate matter

Investment	Policy reform
Provide finance and support for in-situ crop residue management technologies (for example, mulchers, microbial decomposers, and advanced seeders).	*Establish air quality standards*, emissions targets, and legal frameworks.
Support clean cooking and heating with investments in cleaner fuels, efficient appliances, and supplier guarantees.	*Ban open burning of crop residues,* and align national agriculture budgets to incentivize low-emission practices.
Invest in mass transport infrastructure.	*Establish efficiency and quality standards for fuels and stoves*, and provide targeted subsidies and behavior change programs.
Enforce high-occupancy vehicle standards like carpool lanes.	Encourage adoption of more *modern fuel and emission standards.*

Water quality—reducing lead contamination	
Investment	**Policy reform**
Replace lead-containing pipes, fixtures, and plumbing with certified lead-free alternatives as part of water system upgrades.	*Set regulatory limits* for lead in drinking water aligned with global standards, and require inventories of lead service lines.
Deploy water treatment technologies to enhance corrosion control and prevent lead leaching, and finance construction of water quality laboratories.	*Mandate phased removal* of leaded infrastructure with time-bound targets and financial support for utilities and households.
Strengthen local capacity through training on lead detection and prevention, and conduct public awareness campaigns to promote safer water and consumer practices.	*Develop procurement policies* that prioritize certified lead-free technologies and components in water systems.

References

Acemoglu, D., P. Aghion, L. Barrage, and D. Hémous. 2023. "Green Innovation and the Transition toward a Clean Economy." Working Paper 23-14, Peterson Institute for International Economics, Washington, DC.

Acemoglu, D., A. Ozdaglar, and J. Siderius. 2024. "A Model of Online Misinformation." *Review of Economic Studies* 91 (6): 3117–50.

Berreby, D. 2024. "As Use of A.I. Soars, So Does the Energy and Water It Requires." Yale E360, February 2, 2024. https://e360.yale.edu/features/artificial-intelligence-climate-energy-emissions.

Crawford, K. 2024. "Generative AI's Environmental Costs Are Soaring—And Mostly Secret." *Nature* 626 (8000): 693.

Damania, R., S. Desbureaux, A-S. Rodella, J. Russ, and E. Zaveri. 2019. *Quality Unknown: The Invisible Water Crisis.* Washington, DC: World Bank. http://hdl.handle.net/10986/32245 License: CC BY 3.0 IGO.

Damania, R., E. Ebadi, K. Mayr, J. Rentschler, J. Russ, and E. Zaveri. 2025. *Nature's Paradox: Stepping Stone or Millstone?* Washington, DC: World Bank. doi:10.1596/978-1-4648-2164-6.

Dercon, S. 2024. "The Political Economy of Economic Policy Advice." *Journal of African Economies* 33 (2): ii26–ii38. doi:10.1093/jae/ejae027.

ESMAP (Energy Sector Management Assistance Program). 2022. "Mini Grids for Half a Billion People: Market Outlook and Handbook for Decision Makers." World Bank, Washington, DC. http://hdl.handle.net/10986/38082.

European Commission. 2021. *Fit for 55: Delivering on the Proposals.* Brussels: European Commission. https://commission.europa.eu/strategy-and-policy/priorities-2019-2024/european-green-deal/delivering-european-green-deal/fit-55-delivering-proposals_en.

Ford, A. T., A. C. Singer, P. Hammond, and J. Woodward. 2025. "Water Industry Strategies to Manufacture Doubt and Deflect Blame for Sewage Pollution in England." *Nature Water* 3: 231–43.

Forseth, I. N., and A. F. Innis. 2004. "Kudzu (Pueraria Montana): History, Physiology, and Ecology Combine to Make a Major Ecosystem Threat." *Critical Reviews in Plant Sciences* 23 (5): 401–13.

Goulder, L. H., and I. W. Parry. 2008. "Instrument Choice in Environmental Policy." *Review of Environmental Economics and Policy* 2 (2): 152–74.

Goulder, L. H., and A. R. Schein. 2013. "Carbon Taxes versus Cap and Trade: A Critical Review." *Climate Change Economics* 4 (3): 1350010.

Green, B. A. 2009. "Lessons from the Montreal Protocol: Guidance for the Next International Climate Change Agreement." *Environmental Law* 39 (1): 253–83.

Greenstone, M., G. He, R. Jia, and T. Liu. 2022. "Can Technology Solve the Principal-Agent Problem? Evidence from China's War on Air Pollution." *American Economic Review: Insights* 4: 54–70.

Hallegatte, S., C. Godinho, J. Rentschler, et al. 2024. *Within Reach: Navigating the Political Economy of Decarbonization.* Climate Change and Development Series. Washington, DC: World Bank. doi:10.1596/978-1-4648-1953-7.

Hodgson, C. 2024. "US Tech Groups' Water Consumption Soars in 'Data Centre Alley.'" *Financial Times*, August 18. https://www.ft.com/content/1d468bd2-6712-4cdd-ac71-21e0ace2d048.

IEA (International Energy Agency). 2023. "Now Is the Time to Climate-Proof Europe's Economy." IEA, Paris. https://www.iea.org/commentaries/now-is-the-time-to-climate-proof-europes -economy.

IEA (International Energy Agency). 2024. *Electricity 2024: Analysis and Forecast to 2026.* Paris: IEA. https://iea.blob.core.windows.net/assets/6b2fd954-2017-408e-bf08-952fdd62118a/Electricity 2024-Analysisandforecastto2026.pdf.

Islam, M., and S. Beg. 2021. "Rule-of-Thumb Instructions to Improve Fertilizer Management: Experimental Evidence from Bangladesh." *Economic Development and Cultural Change* 70 (1): 237–81.

Kao, H. 2013. "Beyond Solyndra: Examining the Department of Energy's Loan Guarantee Program." *William & Mary Environmental Law and Policy Review* 37 (2): 425.

Karnøe, P. 1990. "Technological Innovation and Industrial Organization in the Danish Wind Industry." *Entrepreneurship & Regional Development* 2 (2): 105–24. doi:10.1080/08985629000000008.

Kauffman, R. L. 2024. "The Valley of Death and the Business of Asset Management: Bridging the Funding Gap for Climate Technology Innovation." Working Paper, Stanford Sustainable Finance Initiative, Stanford University. https://sfi.stanford.edu/publications/catalyzing-private -investment/valley-death-and-business-asset-management.

Kriebel, D., J. Tickner, P. Epstein, et al. 2001. "The Precautionary Principle in Environmental Science." *Environmental Health Perspectives* 109 (9): 871–76.

Katovich, E. and J. Rexer. 2025. "Critical Mining Contributes to Economic Growth and Forest Loss in High-Corruption Settings." Available at https://ssrn.com/abstract=5291760. Background paper for this report.

Li, P., J. Yang, M. A. Islam, and S. Ren. 2023. "Making AI Less 'Thirsty': Uncovering and Addressing the Secret Water Footprint of AI Models." arXiv preprint arXiv:2304.03271.

Mekonnen, M. M., and A. Y. Hoekstra. 2016. "Four Billion People Facing Severe Water Scarcity." *Science Advances* 2 (2): e1500323.

Mohammed, S. M. 2004. "Saving the Commons: Community Involvement in the Management of Mangrove and Fisheries Resources of Chwaka Bay, Zanzibar." *Western Indian Ocean Journal of Marine Science* 3 (2): 221–26.

Nature-based Solutions Initiative. n.d. "Community-Led Mangrove Restoration and Sustainable Fishing." Nature-based Solutions Initiative, Oxford, UK. https://casestudies.naturebasedsolutions initiative.org/casestudy/connecting-the-dots-biodiversity-adaptation-food-security-and -livelihoods-mangrove-restoration-by-the-pred-nai-community-in-thailand.

Nepal, M., A. K. Nepal, and R. P. Berrens. 2017. "Where Gathering Firewood Matters: Proximity and Forest Management Effects in Hedonic Pricing Models for Rural Nepal." *Journal of Forest Economics* 27: 28–37.

Parkin, B., and C. Hodgson. 2024. "India Pulls in Tech Giants for Its AI Ambitions." *Financial Times*, June 17. https://www.ft.com/content/414e912f-c50c-4bc8-b3a2-b9ac36c34ebb.

Rainforest Foundation UK. 2021. "DRC: Community-Based Real Time Monitoring." ForestLink Country Briefing. Rainforest Foundation UK, London. https://forestlink.org/wp-content /uploads/2021/11/FAT8736_Country_Briefing_ENG_DRC_V5_EMAIL.pdf.

Rodrik, D. 2008. "Second-Best Institutions." *American Economic Review* 98 (2): 100–104.

Schmalensee, R., and R. N. Stavins. 2013. "The SO_2 Allowance Trading System: The Ironic History of a Grand Policy Experiment." *Journal of Economic Perspectives* 27 (1): 103–22.

Singh, A., P. Bhattacharya, P. Vyas, and S. Roy. 2010. "Contribution of NTFPs in the Livelihood of Mangrove Forest Dwellers of Sundarban." *Journal of Human Ecology* 29 (3): 191–200.

Solar Energy World. 2011. "Solar History: Bell Labs and the First Modern Silicon Solar Cell." Solar Energy World, August 9. https://www.solarenergyworld.com/solar-history-bell-labs-and-the -first-modern-silicon-solar-cell/.

Stechemesser, A., N. Koch, E. Mark, et al. 2024. "Climate Policies That Achieved Major Emission Reductions: Global Evidence from Two Decades." *Science* 385 (6711): 884–92.

World Bank. 2021. "Developing Human Capital in Egypt through Energy Subsidy Reforms: A Case Study." World Bank, Washington, DC. https://openknowledge.worldbank.org/server/api/core /bitstreams/684a9603-758d-5113-ad5c-df4507114f29/content.

WWF (World Wildlife Fund for Nature). 2021. "Turning Harm into Opportunity: A Blueprint for Repurposing Agricultural Subsidies to Transform Food Systems." WWF International, Gland, Switzerland. https://wwfint.awsassets.panda.org/downloads/wwf-turning-harm-into -opportunity---full-final.pdf.

Policy Case Studies: Winning Moves and Costly Mistakes

Introduction

This spotlight provides brief examples of policy successes and challenges in addressing environmental damage and summarizes lessons that can be learned from them. The examples included here succeeded because they took a systems approach or failed because they did not. Additional case studies involving World Bank operations are provided in spotlight 5, on digital solutions.

Policy successes

Integrating land management with economic development. China's Loess Plateau rehabilitation is a landmark example of ecological restoration and poverty alleviation. Its success stemmed from a systems approach that prioritized long-term environmental interactions and cross-sectoral linkages. Decades of unsustainable farming and overgrazing had caused severe soil erosion, devastating agricultural productivity and increasing sedimentation in the Yellow River. Recognizing that isolated interventions would fail, the project combined land-use zoning, terracing, afforestation, and grazing bans to restore ecological balance while improving livelihoods. Vegetation cover doubled (from 17 to 34 percent), sediment runoff declined, and per capita annual earnings rose from $70 to $200 (World Bank 2007). By addressing the interconnected challenges of land degradation, poverty, and water security, the project became a model for large-scale ecosystem restoration that aligns environmental and economic priorities.

Enhancing biodiversity and tourism through private protected area management. African Parks, a nonprofit organization, collaborates with African governments to manage protected areas, using a systems approach that recognizes links between conservation and economic development. Integrating anti-poaching measures, habitat restoration, and scientific monitoring, African Parks has boosted key wildlife populations, including elephants, lions, and bird species (Denny, Englander, and Hunnicutt 2024). These biodiversity gains have fueled growth in nature-based tourism, with visitor numbers rising in several African Parks–managed parks. Tourism revenue is reinvested in conservation, infrastructure, and park management, creating a self-sustaining model that supports long-term ecosystem health. By recognizing the connections between biodiversity protection and economic incentives, African Parks shows how well-managed conservation areas can thrive while also contributing to development.

Maximizing co-benefits through a systems approach to nitrogen management. Integrating on-farm nitrogen use efficiency with off-farm nitrogen removal strategies reduces pollution and delivers economic and environmental benefits. In China, a decade-long training program for smallholder farmers increased yields by 11 percent while reducing nitrogen fertilizer use by one-sixth, saving 1.2 million tonnes and generating an economic return of $12.2 billion (Cui et al. 2018). More recently, China has phased out fertilizer subsidies, redirecting funds toward improved nitrogen and manure management (Ji, Liu, and Shi 2020). In Bangladesh, training in using a simple leaf color chart reduced fertilizer use by 8 percent without lowering yields, saving 180,000 tonnes—worth $80 million or 14 percent of the input subsidy budget (Islam and Beg 2021). Beyond the farm, wetland restoration provides a cost-effective, nature-based solution to nitrogen pollution. In China, restoring 2.3 million hectares of small wetlands could increase nitrogen removal by 21 percent—four times more than restoring large water bodies (Shen et al. 2025). Wetland restoration could also prevent the need for 800 new wastewater treatment plants, which would cost $8 billion annually—twice the 2022 budget for water pollution control (Shen et al. 2025). Unlike treatment plants, which struggle to remove nitrogen at low concentrations, wetlands naturally filter nitrogen while also enhancing biodiversity, water retention, and carbon storage.

Securing Indigenous land tenure to enhance forest conservation and biodiversity protection. Granting land tenure rights to Indigenous communities has been highly effective in reducing deforestation and preserving biodiversity. Across Latin America, Indigenous-managed lands have experienced lower rates of forest loss than non-Indigenous areas, even under high pressure from agriculture and logging (Ding et al. 2016). When Indigenous land rights are formally recognized, these communities have strong incentives and traditional ecological knowledge to manage forests sustainably (Blackman and Veit 2018). In the Amazon basin, Indigenous groups manage more than 230 million hectares, contributing to global climate regulation and biodiversity conservation (Walker et al. 2014). These successes highlight the importance of empowering local stewards who depend on forests for their livelihoods. However, secure tenure alone is not enough. Effective governance and financial support—for example, to participate in carbon markets—are also needed to ensure that Indigenous communities can continue safeguarding these ecosystems while benefiting from conservation efforts.

Improving air quality through binding emissions reduction targets. A successful example of regional cooperation on air pollution, the Convention on Long-Range Transboundary Air Pollution was signed in 1999 and amended in 2012. It set binding emissions reduction targets for key pollutants and improved Europe's air quality. Between 1990 and 2010, sulfur dioxide emissions dropped 83 percent, and nitrogen oxides dropped 47 percent, cutting acid rain, ground-level ozone, and particulate matter (IIASA 2017).

These improvements yielded major public health benefits, including fewer pollution-related illnesses and premature deaths. The Gothenburg Protocol shows the power of international cooperation on transboundary pollution. Similarly, China's "war on pollution" not only cut fine particulate matter ($PM_{2.5}$) domestically, but it also lowered levels in neighboring countries by 9.6 micrograms per cubic meter of air, yielding $2.62 billion in annual health benefits (Heo, Ito, and Kotamarthi 2025).

Policy challenges

Local water pollution regulation can have geographic spillovers. Empirical analysis of water pollution in a country with federal environmental laws highlighted challenges in decentralized management. Examining pollution levels as rivers approached administrative borders, Lipscomb and Mobarak (2016) uncovered sharp contamination spikes due to upstream jurisdictions having fewer incentives to control pollution impacting downstream areas. The study highlighted a key drawback: when local governments control environmental policies independently, they may underinvest in pollution control due to weak cross-border accountability. These findings suggest that, without strong interjurisdictional coordination, decentralized governance can worsen externalities, underscoring the need for policies that align incentives across regions for effective pollution management. The study also highlighted that policy makers not only struggle to work across national boundaries, but there is also a need to coordinate policies across subnational boundaries.

Carbon offset programs can increase carbon emissions. Although designed to be flexible market-based tools for emissions reduction, carbon offset programs may fail due to weak verification, inaccurate baselines, and poor enforcement of additionality. Some projects, such as forest preservation efforts, have claimed credits for emissions reductions that would have happened anyway, while others have inadvertently increased emissions. For example, Chen, Ryan, and Xu (2025) found that firms in offset programs increased carbon output by 570,000 tonnes over four years, compared to the year before the offset program began. These challenges highlight the importance of accurate data and technology, as real-time satellite monitoring and dynamic baselines could improve oversight and ensure that offsets reflect actual reductions. To remain effective, offset policies must embrace adaptive standards and evolving methods as new data and technologies emerge.

Legal ivory sales can increase illegal poaching. In 2008, a one-time legal sale of ivory stockpiles to China and Japan was intended to reduce black market activity by satisfying demand through legal channels. However, this policy failed to account for linkages across sectors and geographies, leading to unintended consequences. Following the sale, there was a 66 percent increase in illegal ivory production across Africa and Asia, and a 71 percent rise in ivory smuggling out of Africa (Hsiang and Sekar 2016). The legal sale signaled increased legal ivory availability, stimulating demand and providing cover for illegal operations. This case demonstrates the importance of considering complex market dynamics and cross-border effects when designing policies aimed at curbing illegal wildlife trade.

Renewable fuel standards can have environmental trade-offs. The US Renewable Fuel Standard aimed to reduce greenhouse gas emissions by promoting biofuels, such as corn ethanol, but has had unintended environmental consequences (Lark 2023). Its emphasis on corn-based ethanol increased corn cultivation, leading to higher fertilizer application and associated nutrient runoff, which has contributed to water quality degradation and the expansion of hypoxic zones in water bodies. It has also driven land use change, with implications for carbon emissions and biodiversity. These outcomes show the importance of considering cross-sectoral linkages and natural limits when designing policies to ensure that initiatives in one area do not inadvertently cause harm in another.

Large-scale tree-planting programs can have limited impact on forest cover. Although ambitious tree-planting initiatives have been widely implemented, decades of afforestation efforts in South Asia had minimal impact on increasing forest canopy cover and only modestly influenced rural livelihoods (Coleman et al. 2021). Poor species selection led to low survival rates, with non-native or commercially valuable species prioritized over native trees that were better suited to the local environment. Plantation management failures, such as inadequate long-term maintenance and protection from grazing, also resulted in high tree mortality. Local ecological conditions were overlooked, with some areas with unsuitable soil and water availability targeted for planting. Without careful planning, large-scale planting efforts can often fail to deliver the expected benefits.

Targeted fishing area closures can have unintended spillovers. To protect juvenile fish, Peru introduced a policy that temporarily banned fishing where high numbers of juvenile anchoveta were caught. However, rather than reducing the catch, the policy led to a 48 percent increase in total juvenile fish caught (Englander 2023). The closures failed to account for behavioral responses and spatial linkages. Thus, rather than discouraging overfishing, the bans signaled to fishers where juvenile fish were likely to be abundant, causing intensified fishing efforts nearby or in the same locations post-reopening. Weak enforcement outside the closed zones also allowed vessels simply to shift to other areas rather than reduce overall fishing pressure. Without coordination across time and geography, such policies risk being counterproductive, reinforcing the need for adaptive strategies that anticipate fishers' responses and to enforce sustainable catch limits across entire ecosystems.

References

Blackman, A., and P. Veit. 2018. "Titled Amazon Indigenous Communities Cut Forest Carbon Emissions." *Ecological Economics* 153: 56–67. https://www.sciencedirect.com/science/article/pii/S0921800917309746.

Chen, Q., N. Ryan, and D. Xu. 2025. "Firm Selection and Growth in Carbon Offset Markets: Evidence from the Clean Development Mechanism." Working Paper 33636, National Bureau of Economic Research, Cambridge, MA.

Coleman, E. A., B. Schultz, V. Ramprasad, et al. 2021. "Limited Effects of Tree Planting on Forest Canopy Cover and Rural Livelihoods in Northern India." *Nature Sustainability* 4 (11): 997–1004. https://www.nature.com/articles/s41893-021-00761-z.

Cui, Z., H. Zhang, X. Chen, et al. 2018. "Pursuing Sustainable Productivity with Millions of Smallholder Farmers." *Nature* 555 (7696): 363–66.

Denny, S., G. Englander, and P. Hunnicutt. 2024. "Private Management of African Protected Areas Improves Wildlife and Tourism Outcomes but with Security Concerns in Conflict Regions." *Proceedings of the National Academy of Sciences* 121 (29): e2401814121.

Ding, H., P. Veit, E. Gray, K. Reytar, J. C. Altamirano, and A. Blackman. 2016. "Climate Benefits, Tenure Costs." World Resources Institute, Washington, DC. https://www.wri.org/research/climate-benefits-tenure-costs.

Englander, G. 2023. "Information and Spillovers from Targeting Policy in Peru's Anchoveta Fishery." *American Economic Journal: Economic Policy* 15 (4): 390–427.

Heo, Seonmin Will, Koichiro Ito, Rao Kotamarthi. 2025. "International Spillover Effects of Air Pollution: Evidence from Mortality and Health Data." *The Review of Economics and Statistics* 2025: 1–45. doi: https://doi.org/10.1162/rest_a_01581.

Hsiang, S., and N. Sekar. 2016. "Does Legalization Reduce Black Market Activity? Evidence from a Global Ivory Experiment and Elephant Poaching Data." Working Paper 22314, National Bureau of Economic Research, Cambridge, MA.

IIASA (International Institute for Applied Systems Analysis). 2017. "Reducing Air Pollution Worldwide." IIASA, Laxenburg, Austria (accessed July 2025), https://iiasa.ac.at/impacts/feb-2017/reducing-air-pollution-worldwide.

Islam, M., and S. Beg. 2021. "Rule-of-Thumb Instructions to Improve Fertilizer Management: Experimental Evidence from Bangladesh." *Economic Development and Cultural Change* 70 (1): 237–81.

Ji, Y., H. Liu, and Y. Shi. 2020. "Will China's Fertilizer Use Continue to Decline? Evidence from LMDI Analysis Based on Crops, Regions and Fertilizer Types." *PLoS One* 15 (8): e0237234.

Lark, T. J. 2023. "Interactions between US Biofuels Policy and the Endangered Species Act." *Biological Conservation* 279: 109869.

Lipscomb, M., and A. M. Mobarak. 2016. "Decentralization and Pollution Spillovers: Evidence from the Re-Drawing of County Borders in Brazil." *Review of Economic Studies* 84 (1): 464–502.

Shen, W., L. Zhang, E. A. Ury, S. Li, B. Xia, and N. B. Basu. 2025. "Restoring Small Water Bodies to Improve Lake and River Water Quality in China." *Nature Communications* 16 (1): 294.

Walker, W., A. Baccini, S. Schwartzman, et al. 2014. "Forest Carbon in Amazonia: The Unrecognized Contribution of Indigenous Territories and Protected Natural Areas." *Carbon Management* 5 (5–6): 479–85.

World Bank. 2007. *Restoring China's Loess Plateau.* Feature Story, March 15. World Bank, Washington, DC. https://www.worldbank.org/en/news/feature/2007/03/15/restoring-chinas-loess-plateau.

Digital Solutions for a Livable Planet

Introduction

Environmental challenges such as air and water pollution, biodiversity loss, deforestation, and resource depletion are increasingly complex and interconnected. In this context, traditional approaches and practices often fall short. Digital technologies have emerged as a critical addition to the sustainability toolbox—helping policy makers and practitioners bridge gaps in knowledge and deploy more efficient, integrated, and impactful solutions. They can power real-time environmental data generation and analytics, improve coordination across sectors and regions, and provide the insights and tools needed to help governments and communities operate within natural limits.

Despite their transformative potential, digital solutions have often been treated as add-ons rather than core to environmental policy. Yet with third-generation mobile networks covering 90 percent of the global population and over half the world online, even basic digital tools can play a powerful role in connecting people, governments, and businesses to vital climate information and services.

Developing countries increasingly recognize the potential of digital technologies and data for environmental and climate action. Two-thirds of developing countries now include technology as part of their climate action plans (Nationally Determined Contributions) to help adapt to or mitigate the impacts of climate change (World Bank 2023a). There are roles for governments, private companies, and the broad community of nongovernmental and scientific organizations to play in leveraging digital technologies and data to support environmental sustainability. Some of these efforts are explored further in this spotlight as well as in other parts of this report.

This spotlight highlights how digital infrastructure, data platforms, and emerging technologies have already enhanced environmental practices and policy making. By embedding digital tools into mainstream sustainability efforts, governments and institutions can improve monitoring, optimize resource management, and strengthen response strategies.

For digital technologies and data to accelerate climate mitigation and adaptation efforts, inclusive and sustainable digital foundations first need to be in place. While not an exhaustive list, the following digital building blocks are crucial for seizing this opportunity and should be considered in any comprehensive environmental strategy:

- *Towers, cables, and satellites.* Cell towers connect mobile devices to networks for calls, texts, and internet by sending and receiving signals. Fiber-optic cables, on land and undersea, transmit vast amounts of data globally, supporting high-speed internet. Together, towers, cables, and satellites form the backbone of modern telecommunications. Recently, satellite internet using low Earth orbit has begun expanding broadband access to rural areas.

- *Mobile phones.* Near ubiquitous, around 78 percent of the global population owns a mobile phone (ITU 2023).[1] Smartphone use has grown rapidly—GSMA (2023) reports that over half the planet (54 percent, or 4.3 billion people) owns a smartphone. Despite this progress, millions of people lack access. It is essential to bridge this gap to ensure that everyone can leverage digital technologies and data to build resilience for a livable planet.

- *Cloud and data centers.* Data require a physical facility that organizations can use to host their online applications and data. This includes the digital equipment required to store, share, and process data, such as servers, routers, and switches, as well as rapidly increasing computing power for artificial intelligence (AI)–enabled services and analytics.

- *Digital skills.* The number of digital jobs focused on the environment has increased significantly, as demand for technology-enabled mitigation and adaptation efforts has increased. New and evolving skillsets are required to meet this reality and manage and advance emerging climate tech, particularly science, technology, engineering, and mathematics and sustainability skills that can support a livable planet. Furthermore, existing tech jobs will be greened, requiring expertise to improve sustainability, reduce power usage, and conserve natural resources.

- *Data platforms and AI.* Data platforms are digital systems that collect, store, and analyze large amounts of data from various sources, providing real-time insights and facilitating decision-making by leveraging emerging AI applications. They can help monitor environmental conditions, predict climate-related events, and inform strategies for adaptation and mitigation by enabling governments, businesses, and communities to respond proactively to environmental risks, reduce vulnerabilities, and make sustainable decisions that protect both people and the environment. A cohesive data platform is paramount for processing vast volumes of data at high speeds to produce informed and timely climate insights.

Less visibly critical, but equally essential, is physical infrastructure like roads, energy, and water (especially for cooling). While it is technically possible to deploy a connectivity network without such infrastructure (for example, by airlifting solar towers, as seen in Papua New Guinea), this approach is far more expensive to implement and maintain. As a result, it limits both the scale and speed of building a comprehensive national network.

Beyond digital infrastructure, several foundational elements must be in place for communities to harness the power of digital technologies and data. Without a supportive policy and regulatory environment, the deployment and use of digital solutions would face significant obstacles. The following elements form the basis for a supportive environment:

- A robust legal and regulatory framework that provides clear guidance and enforceable legislation on data collection, privacy, and user consent.

- Healthy market competition to spur innovation, reduce costs, and enhance the quality of digital goods and services.

- Interoperability agreements and technical protocols to ensure seamless connectivity and a user-friendly digital experience across platforms and services.

- An investment-friendly climate that welcomes private sector participation and capital, fostering sustainable growth and innovation.

Despite the growing availability of digital technologies and data, many of the people and countries that need them most still lack access. Nearly 3 billion people remain digitally unconnected, with the vast majority living in developing countries. In addition, less than 20 percent of developing countries have modern data infrastructure, such as data centers and direct access to cloud computing facilities (World Bank 2021).

Given the digital divide, more needs to be done to ensure that digital technologies and investments reach the most environmentally vulnerable people, regions, and countries. Figure S5.1 plots the share of people with internet access—an indicator of how digitally connected a country is—against exposure to extreme weather shocks. It shows a clustering of countries in the top left. These countries have very low adoption of digital technologies and very high vulnerability to climate shocks, compounding the impact on those in the margins.

Digital solutions are not only effective for protecting the environment but also practical for developing countries. The following sections explore how these technologies support pollution control, resource efficiency, and ecosystem monitoring while remaining accessible and scalable. Real-world examples demonstrate how governments and communities—especially in lower-income regions—are already using digital tools to drive environmental progress.

FIGURE S5.1 The digital divide is a resiliency divide

Population at high risk from climate-related hazards (%)

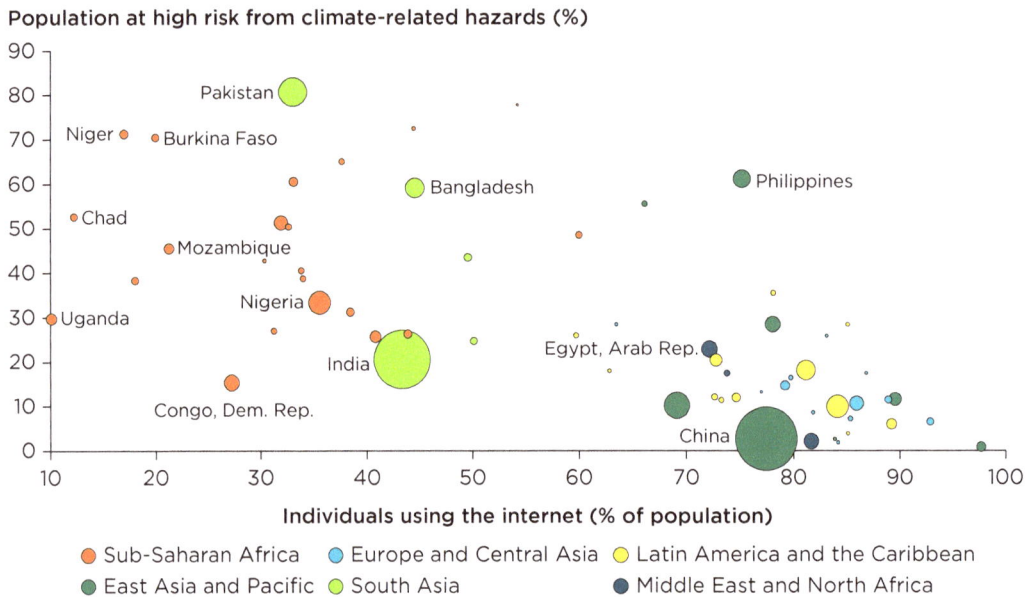

Sources: World Bank 2024a (individuals using the internet (percent of population)); World Bank 2024b (population at high risk from climate-related hazards (percent)).
Note: The figure illustrates the relationship between internet usage and climate vulnerability, highlighting the disparity in digital access among populations at high risk of climate-related hazards. Countries with lower internet penetration tend to have higher percentages of the population vulnerable to climate risks, underscoring the role of digital connectivity in resilience and adaptation efforts.

Digital identity

Digital identity is a digital means of verifying a person's identity using tools like biometrics, cryptographic keys, or government credentials. It enables secure access to services, such as banking, health care, and e-commerce, without physical documents, improving efficiency and reducing fraud. When in place before climate disasters and paired with a robust data governance framework, digital ID and financial systems can deliver aid swiftly, support remote access to services, and share data securely across systems. During COVID-19, such infrastructure allowed some countries to reach more than three times as many people with social support compared to countries without it.

Angola

In Huíla province, Angola, only 31 percent of the population has an ID card (*Bilhete de Identidade*), with a gender gap revealing a bleaker picture for women: only 26 percent have an ID card, compared with 36 percent of men. Without ID documents, public and private service delivery is challenging, especially in rural areas. To improve efficiency and service delivery, a World Bank pilot has been planned to advance access to services by integrating two existing projects in Huíla: the Angola Strengthening Governance for Enhanced Service Delivery Project (NJILA) and the Smallholder Agricultural Transformation Project (MOSAP3).

The NJILA focuses on expanding identity coverage for vulnerable populations, improving fiscal transfers to select municipalities, and strengthening municipal capacity. Complementing this, the MOSAP3 aims to enhance agricultural productivity and resilience among smallholder farmers. A key component is the creation of the National Registry of Agricultural Producers (RNPA), overseen by the Ministry of Agriculture and Forestry. Serving as a functional ID, the RNPA will help identify farmers, ensure secure financial transfers, and promote financial inclusion.

Registered farmers will receive Near Field Communication cards to access inputs via e-vouchers or cash for labor activities, encouraging adoption of climate-resilient agricultural services and practices. The pilot will also facilitate access to the ID card, which is required for obtaining land titles (*Título de Terra*), thereby safeguarding women's land rights and enhancing tenure security. By linking farmers to banking and payment systems, the initiative supports rural credit access and financial independence.

Pakistan

In 2022, an extreme monsoon wreaked havoc in Pakistan, particularly impacting the Balochistan and Sindh provinces. Over three months, Balochistan province received five times its average 30-year rainfall (Center for Disaster Philanthropy 2023), and in Sindh province, rainfall was 5.7 times greater than its 30-year average, with catastrophic effects on people, land, and the economy (Concern Worldwide 2022). For Pakistan's majority rural population (62 percent) (World Bank 2023b), not only were their livelihoods destroyed or damaged, but they were also cut off from crucial supply chains and relief.

Pakistan's civil registry, managed by the National Database and Registration Authority (NADRA) and linked to the National Socio-Economic Registry (NSER), enabled swift identification and cash distribution during the 2022 floods. Using biometric verification and digital payment systems, the Benazir Income Support Program (BISP) delivered emergency transfers to 2.8 million families within days. However, the success of these payments depended on intact telecom infrastructure and physical access to payment points—both of which were disrupted in severely affected areas, limiting cash withdrawal rates despite rapid disbursement.

Leveraging Pakistan's digital infrastructure allowed authorities to triangulate the payment withdrawal data they had with the location data in NSER and other key data sets, such as flood exposure data from the United Nations Institute for Training and Research. This enabled them to pinpoint rural areas with low withdrawal rates. Upon confirming accessibility with BISP field offices, the government coordinated with payment service providers to deploy boats to flood-stricken areas, ensuring that beneficiaries received their payments promptly.

In the face of such crises, the BISP emerged as a lifeline. Established in 2008, the BISP provides crucial support to the poorest families. Evolving from cash to digital transfers, the BISP leverages the NSER to identify eligible beneficiaries swiftly. Amid the COVID-19 pandemic, the BISP gained international recognition for its efficient cash transfers, facilitated by a robust biometric verification system linked to the NADRA, which maintains the country's civil registry.

Digital solutions for early warning systems

Early warning systems (EWS) are integrated frameworks designed to predict and provide timely information about potential hazards, enabling people, communities, and organizations to take proactive measures to mitigate their impact (Yore et al. 2023). The concept of EWS is built on four key components: *risk knowledge and analysis*; *monitoring, forecasting, and warning services*; *dissemination and communication*; *and response capability*. Digital solutions enhance EWS by delivering timely, accurate, and actionable information to at-risk populations. Their effectiveness depends on understanding vulnerabilities and working with communities to tailor information to their needs.

Hydromet services (hydrological and meteorological monitoring) and EWS help predict and communicate weather-related risks, enabling preventive action. The World Bank's Hydromet and EWS unit, through the Global Facility for Disaster Reduction and Recovery,[2] supports countries in modernizing these services, and more than $1 billion has been invested across nearly 120 projects.

In Maldives, Pakistan, and Sri Lanka, the World Bank supported a decision support system for anticipatory action, based on the MOBILISE platform.[3] At a subnational level, this platform offers a suite of digital solutions designed to extract actionable intelligence from diverse data sources, including exposure and vulnerability databases, sensor networks, weather forecasting systems, hazard models, crowdsourced information, and satellite imagery. These data aid decision-makers in reducing climate risks and implementing multi-hazard forecasting and EWS for anticipatory actions. The MOBILISE platform also features secure data-sharing capabilities and interactive visual interfaces and dashboards, facilitating collaboration among stakeholders for effective implementation of risk reduction measures and planning for early warnings and responses.

Sri Lanka faces severe weather impacts and may lose 1.2 percent of its annual gross domestic product by 2050 due to climate change (World Bank 2019). In Kalutara District, the MOBILISE decision support system, part of the Living Lab,[4] supported response efforts during the 2024 monsoon floods and landslides. It provided real-time guidance and collected feedback to improve technical systems and risk communication. The Living Lab served as a platform for coordination among key stakeholders, including government agencies and emergency services. Tools like a mobile app and real-time dashboard helped share early warnings and monitor river levels and disaster incidents. The Living Lab also supported training for local officials, promoting a cost-effective, people-centered early warning system.

Precision agriculture to mitigate environmental impacts

Precision agriculture is the practice of using advanced technologies to optimize crop production and improve productivity, while mitigating environmental impact and strengthening farming practices. It plays a crucial role in enhancing climate resilience for farmers by optimizing resource use, improving productivity, and mitigating environmental impact. These outcomes are achieved through numerous digital tools, such as:

- GPS and mapping for tracking land and its conditions
- Remote sensors and drones for crop and soil health, soil testing, and sampling
- Variable rate technology to automate the management of crop inputs
- Data analytics for informed decision-making around planting and harvest
- Automated machinery for efficient farming.

Through the use of digital technology, precision agriculture can equip farmers with the tools and knowledge needed to adapt to climate change while maintaining productivity and sustainability. By harnessing technology and data, farmers can cultivate resilience in their operations, ensuring food security and economic stability in an increasingly unpredictable climate.

Methane and rice production in Viet Nam

In 2024, rice was the food staple for nearly 50 percent of the global population: around 3.5 billion people depended on it for their primary source of calories, particularly in Asia. However, rice production contributes over 12 percent of global emissions, and 1.5 percent of total greenhouse gas (GHG) emissions.

For Viet Nam, rice is both an essential food and a major export: in the Mekong Delta, over 80 percent of the population is engaged in rice cultivation. Rice also contributes half of the Vietnamese agriculture sector's GHG emissions, and Viet Nam is feeling the impact of extreme rainfall, rising sea levels, and storm surges, increasing the likelihood of flooding (Goh and Lim, 2023; World Bank, 2022).

Several elements contribute to GHG emissions in rice production: inefficient irrigation, high seeding density, ineffective fertilizer application rates, improper disposal (burning) of rice straw and husks, and inefficient energy use in agriculture. Padi requires a large amount of water to grow, but when over flooded, rice systems can emit nitrous oxide (a potent GHG) when fertilized with nitrogen, as well as generating methane-emitting bacteria.

Low-carbon rice aims to cut GHG emissions, boost yields, and improve resilience in rice farming. The World Bank–funded Vietnam Sustainable Agriculture Transformation project used earth observation, sensors, and digital tools to support low-methane rice production. It also established an AI- and satellite-based Measurement, Reporting, and Verification system to track emission reductions. Through the project, 800,000 farmers were trained in the 1 Must Do, 5 Reductions method,[5] promoting certified seeds and reducing the use of inputs like water and fertilizers. This led to higher yields, 28 percent increased income, and a carbon dioxide equivalent reduction of 4 tonnes per hectare per season—equivalent to an average gasoline-powered passenger car driving more than 10,000 miles (US Environmental Protection Agency 2024).

Seismic monitoring and early warning systems in St. Vincent and the Grenadines

To prepare for climate, meteorological, volcanic, and seismic events, the Caribbean Island nation of St. Vincent and the Grenadines has established a monitoring network through the University of the West Indies' (UWI's) Seismic Research Center (SRC). The SRC network utilizes weak-motion seismometers to monitor small to medium-size earthquakes, while accelerometers are used for larger events, detecting earthquake activity, volcanic unrest, tectonic strain, and tsunamis.[6]

The monitoring setup also includes digital tools, such as tiltmeters, digital cameras, and infrasound detectors, alongside continuously operating GPS reference stations (Cross-Origin Resource Sharing) that record crustal strain and deformation. These tools work in conjunction with communication systems supporting Internet Protocol. The UWI SRC leveraged the fiber-optic broadband network, implemented under the World Bank Caribbean Regional Communications Infrastructure Program, to establish terrestrial monitoring stations around the La Soufrière volcano. A portion of the fiber-optic network had to be installed underwater in the northern part of the country due to the dominance of the volcano in that region.

The SRC uses third-party internet, leased satellites, and WiFi radios to support 12 six-channel seismic stations across the Eastern Caribbean, capable of detecting events from micro-earthquakes to magnitude 8–9. These stations are shared with the French West Indies network, forming a key part of the tsunami early warning system, including a station at the Belmont Volcano Observatory in St. Vincent. Volcano monitoring combines radio and internet connections. Remote sites link to access points like schools and health centers, which connect via the government-wide area network (G-WAN) fiber network to the observatory and SRC headquarters in Trinidad. The G-WAN offers high bandwidth and reliability. Plans include adding redundant radio links for greater resilience.

The expansion of high-speed internet has enabled an increase in monitoring stations around the volcano, supporting the UWI SRC team in providing accurate and timely information to disaster management authorities. Consequently, this improved monitoring capability contributed to the successful evacuation of more than 15,000 people from critical hazard zones during the explosive eruption of the La Soufrière volcano in 2021.

Notes

1. Population age 10 and over, as of 2023.
2. Refer to https://www.gfdrr.org.
3. MOBILISE is a cloud-based decision support system developed by the University of Salford in the United Kingdom.
4. Refer to https://kalutara.mobilise-srilanka.org/.
5. The 1 Must Do, 5 Reductions method comes from the Viet Nam Ministry of Agriculture, and refers to "1 must," which is use certified seeds, and "5 reductions," which are reduced: rates of seed, fertilizer, pesticides, water, and postharvest loss.
6. Source: Erouscilla Joseph (Seismic Research Centre, The University of the West Indies), Lloyd Lynch (Seismic Research Centre, The University of the West Indies), and Winston George (Project Coordinator, Caribbean Digital Transformation Project–St. Vincent and the Grenadines), personal communication, December 2024–February 2025.

References

Center for Disaster Philanthropy. 2023. "2022 Pakistan Floods." Center for Disaster Philanthropy, Washington, DC. https://disasterphilanthropy.org/disasters/2022-pakistan-floods/.

Concern Worldwide. 2022. "The Pakistan Floods, Explained." Concern Worldwide, New York. https://concernusa.org/news/pakistan-floods-explained-2022/.

Goh, L., and C. Lim. 2023. "Low Carbon Rice in Vietnam." Presentation slides. World Bank, Washington, DC.

GSMA (Groupe Spécial Mobile Association). 2023. *The State of Mobile Internet Connectivity Report 2023*. London: GSMA. https://www.gsma.com/r/wp-content/uploads/2023/10/The-State-of-Mobile-Internet-Connectivity-Report-2023.pdf.

ITU (International Telecommunication Union). 2023. *Mobile Phone Ownership in 2023*. Geneva: ITU. https://www.itu.int/itu-d/reports/statistics/2023/10/10/ff23-mobile-phone-ownership/.

US Environmental Protection Agency. 2024. *Greenhouse Gas Equivalencies Calculator*. Washington, DC: US Environmental Protection Agency. https://www.epa.gov/energy/greenhouse-gas-equivalencies-calculator.

World Bank. 2019. "Sri Lanka Strengthens Its Climate Resilience." Press Release, June 25. World Bank, Washington, DC. https://www.worldbank.org/en/news/press-release/2019/06/25/sri-lanka-strengthens-its-climate-resilience.

World Bank. 2021. *World Development Report 2021: Data for Better Lives*. Washington, DC: World Bank. https://hdl.handle.net/10986/35218.

World Bank. 2022. *Spearheading Vietnam's Green Agricultural Transformation: Moving to Low-Carbon Rice.* Washington, DC: World Bank.

World Bank. 2023a. *Green Digital Transformation: How to Sustainably Close the Digital Divide and Harness Digital Tools for Climate Action.* Climate Change and Development Series. Washington, DC: World Bank. http://hdl.handle.net/10986/40653.

World Bank. 2023b. "Rural Population (% of Total Population)—Pakistan." World Bank, Washington, DC. https://data.worldbank.org/indicator/SP.RUR.TOTL.ZS?locations=PK.

World Bank. 2024a. "Individuals Using the Internet (% of Population)." Data set. World Bank, Washington, DC. https://data.worldbank.org/indicator/IT.NET.USER.ZS.

World Bank. 2024b. "Population of People at High Risk from Climate-Related Hazards (%)." Data set. World Bank, Washington, DC. https://scorecard.worldbank.org/en/data/indicator-detail/EN_CLM_VULN?orgCode=ALL&refareatype=REGION&refareacode=ACW&age=_T&disability=_T&sex=_T.

Yore, R., C. Fearnley, M. Fordham, and I. Kelman. 2023. "Designing Inclusive, Accessible Early Warning Systems: Good Practices and Entry Points." World Bank, Washington, DC.

CHAPTER 9

Jobs and the Environment: The Green in Greenbacks

"The future of the planet concerns all of us, and all of us should do what we can to protect it. As I told the foresters, and the women, you don't need a diploma to plant a tree."

—*Wangari Maathai*, Kenyan environmentalist and Nobel Peace Prize winner 1940-2011

Key messages

- Jobs and the environment are deeply interconnected, with complex and far-reaching impacts. Environmental conditions shape employment through the effects on human and physical capital, while the green transition reshapes labor markets by shifting jobs from environmentally intensive industries to cleaner alternatives.

- There are at least three ways in which environmental conditions influence jobs: (1) natural endowments such as fertile soil, fisheries, and forests support jobs; (2) pollutants that impact human health and cognitive performance impede labor productivity; and (3) natural disasters disrupt economic activity, leading to job losses.

- The green transition creates job opportunities in growth sectors. Investing in renewable energy and low-pollution sectors creates more jobs but may lead to job losses in other sectors, requiring skill development, and enhanced safety nets.

  ○ Education and training systems can equip workers with the necessary skills to prepare them for the new economy. Education must integrate relevant skills, focusing on foundational and technical competencies. Retraining programs can help workers transition, promoting inclusivity and adaptability as economies evolve.

(continued)

- ○ Social protection systems also help workers affected by the green transition with unemployment benefits, income support, job matching, and social protection to help them adjust to job losses.

- Public works programs are often labor intensive and can create temporary employment opportunities linked to environmental restoration. These programs support vulnerable communities through activities like reforestation and soil conservation, ensuring both social and ecological benefits.

Introduction

In a small village in Kenya during the 1970s, rural women faced a growing crisis. The forests that once shaded their homes and fed their families were vanishing. Rivers were drying up, soil was eroding, and the firewood they relied on for cooking was becoming scarce. These women, already burdened by poverty and the daily toil of survival, felt powerless against the creeping degradation of their land. But when Wangari Maathai arrived, she saw not despair but potential. "You can change this," she told them. "You don't need a diploma to plant a tree."

Maathai's simple yet revolutionary idea gave rise to the Green Belt Movement. Women were taught to plant and nurture tree seedlings, restoring the forests that had been lost. For every tree that survived, the women earned a small wage—enough to buy food, send their children to school, or invest in their homes. As the forests grew back, so did hope. Rivers began to flow again, the soil became more fertile, and families found nourishment with the fruits of the new trees. Beyond restoring the land, the women discovered a new sense of agency and pride, proving to themselves and their communities that they could be the stewards of their own future.

Over the decades, the Green Belt Movement planted millions of trees across Kenya, transforming barren landscapes into verdant forests. Its impact went far beyond the environment. It gave a voice to women long silenced by societal norms and showed that even the smallest acts, like planting a seed, can ripple outward into a movement that changes the world. Wangari Maathai was awarded the Nobel Peace Prize in 2004, in large part due to her work in the Green Belt Movement. She was the first African woman and the first environmentalist to receive the honor. The story of the Green Belt Movement illustrates a powerful truth: environmental stewardship can create jobs and restore livelihoods.

This anecdote illustrates that natural resources can be important in improving the lives and livelihoods of poor people especially in developing countries. It also confirms that when communities have a direct stake, conservation outcomes can be achieved through social trust and participation—even in the absence of strong enforcement.

Effects of the environment on employment and productivity

The links between natural assets and jobs are complex (figure 9.1). Natural endowments, such as fertile soil, fisheries, and forests, are a source of comparative advantage and form the foundation for jobs in agriculture, fisheries, and tourism. Pollutants and other environmental stressors can also impact human health and cognitive performance, impeding productivity and hence adversely impacting labor market outcomes. For instance, while pollution-intensive production may generate employment, the toxic plumes it produces undermine worker productivity and long-term human capital development. Finally, natural disasters disrupt economic activity, leading to job losses during the disasters. Understanding these dynamics is crucial to ensuring that policies support both sustainability and job creation rather than working at cross-purposes.

Jobs, productivity, and natural resource endowments

Natural resources shape the geography of jobs, especially in agriculture, with fertile land, abundant water, and healthy ecosystems serving as economic engines. The agri-food system that depends directly on natural resources alone employs a third of the global workforce, with around 3.2 billion people globally relying on food systems and primary production for their livelihoods (Barbier 2025; Nico and Christiaensen 2023). For the foreseeable future, this is unlikely to change for many emerging markets and developing economies, particularly in poorer countries where food insecurity and hunger remain chronic challenges (Barbier 2025). While farming remains the backbone of employment in many countries, rising productivity and urbanization are driving a transition toward food processing, retail, and services (Nico and Christiaensen 2023).

FIGURE 9.1 **Effects of the environment on employment**

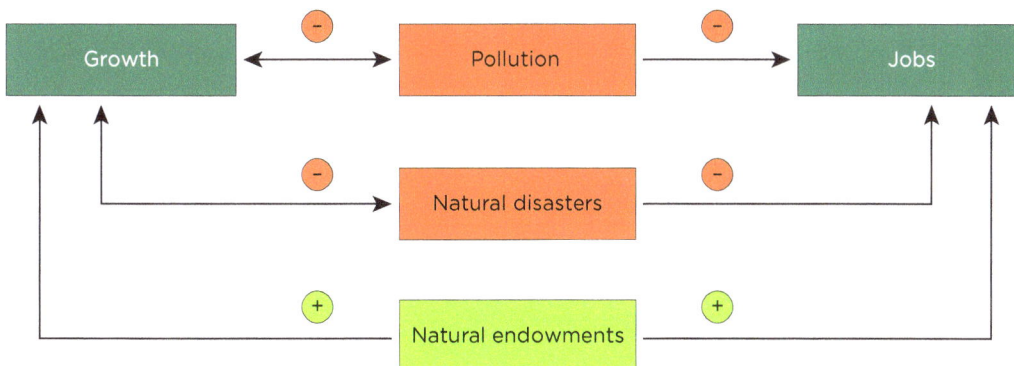

Source: Original figure for this report.

Fisheries and aquaculture are a critical part of the agri-food system, providing not only nutrition and food security but also livelihoods for millions. In 2022, the fishing sector employed 61.8 million people, a slight drop from 62.8 million in 2020, signaling shifts in production methods and labor demand (FAO 2024). However, natural endowments alone are not enough—investment in skills, infrastructure, and markets is critical to translating resource wealth into job opportunities. For instance, expanding cold storage and processing facilities in rural fishing communities can extend the value chain, creating higher-paying jobs beyond primary production. Sustainable land management practices can increase agricultural resilience while supporting employment in eco-friendly agribusiness.

The structure of employment in resource-dependent economies is not static. The shift of labor out of agriculture—known as structural transformation—is perhaps the most universal fact of development. Every country that has experienced sustained economic growth has also seen a drastic fall in the share of its workers in agriculture (Johnston and Mellor 1961; Lewis 1954; Schultz 1953). This question goes to the very heart of development economics and was formalized in the seminal dual sector model developed by Arthur Lewis, whose broader contributions to development theory earned him the Nobel Prize in Economics in 1979. Lewis emphasized the central role of surplus labor moving from traditional, low-productivity agriculture into modern, higher-productivity sectors as a driver of growth. Structural transformation and development move together, but the direction of causality and the geographic scale at which it operates are still unresolved questions in the field.

The structural transformation of the economy is neither linear nor guaranteed. The link between agricultural productivity and labor reallocation varies widely across geographies, shaped by spatial frictions, the tradability of agricultural goods, and the structure of local labor markets (Gollin 2023). In some contexts, gains in farm productivity draw more labor into agriculture, particularly where factor returns rise or when migration is constrained. In others, they catalyze movement into off-farm and urban jobs.

For instance, canal irrigation in India led to sustained agricultural gains, particularly in rural villages, which saw increases in crop yields, population, and indicators of local development (Blakeslee et al. 2023). Yet these gains often reinforced agricultural specialization rather than catalyzing local economic diversification. Over the longer term, however, the productivity shock triggered large-scale population movements: nearly 45 million people relocated to canal-irrigated regions, including about 5 million who settled in nearby towns (Asher et al. 2022). This spatial adjustment—not industrialization of rural areas but urban growth near newly productive agricultural zones—has accounted for an estimated 3–5 percent of India's urbanization since 1950. Structural transformation, in this view, is not just about shifting jobs between sectors but also about people moving across space to access those jobs.

In the short to medium term, agriculture and natural resource sectors will continue to be powerful engines of job creation especially in countries with surplus labor and limited industrialization. Investments in rural value chains, sustainable production practices, and nature-based enterprises can extend the employment benefits of agriculture and fisheries while laying the groundwork for more diversified, resilient economies.

Ultimately, what is needed is a strategy that balances short-run and long-run objectives: boosting farm productivity to generate near-term income, while preparing for long-run constraints like water scarcity and land degradation that could limit agricultural growth. But to realize this strategy, policies must also account for the friction-laden nature of labor mobility, ensuring that opportunities to transition are feasible (Zaveri et al. 2021).

Jobs, productivity, and pollution

As earlier chapters of this report have highlighted, pollution is not just an environmental issue but also an economic one, with direct consequences for worker productivity and wages. Poor air quality saps worker performance across industries, reducing cognitive function, slowing decision-making, and reducing output in both outdoor jobs (Graff-Zivin and Neidell 2012) and indoor jobs (Adhvaryu, Kala, and Nyshadham 2022; Chang et al. 2016, 2019).

Environmental damage also has lasting impacts on labor markets. As figure 9.2 shows, there is a positive correlation between access to improved sanitation, which reduces pollutants linked to poor water quality, and the World Bank's Human Capital Index—a measure of a country's foregone development potential. As sanitation improves, the Human Capital Index increases. Although none of the indicators comprising the Human Capital Index, such as health and education, explicitly includes safe drinking water, sanitation, and hygiene, the figure clearly illustrates the importance of environmental health in improving livability and future productivity (also refer to box 4.2 in chapter 4).

Long-term exposure to pollutants compounds these effects. Exposure to dirty air and contaminated water during fetal development and childhood stunts cognitive and physical growth, lowering human capital formation from infancy, reducing educational attainment, and suppressing lifetime earnings (Almond and Currie 2011; Almond, Currie, and Duque 2018). The damage extends into adulthood—workers born in high-disease areas earn significantly less (Lawson and Spears 2016), and exposure to air pollution in utero is linked to a lower intelligence quotient, weaker workforce participation, and reduced wages (Bharadwaj et al. 2017; Isen, Rossin-Slater, and Walker 2017; Sanders 2012). Even brief pollution spikes leave lasting scars, lowering educational attainment and earnings (Ebenstein, Lavy, and Roth 2016). Conversely, Indian children with better sanitation in their first year scored higher on cognitive tests by age six (Spears and Lamba 2016). Mexico's large-scale municipal program, which disinfected contaminated water, boosted academic achievement for those who were infants during the time of treatment (Bhalotra 2020).

FIGURE 9.2 **Relationship between the Human Capital Index and improved sanitation**

Human Capital Index

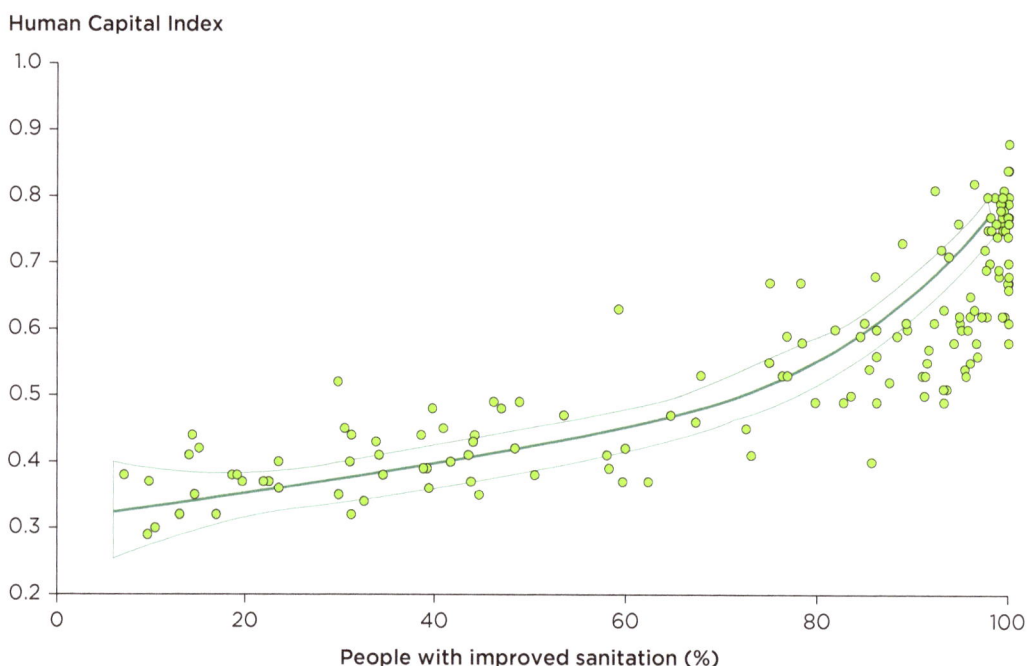

Source: Andres et al. 2018.
Note: Each dot signifies a country included in the World Bank's Human Capital Project.

Pollution also reshapes labor markets by making polluted regions less attractive to workers, driving skilled professionals away. In China, top graduates are far less likely to accept jobs in cities with dirty air (Lai et al. 2021; Zheng et al. 2019), and both China and India have seen a brain drain of executives, innovators, and professionals avoiding polluted environments, weakening firm competitiveness (Wang and Wu 2021; Xue, Zhang, and Zhao 2021). To counteract this, firms in China offer wage premiums of up to 20 percent to lure talent to polluted regions (Lai and Chitravanshi 2019). However, higher wages alone are an economically inefficient and costly way to address the challenges of declining livability, which call for policies to address a public "bad."

Evidence has suggested that when pollution falls, cities and firms reap immediate economic rewards. In Beijing, halving pollution levels could boost average wages by 14.4 percent, driven by an influx of skilled workers and increased productivity among lower-skilled employees (Khanna et al., forthcoming). Without action, polluted cities risk becoming talent deserts, forcing businesses to absorb high retention costs or relocation costs. Increasingly, there is more scope for growth through enhancing the environment than by devastating it.

Jobs, productivity, and natural disasters

Extreme weather events have a direct impact on worker productivity and economic output. In Tanzania, 40 percent of the employees in surveyed firms could not report to work during a major flood in 2018, causing an immediate drop in productivity (Rentschler et al. 2021). Across Latin America and the Middle East, droughts trigger water and power outages that curtail sales and weaken firms (Damania et al. 2017; Desbureaux and Rodella 2019; Islam and Hyland 2019; Zaveri, Gatti, and Islam 2024). The global scale of the problem is staggering—data from 80,000 firms in the World Bank Enterprise Survey program show that businesses exposed to extremely dry weather experience labor productivity drops of up to 50 percent (Gatti et al. 2024) (figure 9.3).

FIGURE 9.3 **Labor productivity falls with extreme weather events**

Log sales per worker

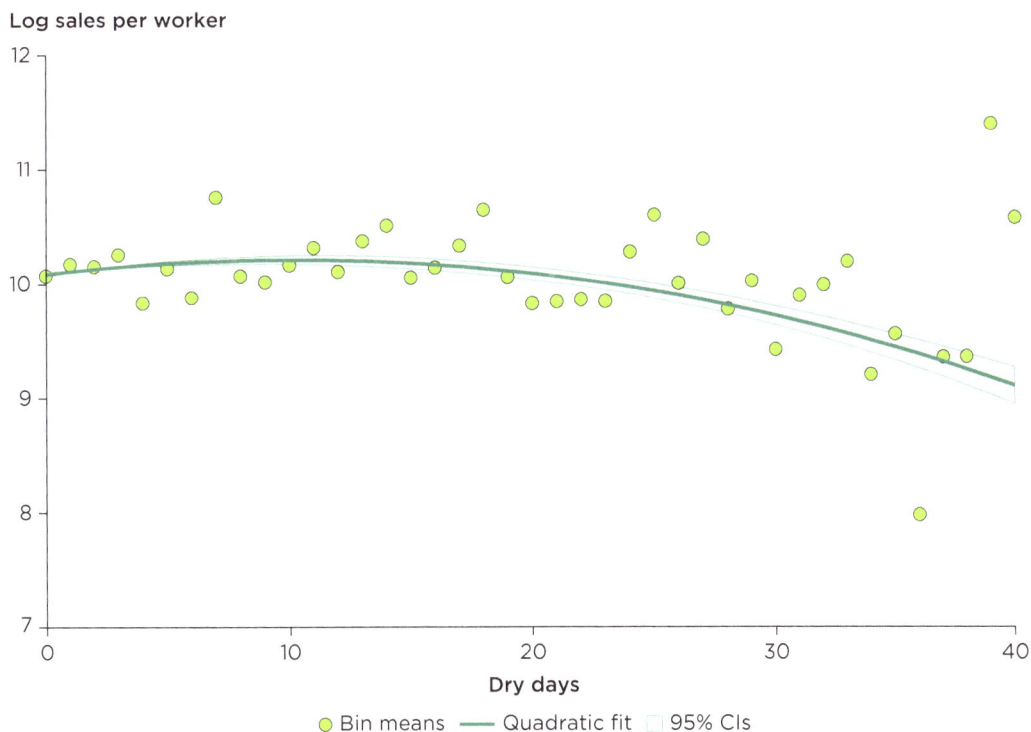

Source: Gatti et al. 2024.
Note: CI = confidence interval.

High temperatures reduce both the quality and quantity of labor, affecting workers across sectors from factory floors to farm fields (Adhvaryu, Kala, and Nyshadham 2020; Garg, Gibson, and Sun 2020; Somanathan et al. 2021). Even indoor manufacturing jobs suffer losses in efficiency on hot days (Adhvaryu, Kala, and Nyshadham 2020;

Somanathan et al. 2021), and the effects are even more pronounced in outdoor manual labor, particularly in agriculture. As extreme weather events and natural disasters intensify, the economic toll will only grow, making resilience strategies critical to protecting jobs and productivity.

What are green jobs and where are they?

New technologies together with the green transition represent a significant shift in the global economic landscape, with profound implications for labor markets. As industries move toward cleaner, more sustainable forms of production, both job creation and job destruction occur. The transition fosters new opportunities in sectors such as renewable energy, energy efficiency, and sustainable agriculture, leading to the emergence of green jobs that contribute to economic growth and environmental sustainability. However, it also poses challenges for workers in traditional industries reliant on fossil fuels and other environmentally harmful practices, necessitating policies for workforce retraining and adaptation. Yet, what is a green job?

Despite the common use of the term "green jobs," there is no consensus on its precise definition. Broadly defined, green jobs can be thought of as those aiming to reduce negative environmental impacts by producing environmentally friendly outputs and promoting environmentally friendly production processes. However, consider a solar panel production plant. The jobs span from production line workers and factory managers to human resources officers and administrative assistants. Should all these occupations be considered green jobs, or only those that directly create or use green technologies, potentially excluding the administrative and human resources staff? Depending on the policy question, either may be correct. This ambiguity highlights the challenge of defining and classifying green jobs within and across various sectors.

Green jobs can be defined in three ways: (1) based on what firms produce, (2) based on how firms produce, or (3) based on what workers do. The first definition would include jobs in firms that produce environmentally friendly goods and services. Thus, the creation of new industries, such as electric vehicles, or the expansion of existing industries, like wind energy, can be expected to see the demand for green jobs increase. The second definition considers jobs in firms that use environmentally friendly production processes, technologies, or inputs. The last definition comes from the workers' occupations. Workers may be carrying out tasks that help improve environmental impacts or use green skills to perform the job, regardless of how green their industry or firm is.[1] For example, someone who works on environmentally friendly methods for disposing of waste at a manufacturing facility might be considered a green worker.

Although information on the topic is incomplete, a variety of economy-wide studies examining "green industries," defined by what firms produce, have found that a modest though reasonable share of jobs are already creating green goods and services. It was estimated that in the United States in 2011, 2.6 percent of total employment (3.4 million jobs) was in the green goods and services sector (BLS 2013). Estimates were in the hundreds of thousands of jobs in Belgium (Nols 2024) and other European countries (OECD 2011). Using similar methodologies, Canada's employment in green industries was 1.6 percent in 2021 (Jiang 2023). A recent study in Brazil found that 4 million Southeastern Brazilians held green jobs in industries ranging from eco-tourism to waste collection (Cirera and Martins Neto 2024).

Turning to how firms produce, a few surveys around the world have found that firms are adopting greener technologies and production processes, with implications for jobs in these firms. A US survey found that 75 percent of firms have adopted green technologies or practices (BLS 2013). In Indonesia, half the firms surveyed had adopted at least one energy efficiency measure and 60 percent had dedicated energy teams or personnel, but the percentage of those involved in green technologies or practices was insignificant (Granata and Posadas 2024). These findings are consistent with the decoupling assessments outlined in spotlight 1, which showed that production externalities are being addressed by efficiency improvements rather than changes in the composition of economic activity.

When focusing on green jobs *within firms* irrespective of sectors, namely the work people do, estimates have suggested that the share of green jobs may be significantly higher, comprising 2-17 percent of all jobs. The estimates differ by country and by methodology. For example, studies have found that 2.3 percent of Indonesian jobs are green (Granata and Posadas 2024), 3.6 percent in Viet Nam (Doan, Luu, and Safir 2023), 5.5 percent in South Africa (Mosomi and Cunningham 2024), and 9 percent in India (Ham, Vazquez, and Yanez Pagans 2024).[2] Turning to developed countries, the shares of green employment were estimated at 17 percent for the United Kingdom and 14 percent for European economies in 2019 (Valero et al. 2021), and up to 20 percent in Latin American and Caribbean countries (Winkler et al. 2024).[3, 4]

There has been rapid job growth in the renewable energy industry, starting from a low base. Kammen (2008) conducted a meta-analysis and concluded that renewable energy generates more jobs than fossil fuel–based portfolios per unit of energy delivered. Blyth et al. (2014) estimated that a shift toward renewable energy and energy-efficient investment could increase the labor intensity of energy by up to one full-time equivalent job per annual gigawatt hour above the previous average of 0.4. The International Renewable Energy Agency estimated that in 2023, 16.2 million people were directly or indirectly employed in the renewable energy sector (IRENA and ILO 2024) (figure 9.4). Renewable energy creates a wide variety of jobs along its value chain, including occupations in biofuels, construction, installation, operation, and maintenance.

FIGURE 9.4 Job distribution in the renewable energy sector, 2023

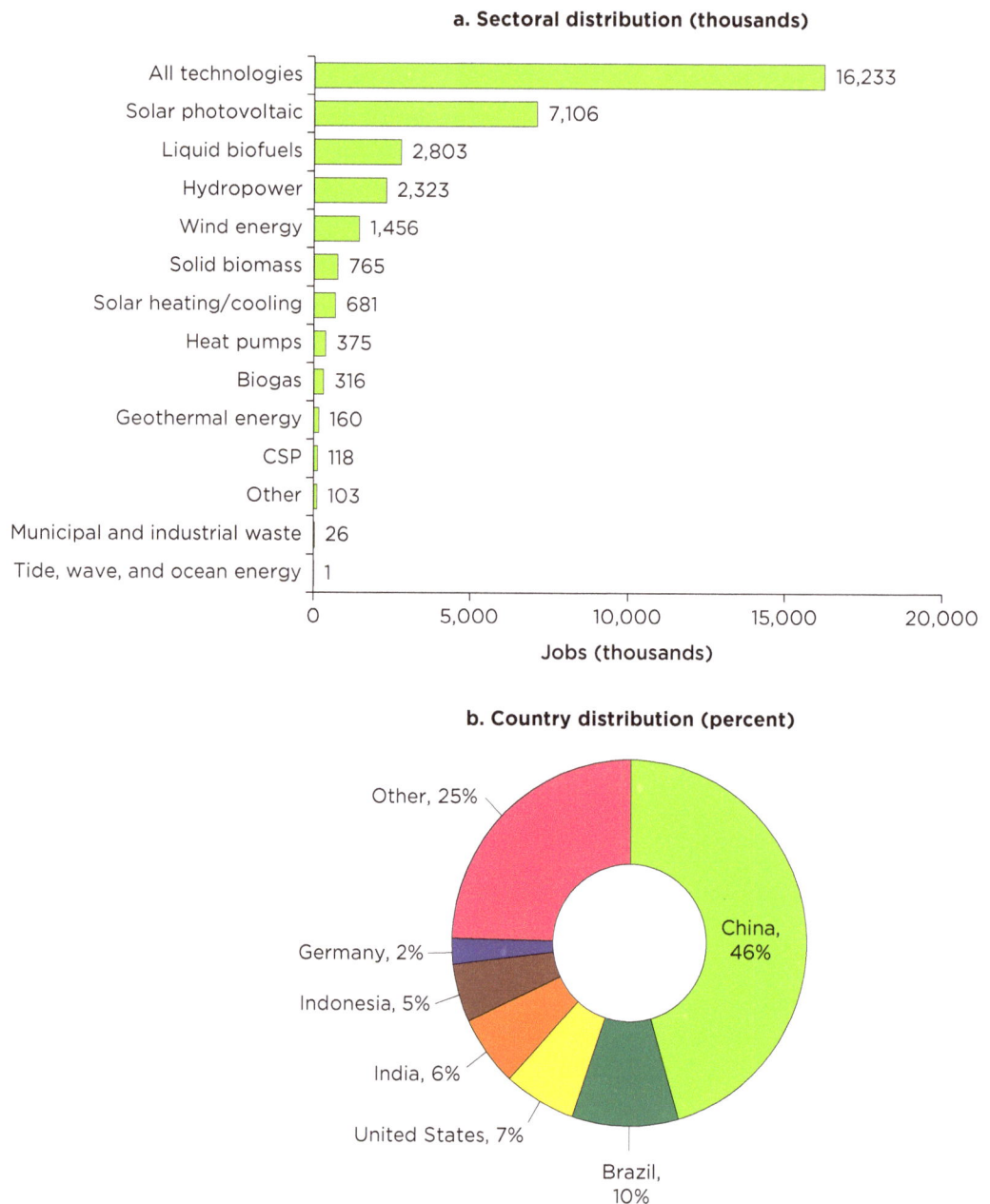

a. Sectoral distribution (thousands)

Technology	Jobs (thousands)
All technologies	16,233
Solar photovoltaic	7,106
Liquid biofuels	2,803
Hydropower	2,323
Wind energy	1,456
Solid biomass	765
Solar heating/cooling	681
Heat pumps	375
Biogas	316
Geothermal energy	160
CSP	118
Other	103
Municipal and industrial waste	26
Tide, wave, and ocean energy	1

Jobs (thousands)

b. Country distribution (percent)

China, 46%
Brazil, 10%
United States, 7%
India, 6%
Indonesia, 5%
Germany, 2%
Other, 25%

Source: Original calculations based on data from IRENA and ILO 2024.
Note: CSP = concentrated solar power.

Contrary to conventional wisdom, not all these occupations require science, technology, engineering, and mathematics (STEM) knowledge and skills. For instance, 47-60 percent of jobs in solar photovoltaics and wind energy are considered "low-skill" (IRENA 2020).

Stimulating job creation

Investments in green sectors, defined by what firms produce, can be a cost-effective way to promote employment, as they tend to be more labor intensive and pay higher wages. Investments in renewable energy and biodiversity conservation create significantly more jobs per dollar spent than, say, fossil fuel sectors (Batini et al. 2022). Renewable energy investments generate up to 1.5 times more economic activity and employment than fossil fuels, partly because clean energy is more labor-intensive and has higher wage multiplier effects. Likewise, every $1 spent on sustainable land use generates nearly $7 more within 5 years, as green land use is more labor-intensive compared to non–green land use (for example, unsustainable agriculture at industrial scales). Furthermore, the output response of non-green spending is exhausted within 5 years, and the green spending response lasts longer. The Great Green Wall in Africa is an example of the significant job creation potential that can be achieved through investment in land restoration and sustainable land management (box 9.1).

BOX 9.1
The Great Green Wall in Africa

The Great Green Wall (GGW) is an African-led initiative launched by the African Union in 2007 to restore and sustainably manage land in the Sahel-Saharan region, addressing land degradation and poverty. Originally envisioned as a 7,000 kilometer (km) long and 15 km wide vegetation barrier, the GGW has evolved into an integrated ecosystem management approach (UNCCD 2020). This includes sustainable dryland management, natural vegetation regeneration, and water conservation measures. The GGW gained significant international support, aiming to transform the lives of 100 million people by restoring 100 million hectares of degraded land, sequestering 250 million tons of carbon, and creating 10 million green jobs by 2030.

Between 2012 and 2020, the World Bank led the Sahel and West Africa Program in Support of the Great Green Wall (SAWAP). The program included $70 million from the Global Environment Facility and $1.25 billion from the International Development Association. SAWAP successfully brought 1.5 million hectares of land under sustainable management and benefited more than 17 million people across nine member countries of the Pan-African Agency of the Great Green Wall: Burkina Faso, Chad, Ethiopia, Mali, Mauritania, Niger, Nigeria, Senegal, and Sudan. The program contributed to the implementation of the GGW by improving landscape resilience and livelihoods, thereby enhancing poverty reduction, food security, and water resource security.

The agroforestry and sustainable land management activities under the GGW initiative have created numerous job opportunities in rural communities, reducing poverty through income-generating activities such as the production and valorization of fruit and nontimber forest products like honey, gum arabic, baobab leaves, fodder, and seedlings. Since 2007, these activities have generated approximately $90 million in revenue across 11 countries. More than 335,000 jobs have been created, primarily in land restoration and the production and sales of nontimber forest products, with additional employment as rangers or nature guards. Overall, more than 500,000 beneficiaries have been directly impacted by GGW activities, and another 10.2 million have benefited from wider regional activities.

Across other sectors, less-polluting industries—defined as those below median pollution levels—generate more jobs compared to more-polluting industries (figure 9.5). Taheripour et al. (2022) estimated that for every $1 million in investment, less-polluting sectors tend to produce significantly more jobs than more-polluting sectors, with forestry leading the way, generating on average more than 38 jobs per $1 million. Less-polluting industries, such as fishing, health, and education, also exhibit high employment multipliers.

There is some evidence that green jobs may be more gender-inclusive. As shown in figure 9.5, less-polluting sectors have tended to have more equal shares of jobs going to men and women, compared to the disproportionately male dirtier sectors. This suggests that investing in less-polluting sectors would not only create more jobs but may attract more women into these jobs. However, under the alternative, task-based definition of green jobs, some evidence has shown that women are considerably less likely to work in green occupations, compared to men (Granata and Posadas 2024; Ham, Vazquez, and Yanez Pagans 2024; Mosomi and Cunningham 2024; Winkler et al. 2024).

Public works programs (PWPs) can be an effective way to generate temporary jobs for low-skill workers. The primary objective of PWPs has been to provide short-term employment earnings to people who are excluded from labor markets. Countries have begun to leverage PWPs to achieve environmental objectives and mitigate the impacts of natural disasters (box 9.2). PWPs include upgrading basic water infrastructure (such as pump maintenance), implementing soil and water conservation techniques, restoring mangroves and coral reefs, and reforesting degraded lands. Ethiopia's Productive Safety Net Program and India's Mahatma Gandhi National Rural Employment Guarantee Scheme have improved groundwater recharge, reduced soil

FIGURE 9.5 **Job multipliers across sectors and by gender**

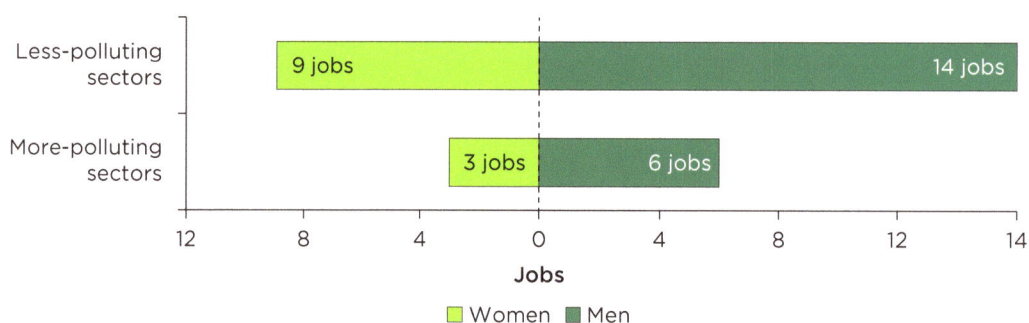

Source: Original calculations based on data from Taheripour et al. 2022.
Note: The figure compares the number of jobs created per $1 million investment across the industries in terms of air pollution, categorized as less- and more-polluting sectors. Industry pollution levels are determined by their average pollution intensities based on Taheripour et al. (2022). Less-polluting industries—including forestry, fishing, health, and education—tend to generate significantly more jobs per dollar invested compared to more-polluting industries such as mining, oil, and transportation. The figure also highlights the gender distribution, showing that less-polluting industries provide more diverse employment opportunities. This analysis included data from more than 141 countries and regions.

erosion, and increased agricultural productivity. Fiji's Jobs for Nature 2.0 and Malawi's Climate-Smart Enhanced Public Works Program and demonstrated how temporary jobs for watershed management and mangrove restoration can bolster community resilience against climate risks while creating short-term jobs. PWPs vary in scale from a few thousand beneficiaries to more than 100 million (Subbarao et al. 2013), positioning them to have transformative impacts on communities and ecosystems when implemented at scale and for multiple years.

Entrepreneurship support programs and hiring incentives can also contribute to clean job creation when targeted to activities and tasks that deliver desirable co-benefits. For example, the Croatian Public Employment Service offers financial grants to unemployed individuals starting businesses that produce environmentally beneficial products or services or provide education and training on green technologies, enhancing public environmental awareness. Slovenia's "Green Jobs" program, launched in 2021, encourages green firms to hire unemployed individuals. It targets employers based on job descriptions, environmental impact, and compliance with green certificates and standards. In lower-income settings, micro-entrepreneurship support can promote green job creation among poor people when the in-kind assets, financial capital, and training interventions are designed to promote environmental conservation and restoration (Sánchez et al. 2024). Indirect measures to foster green innovation and entrepreneurship include structural reforms, such as strengthening the entrepreneurship ecosystem and curriculum reform in schools, as well as quick-return interventions, such as business support services and mentor networks for new entrepreneurs.

BOX 9.2
Designing public works programs for poverty reduction

Public works programs can be a countercyclical tool, providing temporary income support or employment opportunities during periods of economic downturns. Expanding the reach of these programs to additional households when natural disasters occur can also protect people from the negative impacts of natural disasters on their income, assets, or human capital, while simultaneously building or rehabilitating assets that serve the community (Bowen et al. 2020; Weber et al., forthcoming).

Public works programs can improve participants' income and food security when wages are sufficient and predictable over a sustained period. For example, in India, public works participation was associated with a 10 percent increase in monthly food expenditures, increased protein and energy intake, and reduced food insecurity (Deininger and Liu 2019). In the highland regions of Ethiopia, households living in areas that experienced a minimum of two droughts but also received Productive Safety Net Program payments for two or more years did not see their food security decline, and households receiving 4 or 5 years of payments increased their livestock holdings (Coll-Black and Dadzie 2024).

Promoting appropriate skills

Meeting the demand for new jobs will require dedicated policies to train workers to compete in this new environment. The appropriate policy for such workers will differ based on the larger shifts that policy makers are hoping to effect. If the policy objective is to expand industries producing environmentally friendly goods and services, there may be a need for a broad range of skills—technical specialists as well as support services. To enhance green technologies in existing firms, support may be needed to retrain workers in specific skills. For a workforce capable of diverse green tasks across sectors, public and private skills development programs must expand to prepare students for greener workplaces.

Most of the skills needed to meet the demand of an environmentally friendly economy rely on strong foundational and socioemotional skills acquired at school. Workers in the green economy—regardless of how it is defined—require foundational and socioemotional skills and some specialized skills. The top six skills used in the greenest jobs in Organisation for Economic Co-operation and Development countries (using the task-content approach) have been identified as English and communications skills (foundational), teamwork (socioemotional), Microsoft Office (digital), quality assurance (green), and AutoCAD (computer-aided design software) (green/digital). These patterns have been observed in middle-income countries as well (Cao, Elte, and Zhao 2024; Granata and Posadas 2024; Mosomi and Cunningham 2024; Sabarwal et al. 2024). This illustrates the continued need to strengthen foundational literacy and numeracy skills, develop socioemotional skills, and form basic digital skills in pre-primary, primary, and secondary schools.

Some workers will need to acquire specialized knowledge or skills to use green technologies, processes, or information, most of which will not require new or additional education degrees (Sabarwal et al. 2024). Most occupations do not need unique green skills; however, as firms adopt green technologies, new skills will be needed. Low- or medium-skilled jobs often require modest training, like roofers becoming solar installers in six weeks or mechanics learning energy-efficient repairs. Higher-skilled jobs may need extra post-secondary classes. Although most of the skills needed for a green economy will require slight adjustments to curriculum or modest new courses, it is not happening on the required scale (Duncan 2023; LinkedIn 2023). For example, 40 of the 100 occupations that cannot be filled by skilled South Africans are green jobs, pointing to the scarcity of local skills to fill (mostly) high-level green jobs (Department of Home Affairs 2022).

Education systems play a fundamental role in training labor for the demands of twenty-first century economies. Innovators will need STEM skills to invent and implement new technologies and practices (De Ferranti et al. 2002; Sanchez-Reaza, Ambasz, and Djukic 2023). In more advanced economies, STEM skills have increased the likelihood of successfully transitioning into green jobs (LinkedIn 2023). The challenge is to attract more students to these programs, especially women who still cluster in the social fields of study. Institutional changes, such as incentivizing collaboration between university STEM programs and industry and innovation policies can spur innovation and adoption of environmentally sustainable practices.

Supporting workers displaced by the technology transition

Some workers will be displaced by new technology and structural shifts to greener industries. Displaced workers may need assistance to change their professional skills specialization and connect to new job opportunities. Employment services can aid workers in this transition process, helping them identify alternative career paths that are in demand and achievable, prepare for those careers, and swiftly adapt to new jobs (Cunningham and Schmillen 2021).

Demographic factors such as gender, geography, and earnings levels further shape workers' exposure to risks in a changing economy. Men are disproportionately represented in polluting industries, particularly in high-income countries. However, in low-income countries, the gender imbalance is reversed (Alexander et al. 2024). Workers in polluting industries, such as coal mining or oil and gas, face the highest risk of job displacement as economies shift toward clean energy. These workers often have lower educational attainment and are concentrated in rural areas, exacerbating their vulnerability. Workers employed in occupations that are undergoing substantial changes in their tasks are also expected to be impacted by the green transition, as they may need to upskill or reskill to remain relevant.[5] This structural mismatch between workers' existing skills and the demands of the green economy highlights the need for targeted support to address these disparities (box 9.3).

BOX 9.3
Unlocking employment opportunities

Labor market disparities exist in various economies based on gender (Goldin 2014; Ngai and Petrongolo 2017). According to the Arab Barometer 2022, women face significant barriers to entering the labor market in parts of the Middle East and North Africa. Two-thirds of these barriers are economic—such as lack of childcare services, low wages, and limited transportation—and a third stem from discriminatory social norms, like prioritizing men in the labor market (figure B9.3.1). When systemic exclusion prevails, it means that when new economic opportunities arise, vulnerable populations often receive fewer benefits (Damania et al. 2025).

Greening the economy can provide an opportunity to create inclusive employment opportunities while promoting environmental sustainability. Investments in renewable energy create more jobs due to their labor-intensive nature and offer greater workforce inclusivity. Currently, women constitute 32 percent of the global workforce in the renewable energy sector (UN Women 2024). However, off-grid, mini-grid, and stand-alone decentralized renewable energy initiatives in developing economies have created more employment opportunities for women while addressing energy poverty in remote or underserved areas (Baruah 2017).

(continued)

Unlocking employment opportunities *(continued)*

FIGURE B9.3.1 Most challenging barriers to workforce entry among women, Arab Republic of Egypt, 2022

Percent

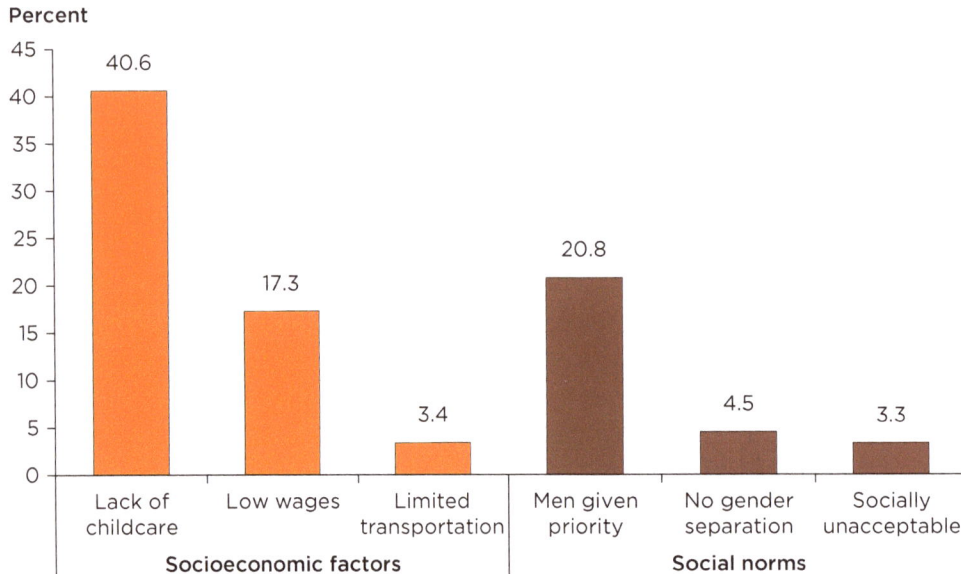

Source: Original calculations based on data from the Arab Barometer Round 7.
Note: The categories do not round to 100 percent because of the unreported "other" category.

As the energy sector shifts to renewable systems, new opportunities for an inclusive workforce are emerging. In renewable energy, hiring women can expand the talent pool, address skill shortages, and enhance product and service design for women. To benefit fully from these initiatives, broader socially progressive policies and transformative changes in societal attitudes toward gender roles are essential (Beides and Maier 2022). More progress in women's leadership is also needed to accompany progress toward gender equality in the labor markets.

Investing in forestry significantly boosts job creation and provides opportunities to leverage Indigenous land stewardship and knowledge, making the investment more efficient. Indigenous land stewardship is crucial for reducing deforestation and protecting biodiversity, as local wisdom aids in forest management (Abas, Aziz, and Awang 2022). Recognizing and enforcing Indigenous rights can ensure effective forest management, benefiting both people and nature (Pratzer et al. 2023). Social certifications, such as Reducing Emissions from Deforestation and Forest Degradation Social and Environmental Standards, are increasingly used to incentivize and reward the creation of additional social and environmental benefits in programs reducing emissions from deforestation and forest degradation. These certifications demonstrate that projects generate net positive impacts on the well-being of vulnerable communities and address any barriers or risks that could prevent benefits from reaching these groups.

Social protection systems are critical for mitigating the risks of an economic transition and supporting workers in vulnerable sectors. Programs such as unemployment benefits, severance payments, and temporary social assistance provide essential income support to displaced workers. These measures are especially important in regions where labor demand is insufficient to absorb displaced workers. For example, in Poland's coal phase-out, some workers received social protection benefits, but many non-mine workers in affected municipalities were excluded, exposing gaps in program coverage (Christiaensen et al. 2022; Honorati and Banaszczyk 2022). Comprehensive social protection and labor programs can address these gaps while enabling workers to adapt to changing labor markets. Social protection programs can also help communities manage environmental shocks, offering both immediate relief and long-term resilience (box 9.4).

BOX 9.4
Social protection programs to help manage environmental shocks

As climate challenges intensify, social protection programs play a pivotal role in supporting vulnerable communities. These programs provide safety nets, ensuring that households can maintain their health, education, and livelihoods during and after climate shocks. Shock-responsive cash transfers are particularly effective, helping households mitigate the impacts. For instance, anticipatory cash transfers in Bangladesh reduced food insecurity by 36 percent in flood-prone households while protecting assets and improving children's nutrition (Pople et al. 2024). Brazil's Bolsa Familia program mitigated the effects of rainfall shocks on school attendance (Fitz and League 2021). Mexico's conditional cash transfers reduced early disadvantages, showing the importance of early-life interventions (Adhvaryu et al. 2024).

Programs combining cash transfers with additional support have demonstrated greater resilience-building impacts. In Niger, regular cash transfers integrated with financial inclusion, skills training, and coaching enhanced households' economic diversification and food security (Bossuroy et al. 2021). In addition, social protection can influence environmentally friendly behaviors, as seen in Indonesia, where conditional cash transfers reduced tree cover loss by 30 percent in rural areas by lowering consumption linked to deforestation (Macours, Premand, and Vakis 2012). These examples highlight the multifaceted benefits of social protection in addressing immediate needs while fostering long-term resilience.

Innovative measures like adaptive social protection enhance the effectiveness of these programs. For example, the Sahel Adaptive Social Protection Program combines cash transfers with investments in human capital and productive inclusion, improving household welfare and resilience to future shocks (Archibald, Bossuroy, and Premand 2021). In Pakistan, the Benazir Income Support Program (BISP), the country's largest government-led cash transfer program, played a key role in the response to the 2022 floods. By scaling up the BISP, 2.8 million families received around $100 in cash support within a week of the floods. To identify them, information from the disaster risk management authorities on affected areas was linked to the geocoded information in the national social registry, which underpins the BISP. These programs underscore the importance of integrating social protection systems with proactive measures to tackle climate-related challenges, including investing in systems and procedures to respond to climate shocks in advance so they can be mobilized rapidly when needed.

During economic transitions, a mix of active labor market policies and temporary income support may be needed. Migration assistance and job mobility programs can further support workers by helping them access employment opportunities in regions with growing labor demand. Integrating education and labor policies with local economic development efforts will be key to fostering sustainable, inclusive growth. The right balance between income support measures (severance payments, unemployment benefits, and temporary social assistance) and active labor market programs (upskilling and reskilling, job, search, and matching assistance) depends on the local labor demand and workers' characteristics.

Notes

1. There are different methodologies to classify green outputs, green technologies, green occupations, green tasks, and green skills. For example, for cross-country comparisons, the World Bank Firm Adoption of Technologies surveys allow estimating the processed-based approach; the US Occupational Information Network Green Economy Program provides a taxonomy of green occupations; text analysis on the International Standard Classification of Occupations (Granata and Posadas 2024) and artificial intelligence–based text analysis provide a taxonomy of green tasks; and LinkedIn and the European Skills, Competences, Qualifications and Occupations provide taxonomies of green skills.
2. These studies broadly use Granata and Posadas's (2024) green dictionary and methodology.
3. Using the Occupational Information Network methodology. Refer to Vona (2021) for details.
4. Refer to Granata and Posadas (2024) for a discussion on crosswalk issues. The higher shares of green jobs in Europe and Latin America were partly due to data challenges that may have introduced significant error in the estimates.
5. For instance, in Latin America and the Caribbean, nearly half of the workers with less than a primary education are employed in non-green sectors (Winkler et al. 2024).

References

Abas, A., A. Aziz, and A. Awang. 2022. "A Systematic Review on the Local Wisdom of Indigenous People in Nature Conservation." *Sustainability* 14 (6): 3415.

Adhvaryu, A., N. Kala, and A. Nyshadham. 2020. "The Light and the Heat: Productivity Co-Benefits of Energy-Saving Technology." *Review of Economics and Statistics* 102 (4): 779–92.

Adhvaryu, A., N. Kala, and A. Nyshadham. 2022. "Management and Shocks to Worker Productivity." *Journal of Political Economy* 130 (1): 1–47.

Adhvaryu, A., T. Molina, A. Nyshadham, and J. Tamayo. 2024. "Helping Children Catch Up: Early Life Shocks and the PROGRESA Experiment." *Economic Journal* 134 (657): 1–22. https://doi.org/10.1093/ej/uead067.

Alexander, N. R., L. Li, J. Mondragon, S. Priano, and M. Tavares. 2024. *The Green Future: Labor Market Implications for Men and Women*. Washington, DC: International Monetary Fund.

Almond, D., and J. Currie. 2011. "Killing Me Softly: The Fetal Origins Hypothesis." *Journal of Economic Perspectives* 25 (3): 153–72.

Almond, D., J. Currie, and V. Duque. 2018. "Childhood Circumstances and Adult Outcomes: Act II." *Journal of Economic Literature* 56 (4): 1360–1446.

Andres, L. A., C. Chase, Y. Chen, et al. 2018. "Water and Human Capital: Impacts across the Lifecycle." World Bank, Washington, DC.

Archibald, E., T. Bossuroy, and P. Premand. 2021. "The State of Economic Inclusion Report 2021, Case Study 1: Productive Inclusion Measures and Adaptive Social Protection in the Sahel." World Bank, Washington, DC.

Asher, S., A. Campion, D. Golli., and P. Novosad. 2022. "The Long-Run Development Impacts of Agricultural Productivity Gains: Evidence from Irrigation Canals in India." STEG Working Paper. Structural Transformation and Economic Growth, London. https://steg.cepr.org/sites/default /files/2022-06/WP004.

Barbier, E.B. 2025. "Greening Agriculture for Rural Development." *World Development* 191: 106974.

Baruah, B. 2017. "Renewable Inequity? Women's Employment in Clean Energy in Industrialized, Emerging and Developing Economies." *Natural Resources Forum* 41 (1): 18–29.

Batini, N., M. Di Serio, M. Fragetta, G. Melina, and A. Waldron. 2022. "Building Back Better: How Big Are Green Spending Multipliers?" *Ecological Economics* 193: 107305.

Beides, H., and E. Maier. 2022. "Getting More Women into the Energy Sector: A RENEW'ed Approach for MENA." World Bank Blogs, July 21, 2022. https://blogs.worldbank.org/en/arabvoices/getting -more-women-energy-sector-renewed-approach-mena.

Bhalotra, S. R. 2020. "Clean Water Programmes Can Improve Cognitive Development." *Advantage Magazine*, Pollution and Climate Change Special, October 20.

Bharadwaj, P., M. Gibson, J. G. Zivin, and C. Neilson. 2017. "Gray Matters: Fetal Pollution Exposure and Human Capital Formation." *Journal of the Association of Environmental and Resource Economists* 4 (2): 505–42.

Blakeslee, D., A. Dar, R. Fishman, S. Malik, H. S. Pellegrina, and K. S. Bagavathinathan. 2023. "Irrigation and the Spatial Pattern of Local Economic Development in India." *Journal of Development Economics* 161: 102997.

BLS (Bureau of Labor Statistics). 2013. "Green Goods and Services Survey: Results and Collection." *Monthly Labor Review* (September).

Blyth, W., R. Gross, J. Speirs, et al. 2014. "Low Carbon Jobs: The Evidence for Net Job Creation from Policy Support for Energy Efficiency and Renewable Energy." Report UKERC/RR/TPA/2014/002, UK Energy Research Centre, London.

Bossuroy, T., M. Goldstein, D. Karlan, et al. 2021. "Pathways Out of Extreme Poverty: Tackling Psychosocial and Capital Constraints with a Multi-faceted Social Protection Program in Niger." Policy Research Working Paper 9562, World Bank, Washington, DC. https://elibrary.worldbank .org/doi/epdf/10.1596/1813-9450-9562.

Bowen, T. V., C. Del Ninno, C. Andrews, et al. 2020. *Adaptive Social Protection: Building Resilience to Shocks*. International Development in Focus. Washington, DC: World Bank. doi:10.1596 /978-1-4648-1575-1.

Cao, X., G. Elte, and Y. Zhao. 2024. "Developing Green Skills for Clean Energy Transition in MENA." Presentation, November 19. World Bank, Washington, DC.

Chang T., J. Graff-Zivin, T. Gross, and M. Neidell. 2016. "Particulate Pollution and the Productivity of Pear Packers." *American Economic Journal: Economic Policy* 8 (3):141–69.

Chang. T. Y., J. Graff-Zivin, T. Gross, and M. Neidell. 2019. "The Effect of Pollution on Worker Productivity: Evidence from Call Center Workers in China." *American Economic Journal: Applied Economics* 11 (1): 151–72.

Christiaensen, L., C. Ferré, T. J. Gajderowicz, M. Honorati, and S. M. Wrona. 2022. "Towards a Just Coal Transition Labor Market Challenges and People's Perspectives from Wielkopolska." World Bank, Washington, DC.

Cirera, X., and A. Martins Neto. 2024. "Measuring Green Jobs." Presentation, September 19. World Bank, Washington, DC.

Coll-Black, S., and C. Dadzie. 2024. "Climate Change Adaptation for the Ethiopian Rural Extreme Poor: The Case Study of the Productive Safety Net Program." Background note for *Rising to the Challenge: Success Stories and Strategies for Achieving Climate Adaptation and Resilience.* Washington, DC: World Bank.

Cunningham, W., and A. Schmillen. 2021. "The Coal Transition: Mitigating Social and Labor Impacts." Social Protection & Jobs Discussion Paper 2105, World Bank, Washington, DC.

Damania, R., S. Desbureaux, M. Hyland, et al. 2017. *Uncharted Waters: The New Economics of Water Scarcity and Variability.* Washington, DC: World Bank.

Damania, R., E. Ebadi, K. Mayr, J. Rentschler, J. Russ, and E. Zaveri. 2025. *Nature's Paradox: Stepping Stone or Millstone?* Washington, DC: World Bank. http://hdl.handle.net/10986/42610.

De Ferranti, D., E. G. Perry, D. Lederman, and W. E. Maloney. 2002. *From Natural Resources to the Knowledge Economy: Trade and Job Quality.* Washington, DC: World Bank.

Deininger, K., and Y. Liu. 2019. "Heterogeneous Welfare Impacts of National Rural Employment Guarantee Scheme: Evidence from Andhra Pradesh, India." *World Development* 117: 98–111.

Department of Home Affairs. 2022. *Government Gazette*, volume 680 (45860). Republic of South Africa.

Desbureaux, S., and A. S. Rodella. 2019. "Drought in the City: The Economic Impact of Water Scarcity in Latin American Metropolitan Areas." *World Development* 114: 13–27.

Doan, D., T. Luu, and A. Safir. 2023. "Green Jobs: Upskilling and Reskilling Vietnam's Workforce for a Greener Economy." World Bank, Washington, DC.

Duncan, K. 2023. "A Review of Green Skills Training Programmes in South Africa." Africa Climate Resilient Investment Facility, World Bank, Washington, DC.

Ebenstein, A., V. Lavy, and S. Roth. 2016. "The Long-Run Economic Consequences of High-Stakes Examinations: Evidence from Transitory Variation in Pollution." *American Economic Journal: Applied Economics* 8 (4): 36–65.

FAO (Food and Agriculture Organization). 2024. *The State of World Fisheries and Aquaculture 2024: Blue Transformation in Action.* Rome: FAO. https://doi.org/10.4060/cd0683en.

Fitz, D., and R. League. 2021. "School, Shocks, and Safety Nets: Can Conditional Cash Transfers Protect Human Capital Investments during Rainfall Shocks?" *Journal of Development Studies* 57 (12): 2002–26. https://doi.org/10.1080/00220388.2021.1928640.

Garg, T., M. Gibson, and F. Sun. 2020. "Extreme Temperatures and Time Use in China." *Journal of Economic Behavior & Organization* 180: 309–24.

Gatti, R., A. M. Islam, C. Maue, and E. Zaveri. 2024. "Thirsty Business: A Global Analysis of Extreme Weather Shocks on Firms." Policy Research Working Paper 10923, World Bank, Washington, DC.

Goldin, C. 2014. "A Grand Gender Convergence: Its Last Chapter." *American Economic Review* 104 (4): 1091–1119.

Gollin, D. 2023. "Agricultural Productivity and Structural Transformation: Evidence and Questions for African Development." *Oxford Development Studies* 51 (4): 375–96.

Graff-Zivin, J., and M. Neidell. 2012. "The Impact of Pollution on Worker Productivity." *American Economic Review* 102 (7): 3652–73.

Granata, J., and J. Posadas. 2024. "Why Look at Tasks When Designing Skills Policy for the Green Transition?" Policy Research Working Paper 10753, World Bank, Washington, DC.

Ham, A., E. Vazquez, and M. Yanez Pagans. 2024. "Characterizing Green and Carbon-Intensive Employment in India." Policy Research Working Paper 10927, World Bank, Washington, DC.

Honorati, M., and A. Banaszczyk. 2022. "Options to Support Workers through a Transition away from Coal in Eastern Wielkopolska." Jobs Group, World Bank, Washington, DC.

IRENA (International Renewable Energy Agency). 2020. "The Post-COVID Recovery: An Agenda for Resilience, Development and Equality." IRENA, Abu Dhabi, United Arab Emirates.

IRENA and ILO (International Renewable Energy Agency and International Labour Organization). 2024. *Renewable Energy and Jobs: Annual Review 2024.* International Renewable Energy Agency, Abu Dhabi: IRENA; Geneva: ILO. https://www.irena.org/-/media/Files/IRENA/Agency /Publication/2024/Oct/IRENA_Renewable_energy_and_jobs_2024.pdf.

Isen, A., M. Rossin-Slater, and W. R. Walker. 2017. "Every Breath You Take—Every Dollar You'll Make: The Long-Term Consequences of the Clean Air Act of 1970." *Journal of Political Economy* 125 (3): 848–902.

Islam, A., and M. Hyland. 2019. "The Drivers and Impacts of Water Infrastructure Reliability: A Global Analysis of Manufacturing Firms." *Ecological Economics* 163: 143–57.

Jiang, K. 2023. "Canada's Environmental and Clean Technology Sector (2021)." Office of the Chief Economist, Government of Canada, Ottawa.

Johnston, B., and J. Mellor. 1961. "The Role of Agriculture in Economic Development." *American Economic Review* 51: 566–93.

Kammen, D. M. 2008. *Putting Renewables to Work: How Many Jobs Can the Clean Energy Industry Generate?* University of California, Berkeley: DIANE Publishing.

Khanna, G., W. Liang, A. Mobarak, and R. Song. forthcoming. "The Productivity Consequences of Pollution-Induced Migration in China." *American Economic Journal: Applied Economics.*

Lai, C., and R. Chitravanshi. 2019. "Asia's Pollution Exodus: Firms Struggle to Woo Top Talent." PHYS. org, March 31. https://phys.org/news/2019-03-asia-pollution-exodus-firms-struggle.html.

Lai, W., H. Song, C. Wang, and H. Wang. 2021. "Air Pollution and Brain Drain: Evidence from College Graduates in China." *China Economic Review* 68: 101624.

Lawson, N., and D. Spears. 2016. "What Doesn't Kill You Makes You Poorer: Adult Wages and Early-Life Mortality in India." *Economics and Human Biology* 21: 1–16.

Lewis, W. A. 1954 "Economic Development with Unlimited Supplies of Labour." *The Manchester School* 22: 139–91.

LinkedIn. 2023. "Global Green Skills Report 2023." LinkedIn Economic Graph. LinkedIn, Sunnyvale, CA.

Macours, K., P. Premand, and R. Vakis. 2012. "Transfers, Diversification and Household Risk Strategies: Experimental Evidence with Lessons for Climate Change Adaptation." Policy Research Working Paper 6053, World Bank, Washington, DC. http://documents.worldbank.org/curated/en /275241468340175496/pdf/WPS6053.pdf.

Mosomi, J., and W. Cunningham. 2024. "Profiling Green Jobs and Workers in South Africa: An Occupational Tasks Approach." Policy Research Working Paper 10779, World Bank, Washington, DC.

Ngai, L. R., and B. Petrongolo. 2017. "Gender Gaps and the Rise of the Service Economy." *American Economic Journal: Macroeconomics* 9 (4): 1–44.

Nico, G., and L. Christiaensen. 2023. "Jobs, Food and Greening: Exploring Implications of the Green Transition for Jobs in the Agri-Food System." Jobs Working Paper 75, World Bank, Washington, DC.

Nols, L. 2024. "The Environmental Goods and Services Sector in Belgium 2010-2011." Federal Planning Bureau, Belgium.

OECD (Organisation for Economic Co-operation and Development). 2011. "The Environmental Goods and Services Sector (EGSS): A New Approach to the Environment Industry." *OECD Statistics Newsletter*, (52), July.

Pople, A., R. Hill, S. Dercon, and B. Brunckhorst. 2024. "The Importance of Being Early: Anticipatory Cash Transfers for Flood-Affected Households." Centre for the Study of African Economies, Oxford, UK.

Pratzer, M., Á. Fernández-Llamazares, P. Meyfroidt, et al. 2023. "Agricultural Intensification, Indigenous Stewardship and Land Sparing in Tropical Dry Forests." *Nature Sustainability* 6 (6): 671–82.

Rentschler, J. E., E. J. Kim, S. F. Thies, S. A. De Vries Robbe, A. E. Erman, and S. Hallegatte. 2021. "Floods and Their Impacts on Firms: Evidence from Tanzania." Policy Research Working Paper 9774, World Bank, Washington, DC.

Sabarwal, S., S. Venegas Marin, M. Spivack, and D. Ambasz. 2024. *Choosing Our Future: Education for Climate Action*. Washington, DC: World Bank.

Sánchez, I. A., J. Heisey, S. Chaudhary, et al. 2024. *The State of Economic Inclusion Report 2024: Pathways to Scale*. Washington, DC: World Bank.

Sanchez-Reaza, J., D. Ambasz, and P. Djukic. 2023. "Making the European Green Deal Work for People: The Role of Human Development in the Green Transition." World Bank, Washington, DC.

Sanders, N. J. 2012. "What Doesn't Kill You Makes You Weaker: Prenatal Pollution Exposure and Educational Outcomes." *Journal of Human Resources* 47 (3): 826–50.

Schultz, T. W. 1953. *The Economic Organization of Agriculture*. McGraw-Hill.

Somanathan, E., R. Somanathan, A. Sudarshan, and M. Tewari. 2021. "The Impact of Temperature on Productivity and Labor Supply: Evidence from Indian Manufacturing." *Journal of Political Economy* 129 (6): 1797–1827.

Spears, D., and S. Lamba. 2016. "Effects of Early-Life Exposure to Sanitation on Childhood Cognitive Skills: Evidence from India's Total Sanitation Campaign." *Journal of Human Resources* 5 (2): 298–327.

Subbarao, K., C. del Ninno, C. Andrews, and C. Rodríguez-Alas. 2013. *Public Works as a Safety Net: Design, Evidence, and Implementation*. Directions in Development. Washington, DC: World Bank.

Taheripour, F., M. Chepeliev, R. Damania, T. Farole, N. L. Gracia, and J. D. Russ. 2022. "Putting the Green Back in Greenbacks: Opportunities for a Truly Green Stimulus." *Environmental Research Letters* 17 (4): 044067.

UN Women. 2024. "Five Ways a Just Energy Transition Can Drive Gender Equality and Social Inclusion." UN Women, New York. https://asiapacific.unwomen.org/en/stories/feature-story/2024/11/five-ways-a-just-energy-transition-can-drive-genderequality-and-social-inclusion.

UNCCD (United Nations Convention to Combat Desertification). 2020. *Great Green Wall: Implementation Status and Way Ahead to 2030*. Bonn, Germany: UNCCD. https://www.unccd.int/resources/publications/great-green-wall-implementation-status-and-way-ahead-2030.

Valero, A., J. Li, S. Muller, C. Riom, V. Nguyen-Tien, and M. Draca. 2021. *Are 'Green' Jobs Good Jobs? How Lessons from the Experience To-Date Can Inform Labour Market Transitions of the Future*. London: Grantham Research Institute on Climate Change and the Environment and Centre for Economic Performance, London School of Economics and Political Science.

Vona, F. 2021. *Labour Markets and the Green Transition: A Practitioner's Guide to the Task-Based Approach*. Luxembourg: Publications Office of the European Union. doi:10.2760/65924, JRC126681.

Wang, F., and M. Wu. 2021. "Does Air Pollution Affect the Accumulation of Technological Innovative Human Capital? Empirical Evidence from China and India." *Journal of Cleaner Production* 285: 124818.

Winkler, H., V. Di Maro, K. Montoya, A. Olivieri, and E. Vazquez. 2024. "Measuring Green Jobs: A New Database for Latin America and Other Regions." Policy Research Working Paper 10794, World Bank, Washington, DC.

Xue, S., B. Zhang, and X. Zhao. 2021. "Brain Drain: The Impact of Air Pollution on Firm Performance." *Journal of Environmental Economics and Management* 110: 102546.

Zaveri, E., R. Gatti, and A. M. Islam. 2024. "Extreme Weather Shocks and Firms in the Middle East and North Africa." Policy Research Working Paper 11004, World Bank, Washington, DC.

Zaveri, E., J. Russ, A. Khan, R. Damania, E. Borgomeo, and A. Jägerskog. 2021. *Ebb and Flow, Volume 1: Water, Migration, and Development*. Washington, DC: World Bank. http://hdl.handle.net/10986/36089.

Zheng, S., X. Zhang, W. Sun, and C. Lin. 2019. "Air Pollution and Elite College Graduates' Job Location Choice: Evidence from China." *Annals of Regional Science* 63: 295–316.

Transition Minerals: Bedrock of the Future

The curiosities beneath our feet

The origins of the green transition trace back to a tiny island in the Swedish archipelago. In the late 18th century, Ytterby was a quiet fishing village sitting on a speck of an island called Resarö. In 1787, while scouting the area for a military fortification, Lieutenant Carl Axel Arrhenhuis—an amateur chemist—found a mysteriously heavy black rock he suspected might contain tungsten. He sent it to Finnish chemist Johan Gadolin, who in 1794 identified it as a new substance: yttria, named after Ytterby. This discovery marked the birth of a new class of materials, known today as *rare earth metals*. Amazingly, that same rock was later found to contain gadolinite, and eventually led to the discovery of four new rare earth elements—erbium, tantalum, terbium, and ytterbium—all named after the same small Swedish village. The same mine would later become the discovery site of four more rare earth metals (gadolinium, holmium, scandium, and thulium), making that small island one of the most significant places in the field of chemistry.

Today, transition minerals like cobalt, copper, lithium, nickel, and rare earth metals like yttria are essential for clean energy technologies, including solar panels, wind turbines, and artificial intelligence data centers. Global demand for these materials has surged, triggering a wave of new mining around the world, and raising concerns about environmental and social risk. This spotlight summarizes a background paper for this report by Katovich and Rexer (2025), which documented worldwide growth in transition mining between 2000 and 2022, evaluated exposure to socioeconomic and environmental risks, and offered policy recommendations for both producer and importer countries.

This spotlight argues that these transition minerals are essential for the net-zero transition, and their environmental footprint can be significantly shaped by the quality of governance and commitment to long-term stewardship. The analysis does not pose a binary choice between mining and preservation, but rather emphasizes the importance of ensuring that essential extraction is done responsibly, transparently, and with damage avoidance embedded at every stage. In addition, the scale and scope of impact are orders of magnitude smaller compared to fossil fuel extraction. A recent estimate suggested that the total material volume required for clean energy deployment in a net-zero pathway would be 500 to 1,000 times lower than that of cumulative fossil fuel extraction (Ritchie 2023a).

There is a further difference in the resource dynamics of clean versus fossil-based energy systems. Fossil fuels—such as coal, oil, and natural gas—are consumed and dissipated when used, leaving behind emissions and waste. In contrast, the transition minerals used in clean energy technologies—such as cobalt, copper, lithium, and rare earths—are incorporated into long-lived assets like batteries, wind turbines, and solar panels. These materials can often be recycled, creating potential to build long-term resource stockpiles that can be reused and remanufactured in future cycles of production. This closed-loop potential stands in stark contrast to the linear and dissipative nature of fossil fuel consumption (Ritchie 2023a, 2023b).

Attention must be paid to the full lifecycle of mining, especially the post-closure phases. Poorly managed mine closures can result in long-term environmental degradation. However, when done right, end-of-life care—such as reforestation, afforestation offsets, soil restoration, and protection of adjacent biodiversity corridors—can mitigate some of the less severe impacts (Cooke and Johnson 2002). Best practice approaches emphasize the importance of fully avoiding damage to irreplaceable ecosystems and threats to endangered species and recognizing that certain biodiversity losses—especially in endemic or highly fragile habitats—are often irreversible (Bull et al. 2020). While minerals are found in many locations across the world, the places that shelter endangered species and irreplaceable habitats are few, fragile, and once lost, cannot be fully restored. As highlighted in chapter 1, where ecosystem services are synergistic and super-additive, piecemeal cost-benefit assessments will be misleading.

Unearthing potential: Managing transition minerals for a sustainable future

The demand for minerals—including both metal ores and non-metallic minerals—has surged more sharply than the demand for any other natural resource over the past two decades. The use of metal and non-metallic minerals over the past 20 years has increased faster than gross domestic product (GDP) growth, outpacing even fossil fuels. This reflects the increasing material intensity of modern economies, particularly as infrastructure expansion, urbanization, and the green transition drive demand for transition inputs. With mineral consumption rising at an unprecedented rate, ensuring a stable and sustainable supply will be a key challenge for policy makers and industries in the years ahead.

Rising demand for transition minerals has spurred a substantial increase in new mines between 2000 and 2022 across a variety of countries. Katovich and Rexer (2025) examined 9,836 transition mineral and metal mines, showing that 58 percent of them were opened in Australia, Canada, the United States, and other high-income countries (HICs) (box S6.1). These commodities are of global importance as they have unique properties—such as high conductivity, magnetic potential, and lightweight durability—that are essential for decarbonization and green technologies. The analysis presented here includes only formally registered commercial transition mineral and metal mines. The mine location data are more complete for HICs, which have higher rates of formalization compared to low-income countries.[1]

BOX S6.1
Transition mineral and metal mining: Global trends and local impacts

To explore mining trends and impacts, Katovich and Rexer (2025) combined data on mine openings (2000–22) with environmental and socioeconomic indicators. They showed how price shocks—driven by surging demand for transition minerals like cobalt, lithium, and rare earths—affected local outcomes (gross domestic product [GDP], population, forest cover, air pollution, and conflict). Compared to non-transition minerals (such as coal and iron), transition mines exhibited a sharper environment-development trade-off.

The study drew from several spatial data sets: mine data from S&P Global Market Intelligence (2023), forest cover (Copernicus Land Monitoring Service 2024), population (NASA 2024), GDP via nighttime lights (Chen et al. 2022), fine particulate matter ($PM_{2.5}$) air pollution (Shen et al. 2024), and conflict events (Uppsala University 2023).

There is no universal definition of transition minerals. This study defines such minerals and metals to include alumina, antimony, bauxite (aluminum ore), chromite (chromium ore), chromium, cobalt, copper, graphite, heavy mineral sands, ilmenite (titanium ore), lanthanides, lithium, manganese, molybdenum, nickel, niobium, palladium, platinum, rutile (titanium ore), scandium, tantalum, tin, titanium, tungsten, vanadium, yttrium, zinc, and zircon. These minerals correspond to those available in the S&P Global database that also appear on the US government's official 2022 list, as determined by the US Energy Act of 2020, which covers nonfuel minerals that serve as inputs into energy technologies. Copper was notably absent from the US government's list; however, it was included in this study in consideration of copper's essential role in electricity infrastructure.

By using global commodity price fluctuations as exogenous shocks, the study estimated the plausibly causal effects of increased mining intensity on nearby environments and communities. It found that global demand drives local extraction, producing both economic benefits and environmental degradation around transition mineral sites.

Transition mines—along with non-transition mines like iron and coal—tend to be in areas with above-average population and GDP, reflecting both their concentration in HICs and their large local economic impact. Some transition mines tend to be located in conflict zones, where death rates from conflict are higher than average. These mines also tend to be in areas with greater forest cover and are more likely to fall within biodiversity hotspots—20.3 percent of transition mines are in such hotspots, which make up just 2.5 percent of global area.

The majority of new transition mines (approximately 98 percent) are in low-risk areas, and the remaining 2 percent are in areas that feature environmental or social conflict risks. The study assessed risks across five dimensions—deforestation, poverty, violent conflict, biodiversity loss, and Indigenous lands—with some of the mines falling into multiple categories:

- Ninety-eight mines are in areas at risk of deforestation—places with substantial baseline forest cover in 2000 that have since seen significant loss. These span diverse biomes, including ponderosa pine forests, boreal forests in North America and Scandinavia, and rainforests in the Amazon, Congo, and Southeast Asia.

- Twenty-nine mines are in regions where GDP per capita is below $10 per day, mostly in Africa and South Asia.[2] However, average GDP per capita in 2019 within 20 kilometers of all transition mine locations was high, at $32,400, largely because 59 percent of transition mines are in Australia, North America, and other HICs.

- Eleven mines are in areas that experienced at least one violent conflict during 2000–22. Notable hotspots included sites in Colombia, the Democratic Republic of Congo, the Philippines, and Uganda. However, over 99 percent of the transition mines experienced no violent conflict within a 20-kilometer radius during this period.

- Seventeen mines are located in biodiversity hotspots with 20 or more threatened species. These are concentrated in tropical regions like the Brazilian Amazon, Cambodia, India, and Indonesia.

- Ninety-seven mines are situated on lands officially recognized as belonging to Indigenous or traditional peoples. The vast majority (82) are in Australia, with others in Botswana, Greenland, Mexico, Namibia, Peru, Spain, the United States, and Zambia. This number likely undercounts true overlap due to limited global registry coverage (Owen et al. 2023). Whether these mines pose risks or offer opportunities depends on local governance and community engagement.

Map S6.1 shows the transition mines identified as high risk along one or more of the risk categories described above, and provides detailed definitions of each risk category. Importantly, these risks represent correlations between transition mining and various risk factors, rather than realized causal effects of mining.

MAP S6.1 Transition mines registered since 2000

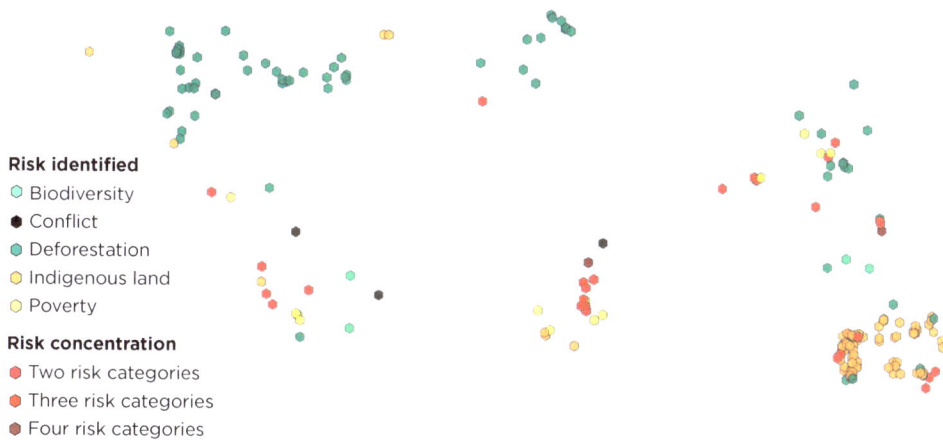

Risk identified
○ Biodiversity
● Conflict
● Deforestation
○ Indigenous land
○ Poverty

Risk concentration
● Two risk categories
● Three risk categories
● Four risk categories

Source: Katovich and Rexer 2025.
Note: The map shows transition mineral and metal mines registered since 2000 that have been identified as being at high risk for at least one of the following five dimensions: *Deforestation:* Had at least 10 percent of their area under forest cover in 2000 and lost 10 percent or more of this by 2020. *Poverty:* GDP per capita within a 20-kilometer radius is below $10/day in 2019 (constant 2017 dollars), based on GDP derived from nighttime lights. *Conflict:* One or more violent conflict events were recorded within a 20-kilometer radius of the mine since 2000. *Biodiversity:* The mine is located within a 10 × 10-kilometer grid square that hosts 20 or more threatened bird, amphibian, or mammal species. *Indigenous land:* The mine is located within lands officially held by Indigenous or traditional communities. Several mines meet multiple high-risk criteria. Selected mines are annotated on the map for illustrative purposes. DRC = Democratic Republic of Congo; GDP = gross domestic product.

The socioeconomic benefits of transition mineral booms are clearly positive. A price increase of 10 percent raises local GDP by 0.9 percent and population by 2.6 percent, likely from in-migration (Alcott and Keniston 2014). Over 2000–22, this translated into a 5 percent rise in economic activity near transition mines. In contrast, no significant local economic effects were observed for non-transition resources.[3] Increased transition mineral prices also led to a slight decrease in the probability of conflict, likely because the large local economic benefits rendered the opportunity cost of fighting higher, reducing the pool of recruits for armed groups, although these effects are not statistically significant (Blair, Christensen, and Rudkin 2021). This finding might be consistent with evidence of an ecological fallacy. For example, in the Democratic Republic of Congo, mining concessions have been found to have no effect on the number of conflicts at the territory level (the lowest administrative unit); however, it has been found that they foster violence at the district level (the higher administrative unit) (Maystadt et al. 2014).

Although most mining affects surface areas, transition mineral mines cause statistically greater deforestation than non-transition mines. Katovich and Rexer (2025) showed that rising mineral prices can drive deforestation for both categories, but the effect is 2.3 times stronger for transition minerals. They found that a price increase of 10 percent reduced forest cover by 0.3 percentage points around transition mines. Cumulatively, price increases during 2000–22 accounted for a loss of 3.6 percent of pre-mining forest cover in tropical areas near transition mines. Appropriate environmental safeguards and improved governance can partly ameliorate these impacts.

The opposite trend was observed for air pollution: non-transition mineral prices were significantly associated with increased fine particulate matter ($PM_{2.5}$) emissions, and transition mineral prices were not. This differential response was likely driven by the composition of non-transition mineral outputs, which were heavily weighted toward pollution-intensive commodities, such as coal and iron ore (Hendryx et al. 2020). The relationship between non-transition mineral prices and $PM_{2.5}$, while statistically significant, was quantitatively small, with an elasticity of just 0.05.

Governance quality shapes the impacts of transition mining. Deforestation impacts from transition mining are worse in poorly governed areas, and economic gains are larger in regions with better governance. This finding underscores the need for governance reforms to balance development with sustainability.

Toward a transition

Countries that export transition minerals can capture significant fiscal benefits from rising global demand; however, doing so requires clear regulatory frameworks and well-designed revenue instruments. Setting extraction taxes high enough to derive fiscal benefits from transition mining would allow countries to fund public investments, provide public goods, and ameliorate mining impacts, but it is important to avoid complex tax structures that are difficult to monitor and enforce. Higher taxes on minerals also need to be accompanied by mechanisms that assure equitable sharing of benefits, especially focused on addressing negative externalities. Enforcing social and environmental safeguards—such as standards for mine decommissioning and pollution control—can also protect local communities (box S6.2).

Transparency and accountability are critical to ensure that the benefits of transition mineral extraction are equitably distributed. The Extractive Industries Transparency Initiative (EITI) offers a valuable governance tool for mineral-rich countries to strengthen oversight, build public trust, and reduce corruption risks. By requiring disclosure of revenues, contracts, and ownership, EITI can strengthen fiscal governance. Evidence has shown that EITI is associated with lower levels of violent conflict and higher government revenues from mining, particularly when paired with strong institutions and civil society engagement (Berman et al. 2017; Pafadnam 2024). Integrating EITI principles into national strategies can help align transition mineral development with inclusive, sustainable growth.

BOX S6.2
Social and environmental safeguards for the green transition

As the global demand for transition minerals surges to support the green transition, the World Bank emphasizes the importance of robust environmental and social safeguards in mining operations. Such safeguards are designed to mitigate the adverse impacts of mining activities, ensuring that transition mineral extraction is not at the expense of environmental sustainability or social well-being. The World Bank's safeguard policies—including the operation manuals (OPs) for environmental assessment (OP 4.01), natural habitats (OP 4.04), and involuntary resettlement (OP 4.12)—provide a comprehensive framework for assessing and managing the environmental and social risks of mining projects. By adhering to these policies, projects can minimize deforestation, protect natural habitats, and ensure the adequate compensation and resettlement of affected communities.

The World Bank's commitment to sustainable mining practices is further underscored by its focus on stakeholder engagement and transparency. Projects must involve local communities in decision making, to ensure that their voices are heard and their concerns addressed. The World Bank also promotes the use of advanced technologies and best practices—such as minimizing water usage, reducing greenhouse gas emissions, and ensuring responsible waste management—to reduce the environmental footprint of mining operations. By implementing these safeguards, the World Bank aims to balance the urgent need for transition minerals with the imperative to protect the environment and uphold social justice, thereby supporting a truly sustainable green transition.

To address local impacts, national governments can adopt transparent revenue-sharing systems that allocate mining revenues to affected communities. However, concentrating too much in local budgets can exacerbate Dutch disease, fiscal instability, and other governance challenges. Revenue-smoothing tools, like sovereign wealth or countercyclical investment funds, may be more effective than fiscal transfers, as they can buffer communities from volatile mineral prices. Use of these tools requires strong governance and fund management capacity.

For countries importing transition minerals, supply chain security is a top priority. Tools like tax credits, subsidies, and streamlined permitting can promote domestic mining, while partnerships with trusted suppliers and diversification strategies reduce the risk of supply disruptions. Yet, easing environmental or social standards to accelerate production can backfire, undermining public trust and long-term sustainability.

Importing countries can also raise global standards by holding companies accountable beyond their borders. Requiring compliance with home-country anti-corruption, environmental, and social standards in overseas operations can promote responsible mining. Joint efforts to create traceable, transparent supply chains help avoid a "race to the bottom" and ensure that transition mineral imports truly support a cleaner and more equitable global future.

The way forward in the green transition hinges on a nuanced understanding of transition minerals and their role in shaping sustainable development. While these resources are essential for green technologies, their extraction and use pose significant environmental and social challenges. Developing countries, which often rely on resource extraction, face unique opportunities and risks as they navigate this transition. At the same time, international collaboration and robust governance frameworks are needed to mitigate the negative impacts of mining and ensure equitable benefits for all. The green transition is not just a technological shift; it is an opportunity to redefine the foundations of global economic and environmental systems, fostering a future that is both sustainable and inclusive.

Notes

1. Informality is highest for artisanal and small-scale mining (ASM), which accounts for 15–35 percent of cobalt mining in the Democratic Republic of Congo and 26 percent of global tantalum production. Although approximately 70–80 percent of ASM mining is estimated to be informal (IGF 2022), commercial mines tend to be much larger than ASM mines, so map S6.1 likely captures most global transition mineral and metal output.
2. GDP per capita values were computed by dividing GDP proxied by nighttime lights by population, which may overstate local incomes—especially in sparsely populated areas with large, light-intensive mines. A higher threshold of $10/day was used to define poverty to account for this.
3. This aggregate zero likely masks substantial heterogeneity across commodities and locations.

References

Alcott, H., and D. Keniston. 2014. "Dutch Disease or Agglomeration? The Local Economic Effects of Natural Resource Booms in Modern America." Working Paper 20508, National Bureau of Economic Research, Cambridge, MA. https://www.nber.org/system/files/working_papers/w20508/w20508.pdf.

Berman, N., D. Couttenier, D. Rohner, and M. Thoenig. 2017. "This Mine Is Mine! How Minerals Fuel Conflicts in Africa." *American Economic Review* 107 (6): 1564–1610.

Blair, G., D. Christensen, and A. Rudkin. 2021. "Do Commodity Price Shocks Cause Armed Conflict? A Meta-Analysis of Natural Experiments." *American Political Science Review* 115 (2): 709–16. https://doi.org/10.1017/S0003055420000957.

Bull, J. W., E. J. Milner-Gulland, P. F. E. Addison, et al. 2020. "Net Positive Outcomes for Nature." *Nature Ecology & Evolution* 4 (1): 4–7.

Chen, J., M. Gao, S. Cheng, et al. 2022. "Global 1 km × 1 km Gridded Revised Real Gross Domestic Product and Electricity Consumption during 1992–2019 Based on Calibrated Nighttime Light Data." *Scientific Data* 9 (1): 202. https://doi.org/10.1038/s41597-022-01322-5.

Cooke, J. A., and M. S. Johnson. 2002. "Ecological Restoration of Land with Particular Reference to the Mining of Metals and Industrial Minerals: A Review of Theory and Practice." *Environmental Reviews* 10 (1): 41–71.

Copernicus Land Monitoring Service. 2024. "Dynamic Land Cover Database." Copernicus Land Monitoring Service, European Environment Agency, Copenhagen, Denmark. https://land .copernicus.eu/en/products/global-dynamic-land-cover.

Hendryx, M., M. S. Islam, G.-H. Dong, and G. Paul. 2020. "Air Pollution Emissions 2008–2018 from Australian Coal Mining: Implications for Public and Occupational Health." *International Journal of Environmental Research and Public Health* 17 (5): 1570. https://doi.org/10.3390/ijerph17051570.

IGF (Intergovernmental Forum on Mining, Minerals, Metals and Sustainable Development). 2022. *Critical Minerals: A Primer*. Geneva: IGF. https://www.igfmining.org/wp-content/uploads /2022/11/critical-minerals-primer-en-WEB.pdf.

Katovich, E., and J. Rexer. 2025. "Critical Mining Contributes to Economic Growth and Forest Loss in High-Corruption Settings." Available at https://ssrn.com/abstract=5291760. Background paper for this report.

Maystadt, J. F., G. De Luca, P. G. Sekeris, and J. Ulimwengu. 2014. "Mineral Resources and Conflicts in DRC: A Case of Ecological Fallacy?" *Oxford Economic Papers* 66 (3): 721–49. https://doi.org /10.1093/oep/gpt037.

NASA (National Aeronautics and Space Administration). 2024. "Gridded Population of the World Version 4. Socioeconomic Data and Applications Center." NASA, Washington, DC. https://sedac .ciesin.columbia.edu/data/collection/gpw-v4.

Owen, J., D. Kemp, A. Lechner, J. Harris, R. Zhang, and É. Lèbre. 2023. "Energy Transition Minerals and Their Intersection with Land-Connected Peoples." *Nature Sustainability* 6: 203–11. https://www.nature.com/articles/s41893-022-00994-6.

Pafadnam, N. A. R. 2024. "How Does Implementing the Extractive Industries Transparency Initiative (EITI) Affect Economic Growth? Evidence from Developing Countries." *European Journal of Political Economy* 85: 102584.

Peñaloza-Pacheco, L., V. Triantafyllou, and G. Martínez. 2023. "The Non-Green Effects of 'Going Green': Local Environmental and Economic Consequences of Lithium Extraction in Chile." CAF Working Paper 2023/05, Banco de Desarrollo de América Latina y el Caribe, Caracas, Venezuela.

Ritchie, H. 2023a. "Mining Quantities for Low-Carbon Energy is Hundreds to Thousands of Times Lower Than Mining for Fossil Fuels." *Sustainability by Numbers* (accessed July 2025). https://www .sustainabilitybynumbers.com/p/mining-low-carbon-vs-fossil.

Ritchie, H. 2023b. "The Low-Carbon Energy Transition Will Need Less Mining Than Fossil Fuels, Even When Adjusted for Waste Rock." *Sustainability by Numbers* (accessed July 2025), https://www.sustainabilitybynumbers.com/p/energy-transition-materials.

S&P Global Market Intelligence. 2023. "S&P Metals and Mining Database." S&P Global Market Intelligence, New York. https://www.spglobal.com/marketintelligence/en/campaigns/metals -mining.

Shen, S., C. Li, A. Van Donkelaar, N. Jacobs, C. Wang, and R. V. Martin. 2024. "Enhancing Global Estimation of Fine Particulate Matter Concentrations by Including Geophysical *a Priori* Information in Deep Learning." *ACS ES&T Air* 1 (5): 332–45. https://doi.org/10.1021/acsestair .3c00054.

Uppsala University. 2023. "The Uppsala Conflict Data Program." Department of Peace and Conflict Research, Uppsala University, Uppsala, Sweden. https://ucdp.uu.se/.

www.ingramcontent.com/pod-product-compliance
Lightning Source LLC
Chambersburg PA
CBHW050906210326

41597CB00002B/37